Proceedings of the Workshop

Semigroups and Languages

Proceedings of the Workshop

Semigroups and Languages

Lisboa, Portugal 27 – 29 November 2002

Editors

Isabel M. Araújo
Universidade De Lisboa & Universidade De Evora, Portugal

Mário J. J. Branco
Universidade De Lisboa, Portugal

Vítor H. Fernandes
Universidade De Lisboa & Universidade Nova De Lisboa, Portugal

Gracinda M. S. Gomes
Universidade de Lisboa, Portugal

NEW JERSEY • LONDON • SINGAPORE • BEIJING • SHANGHAI • HONG KONG • TAIPEI • CHENNAI

Published by

World Scientific Publishing Co. Pte. Ltd.
5 Toh Tuck Link, Singapore 596224
USA office: Suite 202, 1060 Main Street, River Edge, NJ 07661
UK office: 57 Shelton Street, Covent Garden, London WC2H 9HE

British Library Cataloguing-in-Publication Data
A catalogue record for this book is available from the British Library.

SEMIGROUPS AND LANGUAGES

ISBN 981-238-917-2

Printed in Singapore by World Scientific Printers (S) Pte Ltd

Preface

The Workshop on Semigroups and Languages took place in November 2002 at Center of Algebra of the University of Lisbon. It aimed at bringing together researchers in both areas to find out about recent results of common interest.

Jorge Almeida (University of Porto, Portugal), John Fountain (University of York, U.K.), John Howie (University of St. Andrews, U.K.), Donald McAlister (Northern Illinois University, U.S.A.), Douglas Munn (University of Glasgow, U.K.), Jean-Eric Pin (University Paris 7, France), Stephen Pride (University of Glasgow, U.K.), Nikola Ruškuc (University of St. Andrews, U.K.), and Mikhail Volkov (Ural State University, Russia) were the main speakers. Other researchers presented short talks.

These proceedings include part of the material given in various lectures and some of the papers presented here are extended versions of the respective talks.

This workshop was highly successful: about 57 people from 10 different countries attended it. Masters and Ph.D. students who participated in the meeting found it very useful especially because the lectures were presented in a clear and self-contained way. This is the kind of meeting the organizers hope to hold again in the future due to the importance of the interaction of these two research areas.

We would like to thank all the sponsors of this workshop, namely the Center of Algebra of the University of Lisbon, the Faculty of Sciences of the University of Lisbon, the Luso-American Foundation for Development and the Portuguese Foundation for Science and Technology.

Thanks are also due to the city government of Lisbon who kindly organized a reception at, and a visit to, the Museum of the City of Lisbon. The tile whose picture is on the cover of this book is property of this museum.

Lisbon, April 2004

Isabel M. Araújo
Mário J.J. Branco
Vítor H. Fernandes
Gracinda M.S. Gomes

Contents

THE EQUATIONAL THEORY OF ω-TERMS FOR FINITE $\mathcal{R}$-TRIVIAL SEMIGROUPS*

JORGE ALMEIDA†

Centro de Matemática da Universidade do Porto and
Departamento de Matemática Pura,
Faculdade de Ciências da Universidade do Porto,
Rua do Campo Alegre, 687,
4169-007 Porto, Portugal
E-mail: jalmeida@fc.up.pt

MARC ZEITOUN

LIAFA,
Université Paris 7 et CNRS, Case 7014,
2, place Jussieu,
75251 Paris Cedex 05, France
E-mail: mz@liafa.jussieu.fr

A new topological representation for free profinite $\mathcal{R}$-trivial semigroups in terms of spaces of vertex-labeled complete binary trees is obtained. Such a tree may be naturally folded into a finite automaton if and only if the element it represents is an ω-term. The variety of ω-semigroups generated by all finite $\mathcal{R}$-trivial semigroups, with the usual interpretation of the ω-power, is then studied. A simple infinite basis of identities is exhibited and a linear-time solution of the word problem for relatively free ω-semigroups is presented. This work is also compared with recent work of Bloom and Choffrut on transfinite words.

*This work was started during the visit of both authors to the *Centro Internacional de Matemática*, in Coimbra, Portugal. Financial support of *Fundação Calouste Gulbenkian* (FCG), *Fundação para a Ciência e a Tecnologia* (FCT), *Faculdade de Ciências da Universidade de Lisboa* (FCUL), and *Reitoria da Universidade do Porto* is gratefully acknowledged. The work was also supported by the INTAS grant #99-1224.

†This paper has been written during this author's visit to the LIAFA with the support of the *Université Paris 7*. Work also supported by FCT through the *Centro de Matemática da Universidade do Porto*, and by the project POCTI/32817/MAT/2000, which is partially funded by the European Community Fund FEDER.

1. Introduction

Finite $\mathcal{R}$-trivial semigroups form a pseudovariety R which appears naturally as it is generated by the following classes of transformations of a finite chain: full decreasing transformations [16] or partial decreasing and order-preserving transformations [12]. The corresponding variety of languages also appears naturally: it is the smallest variety of languages containing, over a finite alphabet A, the languages B^+ with $B \subseteq A$, the letters $a \in A$, and which is closed under disjoint union and deterministic product [18].

Each regular $\mathcal{D}$-class of a finite $\mathcal{R}$-trivial semigroup forms a (left-zero) band, which places $\mathcal{R}$-trivial semigroups close enough to the well-known variety of (left regular) bands $[xyx = xy, x^2 = x]$. Syntactical techniques that work for R can often be extended to the pseudovariety DA, of all finite semigroups whose regular $\mathcal{D}$-classes are rectangular bands, which plays an important role not only in finite semigroup theory but also in temporal logic [22, 21].

The underlying question which motivated the work summarized in this paper is whether R is "completely tame" for the *canonical* signature $\kappa = \{\_ \cdot \_, \_^{\omega}\}$. This signature is the most commonly used in finite semigroup theory. The notation comes from [3] with a minor change resulting from the fact that we are only interested here in aperiodic semigroups. For R, the question of complete tameness roughly means the following: whether every finite system of equations with clopen constraints which has a solution in the free profinite semigroup *modulo* R also admits such a solution using only terms of the signature κ. From a simple yet rather useful solution of the word problem for ω-terms over the related pseudovariety J, of all finite $\mathcal{J}$-trivial semigroups, that is the identity problem in the signature κ, it is not hard to show that J is completely tame for the canonical signature [1].

Here we consider the related question of solving the word problem for ω-terms over the pseudovariety R and more generally studying the equational theory of R in the signature κ. With the tameness question in mind, we aim at a good understanding of this equational theory as well as efficient algorithms that may be used to test examples and allow us to deepen our intuition.

One can raise a question of a much more general nature, thus abstracting the word problem to a general pseudovariety and a suitable signature: under what general conditions on a pseudovariety V and a signature σ can one guarantee that the identity problem for V in the signature σ is decidable? This is of course a bit vague so we propose a restricted form of this question.

See [1] for undefined terms.

Question. Is there any recursively enumerable pseudovariety V for which there exists a recursively enumerable signature σ (consisting of computable implicit operations) such that V^σ has an unsolvable identity problem?

The results presented here are the following. We start with a new representation of pseudowords over R, namely as certain binary trees. Such trees are regular, that is they may be folded into finite automata by the identification of isomorphic subtrees, if and only if the pseudowords they represent are ω-terms. The minimal such representation of an ω-term is constructible in $O(mn)$-time, where m is the number of letters and n is the length of the term. This gives rise to an algorithm for solving the word problem for ω-terms over R which works in $O(mn)$-time where m is as above and n is now the length of the longest of two ω-terms whose equality over R is to be tested.

We also describe a basis of identities for the variety of *ω-semigroups* generated by R, where ω-semigroups means semigroups with an extra unary ω-power operation. Since no finite basis can be extracted from our basis, this variety is not finitely based.

This paper is meant as an extended abstract of the forthcoming paper with full details. Rather than presenting all technicalities which are required for the detailed account of the results, we concentrate on making clear the underlying ideas at the expense of not being completely precise. In particular, all proofs will be omitted. The reader interested in more details is referred to the full paper [6].

2. R-trees and R-automata

Elements of free profinite semigroups will be generally called *pseudowords*. We may also speak of *pseudowords over* V for a pseudovariety V to indicate the natural projections of pseudowords in free pro-V semigroups. For an introduction to relatively free profinite semigroups, see [1]. Recall in particular that formal equalities between pseudowords are known as *pseudoidentities* and have full descriptive power for defining pseudovarieties of semigroups. We write $\mathcal{C} \models u = v$ to indicate that the class $\mathcal{C}$ of finite semigroups satisfies the pseudoidentity $u = v$.

The basic ingredient in all our results is the following observation taken from [2].

Lemma 2.1. *Let u, v be pseudowords and suppose $u = u_1 a u_2$, $v = v_1 b v_2$*

where a, b are letters such that $a \notin c(u_1)$, $b \notin c(v_1)$, and $c(u_1) = c(v_1)$. If $\mathsf{R} \models u = v$ then $a = b$ and $\mathsf{R} \models u_i = v_i$ $(i = 1, 2)$.

The same holds for the pseudovariety S of all finite semigroups using Proposition 3.5 of [4] combined with simple arguments of language theory. This leads to the *left basic factorization* of a pseudoword, which is well defined over both S and R:

$$u = u_1 a u_2 \text{ with } c(u) = c(u_1 a),\ a \in c(u) \setminus c(u_1). \tag{1}$$

For instance, the left basic factorization of $(ab^\omega a)^\omega$ is $a \cdot b \cdot b^{\omega-1}a(ab^\omega a)^{\omega-1}$. The idea explored in [5] is to iterate this factorization on both factors u_1 and u_2, keeping in mind the observation that every infinite product of pseudowords over R converges. In [5], this is carried out considering at the same level factorizations going rightwards, leading to (in general) infinite factorization trees of finite height.

For instance, for the ω-term $(ab^\omega a)^\omega$ we have the representation in Figure 1. An alternative representation in terms of labeled ordinals was also

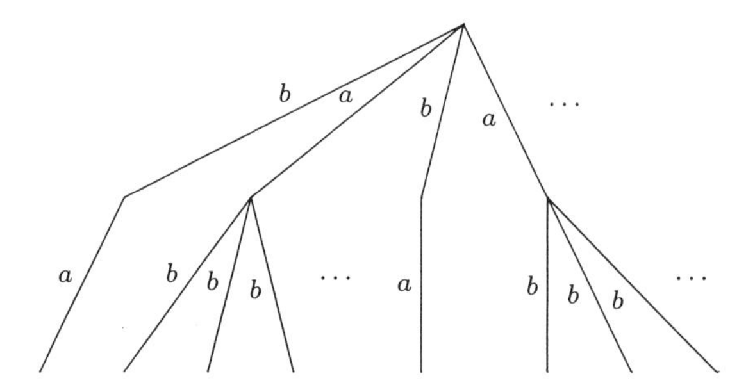

Figure 1. The Almeida-Weil tree representation of $(ab^\omega a)^\omega$

constructed in [5] which, for the same example, can be visualized as follows:

$$a\,\underbrace{bbbb\cdots}_{\omega}\;\underbrace{aa\,\underbrace{bbbb\cdots}_{\omega}\;aa\,\underbrace{bbbb\cdots}_{\omega}\;\cdots}_{\omega}$$

We present here yet another representation, which focuses on the binary flavor of the left basic factorization: each new factor u_i lies either to the left or to the right of the letter marker a which splits the previous factor u according to (1). For the same example, $(ab^\omega a)^\omega$, we obtain the infinite binary tree indicated in Figure 2, where ε stands for the empty word. In view of the left basic factorization of $(ab^\omega a)^\omega = a \cdot b \cdot b^{\omega-1}a(ab^\omega a)^{\omega-1}$, the root is labeled by b, and the process is iterated on the left (with the pseudoword a) and on the right (with $b^{\omega-1}a(ab^{\omega}a)^{\omega-1}$). In this infinite tree

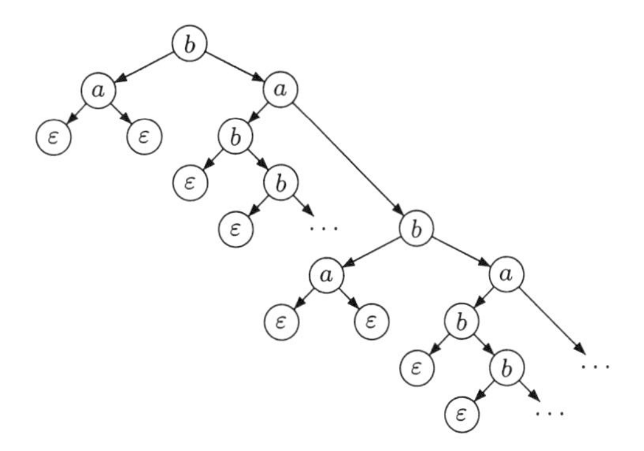

Figure 2. The R-tree of the pseudoword $(ab^{\omega}a)^{\omega}$

one recognizes immediately that certain subtrees are repeated in the sense that isomorphic copies are found several times, where isomorphism stands for topological isomorphism respecting labels. Rather than repeating a subtree T, we may as well point to a node r at which T already appeared with r as a root. This leads to a *folding* of the tree, in our example to the finite graph indicated in Figure 3 where we use the edge labels 0 for left

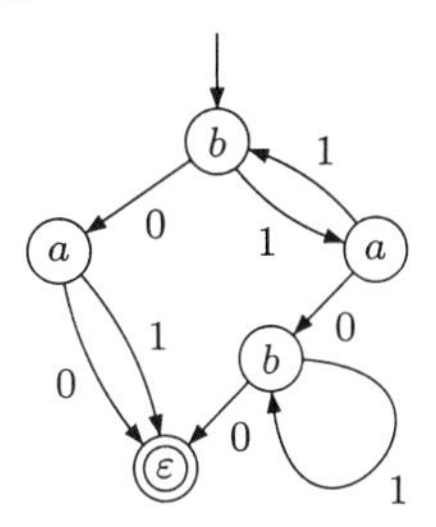

Figure 3. The minimal R-automaton of the pseudoword $(ab^{\omega}a)^{\omega}$

and 1 for right, the incoming arrow at the top represents the node coming from the root and the double circle represents the node coming from the leaves. Another example is given in Section 5.

Examining the procedure leading to the above representations of pseudowords over R, one obtains first a characterization of all vertex-labeled complete binary trees which appear in this way, which we call R-*trees*. More generally, an R-*automaton* over an alphabet A is a deterministic $\{0,1\}$-automaton which is complete, except for the terminal states which have no arrows leaving from them, and which has the states labeled with elements from $A \cup \{\varepsilon\}$ so that the following properties are satisfied:

(1) a state is labeled ε if and only if it is terminal;
(2) for each state v, the labels of states which are accessible from v by a path starting with the edge labeled 0 miss exactly one letter from the labels of all the states accessible from v, namely the label of v.

In particular, R-trees are the acyclic R-automata. An *isomorphism of* R-*automata* is an isomorphism of the underlying automata, in the usual sense of automata theory [14], which respects the labeling of vertices. We will not distinguish between isomorphic R-automata.

The R-tree constructed by iterated left basic factorization from a pseudoword w is denoted $\mathcal{T}(w)$ and is called the R-*tree of* w. From Lemma 2.1 one may easily deduce the characterization of equality of pseudowords over R in terms of equality of their associated R-trees. Furthermore, there is a natural topology on R-trees which is relevant for our characterization: define the distance between two distinct R-trees to be 2^{-r} where r is the smallest integer such that, chopping off all nodes and edges at distance from the root greater than r, we obtain distinct vertex-labeled trees.

Theorem 2.2. *Over a finite alphabet, the correspondence associating to each pseudoword w its* R-*tree $\mathcal{T}(w)$ induces a homeomorphism between the free pro-*R *semigroup and the space of all* R-*trees.*

In particular, to every R-tree T corresponds at least one pseudoword w such that $\mathcal{T}(w) = T$. By Theorem 2.2, the value of such a pseudoword w over R depends only on T. We denote it by $w(T)$.

We may translate an R-automaton $\mathcal{A}$ over an alphabet A into a usual automaton $\mathcal{A}'$ by taking $B = \{0,1\} \times A$ as the alphabet labeling the edges and by transferring the labels of vertices to the labels of edges as second components. This transformation is clearly reversible for B-automata which have properties that are easily identified, such as: no edges leave from terminal states, and all edges leaving from other vertices have labels with the same second component.

The folding procedure may be described as taking the quotient automaton under the congruence $\sim_w$, according to which two nodes are equivalent if the subtrees rooted at these nodes are identical. The resulting automaton will be called the *minimal* R-*automaton* of the pseudoword and denoted $\mathcal{A}(w)$. The language over the alphabet B recognized by the automaton $\mathcal{A}(w)'$ is denoted $L(w)$, and it gives a trace history of how a specific empty factor is eventually obtained by iterated left basic factorization. Note that the R-tree $\mathcal{T}(w)$ may be reconstructed from the R-automaton $\mathcal{A}(w)$ by unfolding it.

Corollary 2.3. *The following conditions are equivalent for a pair u, v of pseudowords:*

(1) $\mathsf{R} \models u = v$;
(2) $\mathcal{T}(u) = \mathcal{T}(v)$;
(3) $\mathcal{A}(u) = \mathcal{A}(v)$;
(4) $L(u) = L(v)$.

In case the congruence $\sim_w$ has finite index, the folding procedure is nothing else than the minimization of the translated automaton followed by the reverse translation, and this is consistent with naming $\mathcal{A}(w)$ the minimal R-automaton of w. The natural question which this observation raises is under what conditions on a pseudoword its minimal R-automaton is finite. For a complete answer to this question, which will be given in the next section, we introduce some pseudowords associated with a given pseudoword.

The subtree T_p of all descendants of a node p of the R-tree $T = \mathcal{T}(w)$ of a pseudoword w is itself an R-tree and therefore it uniquely determines a pseudoword $v = w(T_p)$ over R which we call an R-*factor* of w and may also call the *value of the subtree* T_p and the *value of the node* p. In view of Theorem 2.2, we will usually identify the pseudoword v with the node p and thus use the same notation for pseudowords and nodes in R-trees. Taking into account how $\mathcal{T}(w)$ is constructed by iterated left basic factorization, in case v is a direct right descendant of another node, such pseudowords are called *relative tails* of w.

3. Periodicity and the word problem for ω-terms

By an ω-*term* we mean a well-formed expression in a set of letters using an associative operation of multiplication and a unary operation of ω-power. The following result is a sort of periodicity theorem which answers the question raised at the end of the previous section.

Theorem 3.1. *The following conditions are equivalent for a pseudoword w over* R*:*

(1) w is represented by some ω-term;
(2) the set of R*-factors of w is finite;*

(3) the set of relative tails of w is finite;
(4) the folded R*-automaton of w is finite;*
(5) the language $L(w)$ is rational.

Thus, the word problem for ω-terms over R will be solved if we find a way to compute the folded R-automaton $\mathcal{A}(w)$ for each ω-term w since $\mathcal{A}(w)$ is finite and completely determines w. This may seem however a rather strange way of solving the word problem. A seemingly more natural approach would at first sight appear to be devising a finite confluent set of reduction rules which, applied repeatedly in any order until no further reduction is possible, leads to a canonical form of a given ω-term such that two ω-terms are equal over R if and only if they have the same canonical form. We have found no such system and we conjecture that there is none.

This motivates the construction of an algorithm that will efficiently produce an R-automaton recognizing the language $L(w)$ for an ω-term w, though not necessarily the minimal automaton $\mathcal{A}(w)$. Once this goal is achieved, the computation of $\mathcal{A}(w)$ can then be concluded by minimization of the R-automaton thus obtained. If the computed R-automaton is close to being minimal and one can further optimize the minimization procedure, then the computation of the minimal R-automaton may be close to being optimal. One can then invoke Corollary 2.3 to deduce an efficient solution of our word problem.

So, let us consider an ω-term w. Note that, unless w is a finite word, $\mathcal{T}(w)$ is an infinite tree and of course we do not want to use the word problem to fold it by comparing the pseudowords corresponding to its subtrees. The idea is to look closer at the syntactic structure of w and to relate the values $w(T_p)$ of subtrees of $T = \mathcal{T}(w)$ with certain terms that can be syntactically constructed from w.

The essential reason why the above idea works is best explained in a picture which represents one of a few typical cases, as depicted in Figure 4. The other cases are handled similarly. In it we consider a subtree S of $\mathcal{T}(w)$ and we assume that its root is a descendant of a node along its left line which in turn is the direct right descendant of some node. The value of $w(S)$ is then precisely the factor of w that can be found between the positions in w determined by a and b:

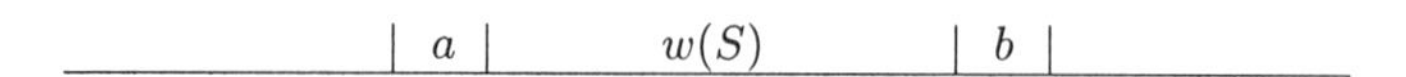

The value of $w(S)$ may be computed syntactically from the given ω-term w since the indicated b is the first occurrence of this letter to the right of

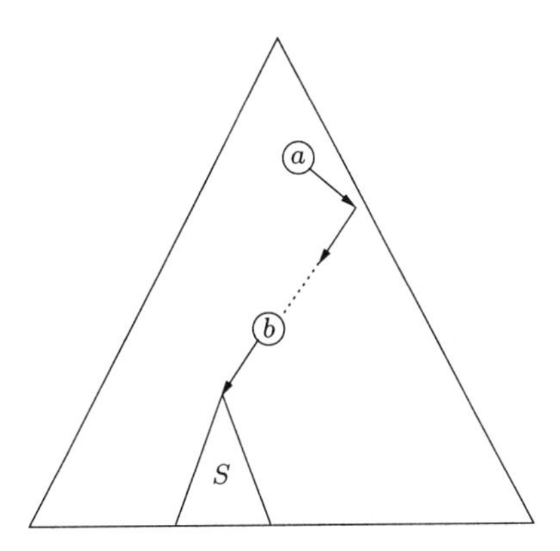

Figure 4. A subtree of $\mathcal{T}(w)$

the indicated occurrence of a. Thus, by canonically splitting ω-powers so as to extract any required factors according to the formula

$$(uvw)^{\omega} = uv(wuv)^{\omega},$$

which is valid in R, one can compute the successive values of subtrees until no new values are obtained. More precisely, one may compute, for each position i of a letter in w and each letter b which either occurs to the right of that position in w, or occurs within the same base of an ω-power as the position i, an ω-term $w(i,b)$ corresponding to the value of a subtree. Moreover every such value may be obtained in this way. It is clear that there are at most nm such ω-terms, where $n = |w|$ is the length of w disregarding ω-powers and $m = |c(w)|$ is the number of letters occurring in w.

By arranging appropriately the order in which the calculation of these ω-terms derived from w is performed, one may thus achieve the computation of all values of subtrees of $\mathcal{T}(w)$ in $O(mn)$-time. Moreover, from the computation it is clear that the value of the left descendant of any node v in $\mathcal{T}(w)$ with value $w(i,b)$ is $w(i,c)$, where c is the only letter of $w(i,b)$ such that every other letter in $w(i,b)$ also occurs in $w(i,c)$; if j is the position in w corresponding to the first occurrence of c in $w(i,b)$, the value of the right descendant of v is $w(j,b)$; finally the label of v is c. Thus basically, for each new encountered pair (i,a), one computes the list of letters occurring in $w(i,a)$ and the reverse order in which they are found in terms of the left basic factorization iterated on the left factors. For convenience of calculation, one adds a position 0 corresponding to the beginning of w and a letter # to mark the end of w.

Theorem 3.2. *There is an algorithm which computes in time $O(|w|\cdot|c(w)|)$ an* R*-automaton recognizing $L(w)$ for any given ω-term w.*

In our earlier example, $w = (ab^\omega a)^\omega$, one starts with the initial state $w(0,\#)$ and looks for the last letter to occur for the first time, namely b. Its position is 2, so the state $w(0,\#)$ is labeled b and it has an edge labeled 0 leading to $w(0,b)$ and an edge labeled 1 leading to $w(2,\#)$. More systematically, one computes successively each row in the next table, where the last three columns are computed immediately from the previous two and are only indicated to facilitate understanding the procedure. In the second and third columns of a row (i,a) we indicate, for each letter, the first position in w "after" position i and before the first occurrence of a to the right of i, where the letter is found, if such a position exists. The term "after" here has to be understood in the sense that if the position i is found within the base of an ω-power, then we are allowed to read again the whole base before leaving the ω-power. Identifying all states labeled ε,

(position, letter)	a	b	label	left	right
$(0,\#)$	1	2	b	$(0,b)$	$(2,\#)$
$(0,b)$	1	–	a	$(0,a)$	$(1,b)$
$(2,\#)$	3	2	a	$(2,a)$	$(3,\#)$
$(0,a)$	–	–	ε	–	–
$(1,b)$	–	–	ε	–	–
$(2,a)$	–	2	b	$(2,b)$	$(2,a)$
$(3,\#)$	1	2	b	$(3,b)$	$(2,\#)$
$(2,b)$	–	–	ε	–	–
$(3,b)$	1	–	a	$(3,a)$	$(1,b)$
$(3,a)$	–	–	ε	–	–

the resulting R-automaton is given in Figure 5 where we add as an index to the state-label the pair (i,a) which determines the state. The minimization of this automaton identifies two pairs of states according to the equalities $w(0,b) = w(3,b)$, $w(0,\#) = w(3,\#)$ and produces exactly the R-automaton of Figure 3. We stress that the fact that we do get these equalities of ω-terms over R comes from automata manipulations rather than by invoking again the word problem.

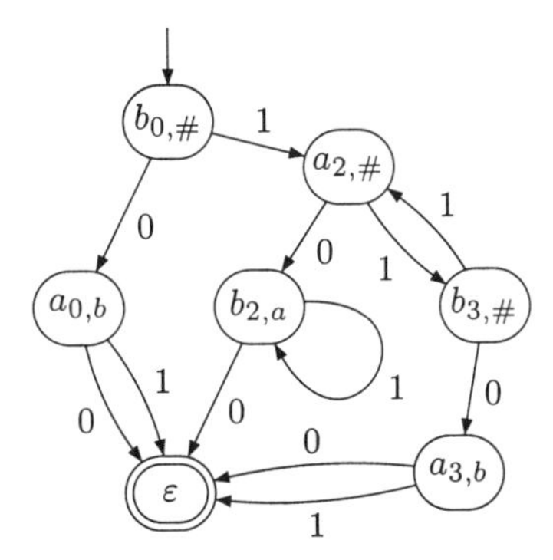

Figure 5. The computed R-automaton of the pseudoword $(ab^{\omega}a)^{\omega}$

For the minimization procedure for finite deterministic automata, there is an algorithm due to Hopcroft [13] which works in $O(\ell s \log s)$-time (see [15] for a recent complete complexity analysis), where ℓ is the number of letters and s stands for the number of states. Combining with Theorem 3.2, we obtain the following result.

Theorem 3.3. *There is an algorithm which computes in time $O(m^2 n \log(mn))$ the minimal* R*-automaton recognizing $L(w)$ for any given ω-term w, where $m = |c(w)|$ and $n = |w|$.*

Since R-automata are so special one should be able to do better. The R-automaton of a word is acyclic, and in this case, we already have a linear algorithm due to Revuz [17]. If there are cycles (that is, if we started from a term involving at least one ω-power) then we will adapt this algorithm to completely fold a finite R-automaton in time $O(mn)$.

The additional difficulty for minimizing R-automata comes from the cycles: in the acyclic version, a height function measuring the longest path starting from each state is computed at the beginning of the algorithm. The situation is then simple, in that the minimization can only identify states having the same height.

If we do have cycles, such paths can be infinite. But, due to the specific stucture of R-automata, all cycles are disjoint and 1-labeled. For that reason we can, after a preprocessing phase, treat separately the states belonging to cycles and the other states.

The analog of Revuz's height function is the level, obtained by letting edges in cycles have weight zero. Proceeding then bottom-up, the preprocessing stage consists in rolling paths coming to a cycle, if this does not change the language. Consider for example a usual automaton with a single initial state q_0, one simple path from q_0 to q_1 labeled v and one cycle around q_1 labeled u, as pictured in Figure 6. If $v = u'u^r$ with $r \geq 0$ and u' a suffix

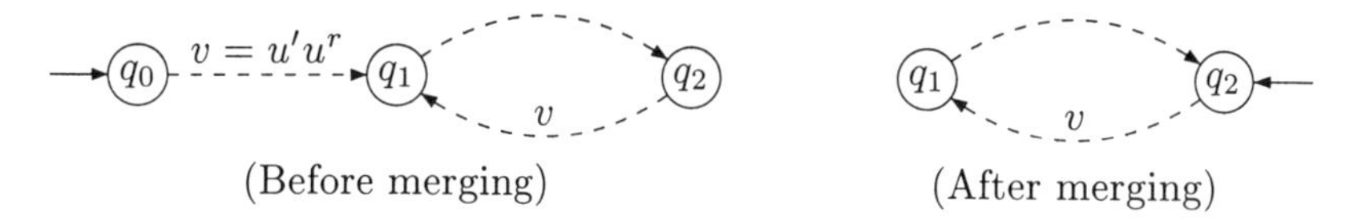

Figure 6. Merging a path ending in a cycle

of u, then we do not change the language by rolling the simple path around the cycle, that is, by only retaining the cycle and choosing as the new initial state the unique state q_2 of the cycle such that $q_2 \cdot v = q_1$. The next step for cycles is to minimize them one by one, *i.e.*, to find the primitive roots of their labels. We then identify, at one level, all equal cycles and all states which do not belong to a cycle. This is essentially Revuz's algorithm.

Finally, notice that rolling paths around cycles may change the level of states that lie above them. We therefore have to recompute this level. The recomputation is local, we just update correctly levels of states we are about to treat.

Here is a more detailed, but informal, commented sketch of the algorithm (see [7] which will be devoted to a detailed presentation and a correctness proof):

(1) Given a finite R-automaton $\mathcal{A}$, compute its cycle structure. For this one can use for instance Tarjan's algorithm [20] which computes the strongly connected components of a graph.
(2) Compute an initial *level* function that measures, for each state, the maximum weight of a path to the terminal state, assigning weight 0 to edges in cycles and weight 1 to all other edges. This can be done efficiently by a simple traversal of the graph that is further used to assign a level value to each edge that is not in a cycle, a value which is initialized to the level of the end state plus 1. Both these level functions will be updated in the main loop of the algorithm as a result of rolling paths with all edges labeled 1 around cycles to which they lead. The level of edges serves as a mechanism to propagate to higher levels changes coming from identifications done at lower levels.
(3) From this point on, we construct successive equivalence relations on sets of states which are approximations to the congruence on $\mathcal{A}$

whose quotient determines the minimized R-automaton. We do so level by level, at each stage suitably joining elements into equivalence classes. The first step consists in putting the final state into its own class.

(4) This is the main cycle in the algorithm. Proceed by increasing level $n \geq 1$, as in the following loop. At the end of level n, all vertices processed in it will have level-value n and they will all be assigned to an equivalence class, which remains unaltered at higher levels.

For each non-terminal state v, denote by $0_v, 1_v$ the edges starting from v labeled $0, 1$, respectively. If $\mathsf{level}(0_v) \leq n$, then let $\zeta(v)$ denote the pair consisting of label $\lambda(v)$ of the state v and the class $[v0]$ containing the state at the end of the edge 0_v.

(a) Call subroutine $\mathsf{Level}(n)$ which returns the list S of states whose current level-value is n.
(b) For each vertex in S which lies in a cycle, put it in its own singleton class.
(c) Roll 1-labeled paths leading to cycles in S around the corresponding cycles by testing for each successive vertex v away from the cycle whether $\zeta(v)$ is defined and whether it coincides with $\zeta(w)$, where w is the unique state in the cycle such that for all sufficiently large k, $v1^k = w1^k$. In the negative case, do not proceed with the test for states u such that $u1 = v$. In the affirmative case, add v to the class of w, as a result of which the edge 1_v becomes a cycle-edge and thus no longer contributes to the level function; this leads us to reduce $\mathsf{level}(v)$ to n and $\mathsf{level}(e)$ to $n+1$ for every edge e which ends at state v.
(d) Since the previous step may change the level functions, lowering to level n states that were previously considered at higher levels, we call subroutine $\mathsf{Level}(n)$ again. This will return an updated value for S which contains the previous value since the previous step only affects the level-values of states at higher levels.
(e) For each cycle C in S, do the following steps which suitably merges all equivalence classes of states in the cycle according to their identification in the minimized R-automaton:

i. compute the (circular) word W_C whose letters are the successive $\zeta(w)$ with w in C;

ii. compute the primitive W'_C root of W_C; this can be done by computing the shortest border[a] u of W_C such $u^{-1}W_C$ is also a border; that this computation can be performed in linear time in terms of the length of W_C follows from the fact that the list of all borders can be computed within this time-complexity [11];

iii. compute the minimal conjugate V_C of W'_C; this can be done in linear time in terms of the length W'_C [9, 19];

iv. merge classes of states in C according to the periodic repetition of V_C in W_C.

(f) To merge classes of states in different cycles C of S start by lexicographically sorting the words V_C using bucket sort [10]. This determines in particular which cycles have the same words $V = V_C$ and their classes associated with corresponding positions in V are merged.

(g) To merge the remaining states v in S into classes start by lexicographically sorting (by a bucket sort) the associated triples $(\lambda(v), [v_1], [v_2])$, where v_1, v_2 denote the ends of the edges $0_v, 1_v$, respectively. As in the previous step, this determines in particular which states have the same associated triples, and those that do are merged into the same class.

(h) Increment n by 1 and proceed until a subroutine call returns the empty list.

To complete the description of the algorithm, it remains to indicate what the subroutine $\mathsf{Level}(n)$ does. It starts by updating the level-value of the beginning state v of each edge e such that $\mathsf{level}(e) = n$ according to the formula

$$\mathsf{level}(v) = \begin{cases} \max\{\mathsf{level}(e), \mathsf{level}(f)\} & \text{if } e \text{ is not in a cycle} \\ & \quad \text{and } \{e, f\} = \{0_v, 1_v\} \\ \max_x \mathsf{level}(x) & \text{otherwise} \end{cases}$$

where the second maximum runs over all edges x with label 0 which start in the cycle that contains v. Then return all states for which the new level-value is n.

Theorem 3.4. *The above algorithm minimizes a given* R*-automaton with* s *states in time* $O(s)$.

[a] A border of a word w is a word which is both a prefix and a suffix of w.

Taking into account Corollary 2.3, this gives our solution of the word problem:

Theorem 3.5. *The word problem for ω-terms over R can be solved in time $O(mn)$, where m is the number of letters involved and n is the maximum of the lengths of the ω-terms to be tested.*

4. Equations for ω-terms

Consider the set Σ consisting of the following identities:

$$(xy)^\omega x^\omega = (xy)^\omega x = x(yx)^\omega = (xy)^\omega$$
$$(x^\omega)^\omega = x^\omega$$
$$(x^r)^\omega = x^\omega \quad (r \geq 2)$$

It is immediately checked that R satisfies these identities (viewed as pseudoidentities).

Theorem 4.1. *The set Σ is a basis for the variety R^κ of ω-semigroups generated by R.*

The proof of this result depends on the construction of canonical forms presented in the next section. There we will show that the canonical form $\mathsf{cf}(w)$ of an ω-term w uniquely determines the value of w over R and that the equality $w = \mathsf{cf}(w)$ can be formally deduced from Σ. Hence, if two ω-terms v and w have the same value over R, then they have the same canonical form and therefore the identity $v = w$ may be formally deduced from Σ. This will show that indeed Σ is a basis of identities for R^κ.

Using the completeness theorem for equational logic, it is now easy to deduce that

Corollary 4.2. *The variety R^κ is not finitely based.*

The proof of Corollary 4.2 consists in verifying that the semigroup

$$S_p = \langle a, e, f : a^p = 1,\ ea = ef = e^2 = e,\ fa = fe = f^2 = f,$$
$$ae = e,\ af = f\rangle,$$

where p is a positive integer, with the unary operation σ defined by taking

$$\sigma(e) = e,\ \sigma(f) = f,\ \sigma(1) = e,\ \sigma(a^k) = f\ (k \in \mathbb{Z} \setminus p\mathbb{Z})$$

as the ω-power, satisfies the identities in the basis Σ for r relatively prime with p, but fails $(x^p)^\omega = x^\omega$.

5. Canonical forms

From a finite R-automaton one can read off an ω-term in a canonical way. The definition is recursive in the number of states:

- for the trivial R-automaton, in which there is only one state, take the empty pseudoword 1;
- for an R-automaton in which the initial state r, labeled a, is not the end of an edge labeled 1, take uav where u and v are respectively the ω-terms canonically associated with the R-subautomata obtained by changing the initial state respectively to $r0$ and $r1$;
- for an R-automaton in which the initial state r, labeled a, is the end of an edge labeled 1, take $(uav)^\omega$ where u and v are respectively the ω-terms canonically associated with the R-subautomata obtained by changing the initial state respectively to $r0$ and $r1$ and, in the second case, changing the end of the edge labeled 1 pointing to r to a terminal state.

We call the resulting ω-term for $\mathcal{A}(w)$ the *canonical form* of a given ω-term w and denote it $\mathsf{cf}(w)$. For example, since the minimal R-automaton of the ω-term $w = (ab^\omega a)^\omega$ is that given by Figure 3, the canonical form is $\mathsf{cf}(w) = (abb^\omega a)^\omega$.

A factor u of an ω-term w is *fringy* if there is a factorization $u = va$ with $c(u) = c(w)$ and a a letter which is not in $c(v)$. A Σ-*fringy decomposition* of an ω-term w is a finite sequence $w_1, \ldots, w_n$ of ω-terms such that each w_i is a fringy factor of $w_1 \cdots w_n$ and the identity $w = w_1 \cdots w_n$ is a consequence of Σ. The following technical result is used to compute the canonical form.

Proposition 5.1. *Let w be an ω-term.*

(1) *If $w = uav$, a is a letter, and the factor ua is fringy, then it is possible to deduce the identity $\mathsf{cf}(w) = \mathsf{cf}(u)\, a\, \mathsf{cf}(v)$ from Σ.*

(2) *If w admits a Σ-fringy decomposition, then the identity $\mathsf{cf}(w^\omega) = (\mathsf{cf}(w))^\omega$ is a consequence of Σ.*

(3) *Let w_0 and w be ω-terms such that*
   - $\mathsf{R} \not\models w = w^2$,
   - $c(w_0) \subseteq c(w)$,
   - *either* $c(w_0) \subsetneq c(w)$, *or* $\mathsf{R} \not\models w_0 = w_0^2$.

   Then, there are ω-terms u, v such that $uv = w$ is a consequence of Σ and there are positive integers r, s such that $w_0 w^r u$ and $(vu)^s$ admit Σ-fringy decompositions.

Rules (1) and (2) allow to "push" cf operators downwards in the term tree of an ω-term, while preserving equality modulo Σ. However, they only can be used starting from particular forms, namely where fringy factors occur. To make fringy factors appear, we use item (3). For instance, we may deduce from Σ the following identities:

$$\begin{aligned}
\mathsf{cf}((ab^\omega a)^\omega) &= \mathsf{cf}\big((ab \cdot b^\omega a)^\omega\big) && \text{(note that } ab \text{ and } b^\omega a \text{ are fringy)}\\
&= \big(\mathsf{cf}(ab \cdot b^\omega a)\big)^\omega && \text{by (2)}\\
&= \big(\mathsf{cf}(a) \cdot b \cdot \mathsf{cf}(b^\omega a)\big)^\omega && \text{by (1)}\\
&= (ab\,\mathsf{cf}(b^\omega)a)^\omega && \text{by (1)}\\
&= (ab\,\mathsf{cf}(b)^\omega a)^\omega && \text{by (2)}\\
&= (abb^\omega a)^\omega = (ab^\omega a)^\omega
\end{aligned}$$

In this example, fringy factors needed to apply Proposition 5.1 appear with almost no extra work. As another example, consider $w = (aabb)^\omega$, whose automaton and wrapped automaton computed by our algorithms are shown in Figure 7. Using $\mathcal{A}((aabb)^\omega)$ and the definition of the canonical form, we

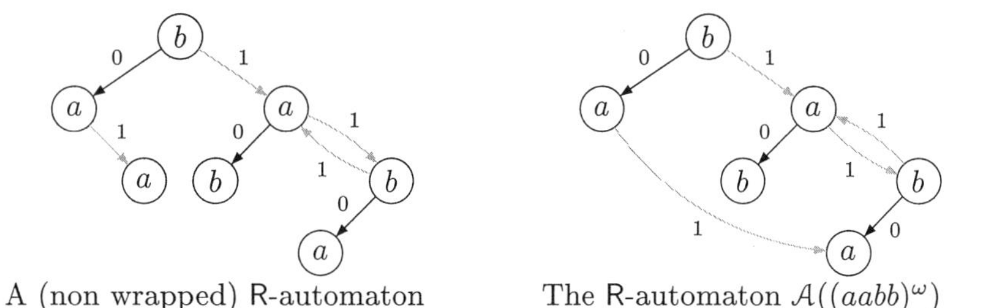

Figure 7. R-automata associated to $(aabb)^\omega$

get $\mathsf{cf}((aabb)^\omega) = aab(baab)^\omega$. On the other hand, one deduces from Σ, using Proposition 5.1:

$$\begin{aligned}
\mathsf{cf}((aabb)^\omega) &= \mathsf{cf}\big(aab(ba \cdot ab)^\omega\big) && \text{using } x(yx)^\omega = (xy)^\omega\\
&= aab \cdot \mathsf{cf}\big((ba \cdot ab)^\omega\big) && \text{by Prop. 5.1 (1)}\\
&= aab \cdot \big(\mathsf{cf}(ba \cdot ab)\big)^\omega && \text{by Prop. 5.1 (2)}\\
&= aab(baab)^\omega
\end{aligned}$$

The first equality $(aabb)^\omega = aab(ba \cdot ab)^\omega$ deduced by Σ makes fringy factors appear. Proposition 5.1 (3) essentially states that one can always produce these fringy factors when they are needed.

These examples illustrate how Proposition 5.1 may be used to prove the following theorem already announced in Section 4.

To prove that Σ implies $\mathsf{cf}(w) = w$, we start with $\mathsf{cf}(w)$. Then, using Σ, one may split ω-powers by bringing out of them fringy factors and push inside the canonical form "calculation" at the cost of taking a power of a conjugate of the original base of the ω-power, and thus reduce the number of nested ω-powers. For fringy factors, we also use Proposition 5.1 to bring out the last letter and thus reduce the content. In either type of step, a non-negative integer parameter is reduced and therefore this process terminates in a finite number of steps. Once no more canonical forms remain to be "computed", all transformations can be traced back without the cf operator using Σ to recover the ω-term w.

Theorem 5.2. *For every ω-term w, the identity $\mathsf{cf}(w) = w$ is a consequence of Σ.*

Note that we do not claim that the above procedure serves to compute the canonical form of a given ω-term. But, even if it would serve that purpose, the algorithm to solve the word problem obtained by such a computation followed by a test of graphical equality of the canonical forms would not be efficient. This is shown by the following example in which the canonical form computation has to take too long to perform.

Indeed the canonical form of an ω-term may be exponentially longer than the given ω-term. This is so for the sequence of ω-terms

$$w_k = (\cdots((a_1a_1b_1b_1)^\omega a_2a_2b_2b_2)^\omega \cdots a_ka_kb_kb_k)^\omega.$$

For instance, $k = 1$ corresponds to the ω-term $(aabb)^\omega$, whose canonical form was previously computed using $\mathcal{A}((aabb)^\omega)$. For $k = 2$, the automaton is more involved. By the same method, one can compute its canonical form:

$$\underbrace{aab(baab)^\omega ccd}_{u_0} \cdot \Big(\underbrace{daab(baab)^\omega c}_{u_1} \cdot \underbrace{cddaab}_{u_2} \cdot \underbrace{(baab)^\omega ccd}_{u_3}\Big)^\omega$$

where $a = a_1$, $b = b_1$, $c = a_2$, $d = b_2$. Here, u_0 is a fringy factor of this factorization, and u_1, u_2, u_3 are fringy factors of the external ω-power.

An ω-term w is *reduced* if it has no subterm of the form $(x^r)^\omega$ with $r > 1$, no subterm of the form $y^\omega z$, with $c(z) \subseteq c(y)$, and no subterm of the form $(xy^\omega z)^\omega$, where x, z may be empty and $c(xz) \subseteq c(y)$. Note that Σ may be used to deduce the equality of any ω-term with a shorter one in reduced form. Surprisingly, a canonical factorization is not necessarily reduced. For instance, the ω-term inside the ω-power of $\mathsf{cf}(w_2)$ is a square.

Coming back to the calculation of $\mathsf{cf}(w_k)$, one can show that, in general:

$$|w_k| = 4k, \ |c(w_k)| = 2k, \ |\mathit{cf}(w_k)| = (25 \cdot 3^{k-1} - 11)/2.$$

Another example is given by the sequence v_k defined by

$$v_k = (a_k(\cdots(a_2(a_1)^\omega)^\omega\cdots)^\omega.$$

Here we can compute $|v_k|$ to be k, and the number of states of $\mathcal{A}(v_k)$ to be $(k+2)(k+3)/2 - 5$, while the number of states of the R-automaton of v_k computed by our algorithm is $(k+2)(k+3)/2 - 3$. This shows that when we allow m and n to grow at equal rates, both the algorithm of Section 3 to compute a finite R-automaton for an ω-term and the minimization algorithm have optimal complexity rate.

It is also possible to have an exponential decrease in length in going from an ω-term to its canonical form, even for ω-terms for which there are no obvious reductions, as shown by Proposition 5.3 below.

Let us examine next why the first strategy that comes to mind to reduce ω-terms does not work. It would consist in taking rules that result from our proposed basis of identities for the variety of ω-semigroups generated by R by orienting the identities in the length-reducing direction:

r1. $(xy)^\omega x \to (xy)^\omega$
r2. $(xy)^\omega x^\omega \to (xy)^\omega$
r3. $(x^\omega)^\omega \to x^\omega$
r4. $(x^r)^\omega \to x^\omega \ (r \geq 2)$
r5. $x(yx)^\omega \to (xy)^\omega$

In general, for any set of reduction rules, its use is the following: whenever in a term w we find a subterm u which matches the pattern of the left side of a rule $\ell \to r$, then we may replace in w an occurrence of u by the resulting value of applying the rule to u to obtain a new ω-term w'; we then write $w \Rightarrow w'$. If none of the above reduction rules may be applied to an ω-term w then we say that w is *irreducible*.

Proposition 5.3. *Consider the sequence z_n, defined recursively by $z_0 = 1$, $z_{n+1} = (z_n a_n z_n)^\omega$, where the a_n are distinct letters. Then the z_n are irreducible ω-terms such that $|z_n| = 2^n - 1$ and $|\mathsf{cf}(z_n)| = n$.*

In particular, it follows from Proposition 5.3 that the system of reduction rules (r1)–(r5) cannot be confluent since it is length-reducing. One can try to apply the *Knuth-Bendix algorithm* by adding rules to obtain confluence without losing the Noetherian property. But, since our system is essentially infinite (not only because of the parameter $r \geq 2$, which does not bother much here but because x, y stand for arbitrary terms, that is they serve as a means of giving patterns for terms), this quickly becomes an unmanageable task.

6. Comparison with related work of Bloom and Choffrut

Bloom and Choffrut [8] have established results which bare some similarities with those reported in this paper. They consider posets labeled with elements of a finite nonempty alphabet under two operations:

- *series concatenation*: PQ extends the orders of each factor by declaring the elements in P to precede all elements in Q;
- ω-*power*: $P^\omega = \omega \times P$ under lexicographic order.

The finite or countable labeled posets over an alphabet A, considered up to isomorphism, constitute an algebra $P(A)$ which satisfies the following identities:

- $x(yz) = (xy)z$,
- $(xy)^\omega = x(yx)^\omega$,
- $(x^n)^\omega = x^\omega$ for $n \geq 1$.

The subalgebra generated by the singletons labeled with letters a from an alphabet A is denoted $W(A)$. Note that its elements are labeled ordinals of length less than ω^ω.

A *tail* of an ordinal α with labeling function f corresponding to an ordinal $\beta < \alpha$ is the unique ordinal γ such that $\beta + \gamma = \alpha$ with the labeling function $g(\delta) = f(\beta + \delta)$. A labeled ordinal is *tail-finite* if it has only finitely many tails.

The main results in the paper [8] are the following:

(1) A labeled ordinal belongs to $W(A)$ if and only if it has length less than ω^ω and it is tail-finite, if and only if it is defined by an ω-term.
(2) The variety V generated by the $P(A)$ is equal to the variety generated by the $W(A)$ and it is defined by the above identities. Moreover, $W(A)$ is free over A in this variety.
(3) The variety V is not finitely based.
(4) The equality of two ω-terms u_1, u_2 in V can be decided (with the help of *Choueka automata*) in $O(m^2 n_1^2 n_2^2)$-time, where n_i denotes the length of u_i and m the maximum of their heights in terms of nested ω-powers.

Our results are similar, so the natural question is if they are connected by some causal relationship. Our basis of identities for R^κ is obtained from the above basis for V by adding the identities

$$(x^\omega)^\omega = x^\omega, \quad (xy)^\omega x^\omega = (xy)^\omega x = (xy)^\omega.$$

In particular, we have the inclusion $\mathsf{R}^\kappa \subsetneq V$ which does not appear to be obvious taking into account the rather different types of generators considered for the two varieties. That the inclusion is proper is easy to show: for instance, the identity $x^\omega x = x^\omega$ fails in the algebra $P(A)$ since the ordinals $\omega + 1$ and ω labeled with only one letter are distinct.

Finally, since the two varieties are different, the solution of the word problem for the free algebra in one of the varieties is insufficient to solve the word problem for the other. A curious difference in the two algorithms is that the complexity of our algorithm depends on the size of the alphabet whereas that of Bloom and Choffrut does not depend directly on this parameter and instead depends on the height of nested ω-powers. Note however that for reduced ω-terms in the sense of Section 5, the indicated height is bounded by the number of letters which appear in the term.

Acknowledgments

The authors wish to thank the anonymous referees for their careful reading of a preliminary version of the paper and for their comments.

References

1. J. Almeida. Finite semigroups: an introduction to a unified theory of pseudovarieties. In G. M. S. Gomes, J.-E. Pin, and P. V. Silva, editors, *Semigroups, Algorithms, Automata and Languages*, pages 3–64, Singapore, 2002. World Scientific.
2. J. Almeida and A. Azevedo. The join of the pseudovarieties of $\mathcal{R}$–trivial and $\mathcal{L}$–trivial monoids. *J. Pure Appl. Algebra*, 60:129–137, 1989.
3. J. Almeida and B. Steinberg. On the decidability of iterated semidirect products and applications to complexity. *Proc. London Math. Soc.*, 80:50–74, 2000.
4. J. Almeida and P. G. Trotter. Hyperdecidability of pseudovarieties of orthogroups. *Glasgow Math. J.*, 43:67–83, 2001.
5. J. Almeida and P. Weil. Free profinite -trivial monoids. *Int. J. Algebra Comput.*, 7:625–671, 1997.
6. J. Almeida and M. Zeitoun. An automata-theoretic approach of the word problem for ω-terms over R. In preparation.
7. J. Almeida and M. Zeitoun. A linear time minimization algorithm for disjoint loop automata. In preparation.

8. S.L. Bloom and Ch. Choffrut. Long words: The theory of concatenation and ω-power. *Theor. Comp. Sci.*, 259:533–548, 2001.
9. K. S. Booth. Lexicographically least circular substrings. *Inform. Process. Lett.*, 10:240–242, 1980.
10. Thomas H. Cormen, Clifford Stein, Ronald L. Rivest, and Charles E. Leiserson. *Introduction to Algorithms.* McGraw-Hill Higher Education, 2001.
11. M. Crochemore and W. Rytter. *Text Algorithms.* The Clarendon Press Oxford University Press, New York, 1994. With a preface by Zvi Galil.
12. P. M. Higgins. Divisors of semigroups of order-preserving mappings of a finite chain. *Int. J. Algebra Comput.*, 5:725–742, 1995.
13. J. E. Hopcroft. An $n \log n$ algorithm for minimizing states in a finite automaton. In Z. Kohavi, editor, *Theory of machines and computations (Proc. Internat. Sympos., Technion, Haifa, 1971)*, pages 189–196, New York, 1971. Academic Press.
14. J. E. Hopcroft and J. D. Ullman. *Introduction to Automata Theory, Languages, and Computation.* Addison-Wesley, Reading, Mass., 1979.
15. T. Knuutila. Re-describing an algorithm by Hopcroft. *Theor. Comp. Sci.*, 250:333–363, 2001.
16. J.-E. Pin. *Varieties of Formal Languages.* Plenum, London, 1986. English translation.
17. D. Revuz. Minimisation of acyclic deterministic automata in linear time. *Theor. Comp. Sci.*, 92:181–189, 1992.
18. M. P. Schützenberger. Sur le produit de concaténation non ambigu. *Semigroup Forum*, 13:47–75, 1976.
19. Y. Shiloach. Fast canonization of circular strings. *J. Algorithms*, 2:107–121, 1981.
20. R. E. Tarjan. Depth first search and linear graph algorithms. *SIAM Journal on Computing*, 1(2):146–160, June 1972.
21. P. Tesson and D. Thérien. Diamonds are forever: the variety DA. In G. M. S. Gomes, J.-E. Pin, and P. V. Silva, editors, *Semigroups, Algorithms, Automata and Languages*, pages 475–499, Singapore, 2002. World Scientific.
22. D. Thérien and Th. Wilke. Over words, two variables are as powerful as one quantifier alternation. In *STOC '98 (Dallas, TX)*, pages 234–240. ACM, New York, 1999.

SOME RESULTS ON ČERNÝ TYPE PROBLEMS FOR TRANSFORMATION SEMIGROUPS

D. S. ANANICHEV and M. V. VOLKOV*

Department of Mathematics and Mechanics,
Ural State University,
620083 Ekaterinburg, Russia
E-mail: {Dmitry.Ananichev,Mikhail.Volkov}@usu.ru

It is known that several problems related to the longstanding Černý conjecture on synchronizing automata can be conveniently thought of as questions about various types of transformation semigroups. Within this framework, we consider such "Černý type problems" for semigroups of order preserving or order reversing transformations of a finite chain as well as semigroups of orientation preserving or orientation reversing transformations of a finite cycle.

1. Motivation and overview

Let $\mathscr{A} = \langle Q, \Sigma, \delta \rangle$ be a deterministic finite automaton (DFA), where Q denotes the state set, Σ stands for the input alphabet, and $\delta : Q \times \Sigma \to Q$ is the transition function defining an action of the letters in Σ on Q. The action extends in a unique way to an action $Q \times \Sigma^* \to Q$ of the free monoid Σ^* over Σ; the latter action is still denoted by δ. The automaton $\mathscr{A}$ is called *synchronizing* if there exists a word $w \in \Sigma^*$ whose action resets $\mathscr{A}$, that is, leaves the automaton in one particular state no matter which state in Q it started at: $\delta(q, w) = \delta(q', w)$ for all $q, q' \in Q$. Any word w with this property is said to be a *reset* word for the automaton.

The following picture shows an example of a synchronizing automaton with 4 states. The reader can easily verify that the word ab^3ab^3a resets the automaton leaving it in the state 2. With somewhat more effort one

*The authors acknowledge support from the Education Ministry of Russian Federation, grants E02-1.0-143 and 04.01.059, from the Russian Foundation of Basic Research, grant 01-01-00258, from the President Program of Leading Scientific Schools, grant 2227.2003.1, and from the INTAS through Network project 99-1224 'Combinatorial and Geometric Theory of Groups and Semigroups and its Applications to Computer Science'.

can also check that ab^3ab^3a is the shortest reset word for this automaton.

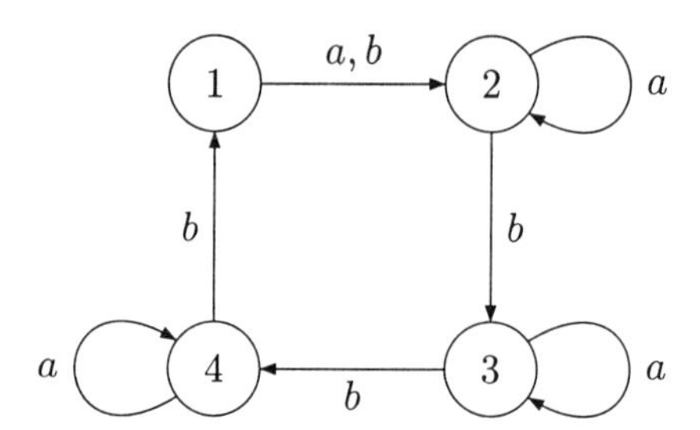

Figure 1. A synchronizing automaton

For a mathematician, the notion of a synchronizing automaton is pretty natural by itself but we would like to mention here that it is also of interest for various applications, for instance, for *robotics* or, more precisely, *robotic manipulation* which deals with part handling problems in industrial automation such as part feeding, fixturing, loading, assembly and packing (and which is therefore of utmost and direct practical importance). Of course, there exists vast literature about the role that synchronizing automata play in these matters (tracing back to Natarajan's pioneering papers [8, 9]) but we prefer to explain the idea of using such automata on the following simple example.

Suppose that one of the parts of a certain device has the following shape:

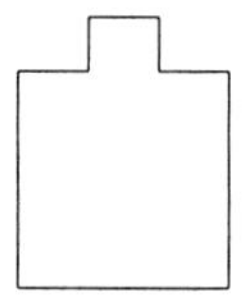

Figure 2. A polygonal detail

Such parts arrive at manufacturing sites in boxes and they need to be sorted and oriented before assembly. For simplicity, assume that only four initial orientations of the part shown on Fig. 2 are possible, namely, the following ones:

Figure 3. Four possible orientations

Further, suppose that prior the assembly the detail should take the "bump-left" orientation (the second one on Fig 3). Thus, one has to construct an orienter which action will put the detail in the prescribed position independently of its initial orientation.

Of course, there are many ways to design such an orienter but practical considerations favor methods which require little or no sensing, employ simple devices, and are as robust as possible. For our particular case, these goals can be achieved as follows. We put details to be oriented on a conveyer belt which takes them to the assembly point and let the stream of the details encounter a series of passive obstacles placed along the belt. We need two type of obstacles: high and low. A high obstacle should be high enough in order that any detail on the belt encounters this obstacle by its rightmost low angle (we assume that the belt is moving from left to right). Being curried by the belt, the detail then is forced to turn 90° clockwise. A low obstacle has the same effect whenever the detail is in the "bump-down" orientation (the first one on Fig. 3); otherwise it does not touch the detail which therefore passes by without changing the orientation.

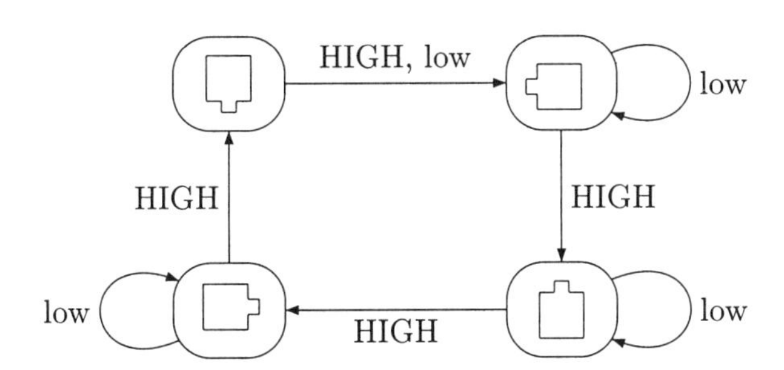

Figure 4. The action of the obstacles

The scheme on Fig. 4 summarizes how the aforementioned obstacles effect the orientation of the detail. The reader immediately recognizes the synchronizing automaton from Fig. 1. Remembering that its shortest reset word is the word ab^3ab^3a, we conclude that the series of obstacles

low–HIGH–HIGH–HIGH–low–HIGH–HIGH–HIGH–low

yields the desired sensorless orienter.

Another, perhaps, even more striking application of synchronizing automata is connected with *biocomputing*. Mastering a simple illustrating example from this area is not that easy (one need to be acquainted at least with some rudiments of molecular biology) so we just refer to recent experiments (see [2, 3]) in which DNA molecules have been used as both hardware and software for constructing finite automata of nanoscaling size. For instance, the authors of [3] have produced a "soup of automata", that is, a solution containing 3×10^{12} identical automata per μl. All these molecular automata can work in parallel on different inputs, thus ending up in different and unpredictable states. In contrast to an electronic computer, one cannot reset such a system by just pressing a button; instead, in order to synchronously bring each automaton to its "ready-to-restart" state, one should spice the soup with (sufficiently many copies of) a DNA molecule whose nucleotide sequence encodes a reset word.

Clearly, from the viewpoint of the above applications (as well as from mathematical point of view) the following questions are of importance:

(1) Given an automaton $\mathscr{A}$, how to determine whether or not $\mathscr{A}$ is synchronizing?
(2) Given an automaton $\mathscr{A}$ with the input alphabet Σ and a word w over Σ, how to check if w is a reset word of the minimum length for $\mathscr{A}$?
(3) Given a positive integer n, how long can be reset words for synchronizing automata with n states?

While there exist polynomial (in fact, cubic of the size of the state set of the automaton under inspection) algorithms that answer the first question, the second question is known to be NP-hard in general, see the recent survey [7] for details. The third question is the most intriguing as it remains open for almost 40 years. In 1964 Černý [4] produced for each n a synchronizing automaton with n states whose shortest reset word has length $(n-1)^2$ (the automaton on Fig. 1 is Černý's example for $n = 4$) and conjectured that these automata represent the worst possible case, that is, every synchronizing automaton with n states can be reset by a word of length $(n-1)^2$. By now this simply looking conjecture is arguably the most longstanding open problem in the theory of finite automata. The best upper bound achieved so far guarantees that for every synchronizing automaton with n states there exists a reset word of length $\frac{n^3-n}{6}$; this result is due to Pin [11] and is based on a combinatorial theorem by Frankl [6].

Observe that Černý's conjecture can be easily restated in terms of the semigroup $\mathcal{T}_n$ of all full single-valued transformations of the set $\{1,\dots,n\}$. Indeed, the fact that an automaton with n states is synchronizing amounts to saying that its transition semigroup (which can be identified with a subsemigroup in $\mathcal{T}_n$) contains a constant mapping, that is, an element of the kernel K of $\mathcal{T}_n$. Therefore Černý's conjecture can be reformulated as follows: for each subset $\Gamma \subseteq \mathcal{T}_n$ such that $\langle\Gamma\rangle \cap K \neq \varnothing$, one has $\Gamma^{(n-1)^2} \cap K \neq \varnothing$. (Here $\langle\Gamma\rangle$ stands for the subsemigroup generated by Γ while Γ^k denotes the set of all products of k elements of Γ.) This point of view naturally suggests placing the original conjecture in a somewhat broader context.

Suppose that we have a series of "natural" finite transformation semigroups $\mathcal{S}_n$ associated with each positive integer n. (In general, transformations in $\mathcal{S}_n$ can be partial and/or multi-valued and they can be defined on a certain structure determined by n—say, on the n-dimensional vector space over a fixed finite field—rather than on the set $\{1,\dots,n\}$ itself.) Let K denote the kernel of $\mathcal{S}_n$. Clearly, for each subset $\Gamma \subseteq \mathcal{S}_n$ such that $\langle\Gamma\rangle \cap K \neq \varnothing$, there exists the least positive integer $k = k(\Gamma)$ such that $\Gamma^k \cap K \neq \varnothing$. By the *Černý type problem* (ČTP, for short) for the series $\mathcal{S}_n$ we understand the problem of determining the maximum value of the numbers $k(\Gamma)$ as a function of n (this function will be called the *Černý function* for the series $\mathcal{S}_n$). For example, we may speak of the ČTP for the semigroups of all binary relations over the set $\{1,\dots,n\}$—it is easy to see that being retold in the automata-theoretic terms, this problem may be naturally thought of as the Černý problem for non-deterministic finite automata, see [10].

In this paper we consider the ČTP for certain series of finite transformation semigroups which have been intensively studied from various viewpoints over the last decade, namely, for semigroups of order preserving (or order reversing) transformations of a finite chain and semigroups of orientation preserving (or orientation reversing) transformations of a finite cycle. It turns out that the compatibility of transformations with a linear or cyclic order on the base set simplifies the situation and allows one to precisely determine the corresponding Černý functions.

2. Semigroups of orientation preserving or orientation reversing transformations

In this section we assume that the domain of all transformations under consideration is the set $\{1, 2, \ldots, n\}$ equipped with the natural cyclic order

$$1 \prec 2 \prec \cdots \prec n-1 \prec n \prec 1 \qquad (1)$$

(here $k \prec \ell$ means that ℓ immediately follows k). If $i_1, i_2, \ldots, i_m$ are numbers in $\{1, 2, \ldots, n\}$, we call the sequence $i_1, i_2, \ldots, i_m$ *cyclic* if, after removal of possible adjacent duplicate numbers, it is a subsequence of a cyclic shift of the sequence $1, 2, \ldots, n$. In a slightly more formal language, we may say that $i_1, i_2, \ldots, i_m$ is a cyclic sequence if there exists no more than one index $t \in \{1, \ldots, m\}$ such that $i_{t+1} < i_t$ where i_{m+1} is understood as i_1 and $<$ stands for the usual strict linear order on $\{1, 2, \ldots, n\}$. This version of the definition makes apparent that every subsequence of a cyclic sequence is cyclic. We say that a number j is *between* two numbers i, k if the sequence i, j, k is cyclic.

A sequence $i_1, i_2, \ldots, i_m$ is called *anti-cyclic* if its reverse $i_m, \ldots, i_2, i_1$ is cyclic; in other words, there exists no more than one index $t \in \{1, \ldots, m\}$ such that $i_t < i_{t+1}$. Of course, a sequence may well be neither cyclic nor anti-cyclic: for instance, any 4-element sequence of the form i, j, i, k with $j, k \neq i$ has neither of the two properties.

A full transformation α of the set $\{1, 2, \ldots, n\}$ is said to be *orientation preserving* if the numbers $1\alpha, 2\alpha, \ldots, n\alpha$ form a cyclic sequence. Similarly, α is *orientation reversing* if the sequence $1\alpha, 2\alpha, \ldots, n\alpha$ is anti-cyclic. All orientation preserving transformations of the set $\{1, 2, \ldots, n\}$ form a subsemigroup in the full transformation semigroup $\mathcal{T}_n$; this subsemigroup is denoted by $\mathcal{OP}_n$. If one adds to $\mathcal{OP}_n$ all orientation reversing transformations of $\{1, 2, \ldots, n\}$, one gets a bigger subsemigroup in $\mathcal{T}_n$ denoted by $\mathcal{OR}_n$.

Now let $\mathscr{A}$ be a DFA with n states. We say that the automaton $\mathscr{A}$ is *orientable* if its states can be indexed by the numbers $1, 2, \ldots, n$ so that the transition semigroup of $\mathscr{A}$ becomes a subsemigroup in $\mathcal{OP}_n$. For instance, the automaton in Fig. 1 is orientable and, more generally, the aforementioned examples from [4] are orientable automata. Therefore, the lower bound for the Černý function for the semigroups $\mathcal{OP}_n$ is $(n-1)^2$ as it is in the case of $\mathcal{T}_n$. Eppstein [5] has proved that, conversely, every orientable synchronizing automaton with n states has a reset word of length $(n-1)^2$. In our terminology, Eppstein's result can be stated as follows:

Theorem 2.1. *The Černý function for the semigroups of all orientation preserving transformations of* $\{1, 2, \ldots, n\}$ *is equal to* $(n-1)^2$.

Eppstein's interest in orientable automata (which he called *monotonic*) was motivated by the robotics applications of synchronizing automata. Indeed, in the problem of sensor-free orientation of polygonal parts one deals with solid bodies whence only those transformations of polygons are physically meaningful that preserve relative location of the faces of these polygons. It was observed already by Natarajan [8] that in the "flat" case (when the polygonal parts do not leave a certain plane, say, the surface of a conveyer belt) this physical requirement leads precisely to orientation preserving transformations. However, in the physical world we are allowed to make spatial transformations as well. Consider, for instance, the detail shown on Fig. 2. We can arrange two conveyer belts so that when the detail slides from one belt onto the other, it turns 180° around its horizontal axis. This adds a new transformation to the ones shown in Fig. 4 so that we get the following picture (in which the new transformation is called *turn*):

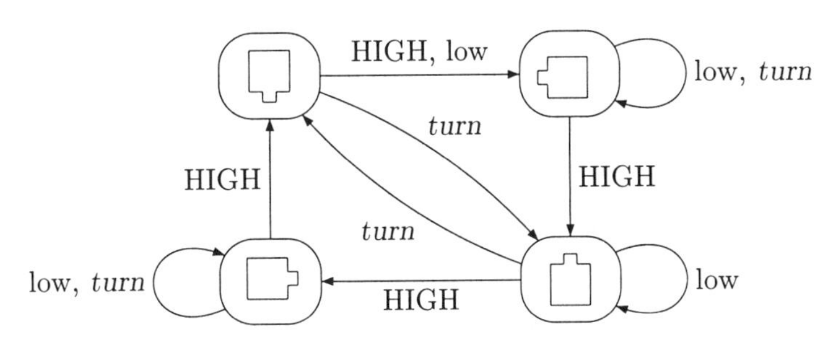

Figure 5. Adding the rotation around the horizontal axis

The reader can readily verify that "upgrading" the synchronizing automaton from Fig. 1 this way leads to an automaton with a shorter reset word, and hence, to a more compact sensorless orienter, namely,

low–HIGH–HIGH–HIGH–low–*turn*–low.

One can show that enhancing orientable automata by adding a bunch of physically meaningful "spatial" transformations precisely corresponds to extending subsemigroups of $\mathcal{OP}_n$ to subsemigroups of $\mathcal{OR}_n$. This suggests considering automata whose transition semigroups are (under a suitable numbering of states) subsemigroups in $\mathcal{OR}_n$. We call such automata *weakly orientable.* (For instance, the automaton shown in Fig. 5 is weakly orientable though not orientable.) Thus, the ČTP for the semigroups $\mathcal{OR}_n$ is nothing but the Černý problem for weakly orientable automata.

Using Eppstein's method, it is not too hard to obtain the following analogue of Theorem 2.1:

Theorem 2.2. *The Černý function for the semigroups of all orientation preserving or orientation reversing transformations of* $\{1, 2, \ldots, n\}$ *is equal to* $(n-1)^2$.

One gets the lower bound $(n-1)^2$ for the Černý function for $\mathcal{OR}_n$ for free because each orientable automaton is also weakly orientable. In contrast, no upper bound for this function can be extracted from Theorem 2.1 because the "orientable reduct" of a weakly orientable synchronizing automaton need not be synchronizing. Here is a simple example:

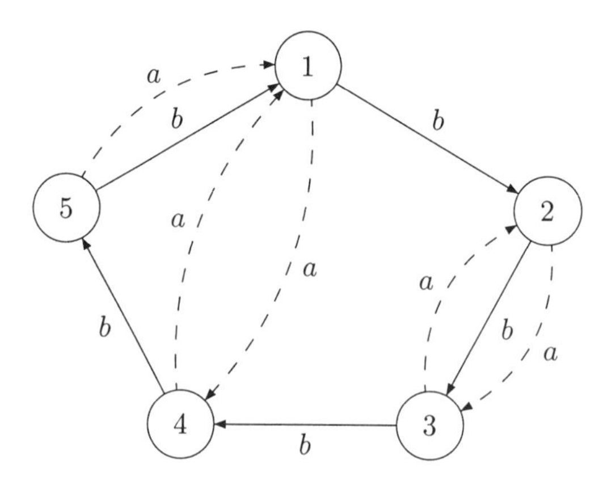

Figure 6. A weakly orientable synchronizing automaton

Indeed, the letter b acts here as an orientation preserving transformation (with respect to the cyclic order defined by the indicated numbering of the states) while the letter a acts as an orientation reversing transformation so the automaton is weakly orientable. It is easy to check that the word $(aba)^3$ resets the automaton but if one omits the letter a then the remaining orientable automaton obviously is not synchronizing.

Our proof that the Černý function for $\mathcal{OR}_n$ does not exceed $(n-1)^2$ closely follows Eppstein's proof for $\mathcal{OP}_n$. However, for the reader's convenience, we provide a self-contained presentation that hopefully can be understood without studying [5].

By an *interval* $[i,j]$ we mean an arbitrary pair of numbers in the set $\{1,2,\ldots,n\}$ while the set $\overline{[i,j]}$ represented by this interval $[i,j]$ consists of all numbers between i and j, including the endpoints i and j. We notice that by the definition $[2,1]$, $[3,2]$, ..., $[n,n-1]$, $[1,n]$ are treated as different intervals even though they all represent the same set $\{1,2,\ldots,n\}$; however, every other set which can be represented as an interval has exactly one such representation.

As usual, for a transformation $\alpha \in \mathcal{T}_n$ and a subset $M \subseteq \{1,2,\ldots,n\}$ we consider the set $M\alpha^{-1} = \{i \mid i\alpha \in M\}$.

Lemma 2.1. *If a subset $M \subseteq \{1,2,\ldots,n\}$ can be represented as an interval and $\alpha \in \mathcal{OR}_n$, then the set $M\alpha^{-1}$ can also be represented as an interval.*

Proof. If $M\alpha^{-1} = \{1,2,\ldots,n\}$, we have nothing to prove. If $M\alpha^{-1}$ is a proper subset of the set $\{1,2,\ldots,n\}$, then so is M, whence, as discussed above, M has a unique representation as $M = \overline{[i,j]}$ for some interval $[i,j]$. Let $[k_1,k_2]$ be a maximal (with respect to the size of the set $\overline{[k_1,k_2]}$) interval such that $\overline{[k_1,k_2]} \subseteq M\alpha^{-1}$. Then $i_1 = k_1\alpha \in M$ and $i_2 = k_2\alpha \in M$ while $j_1 = \ell_1\alpha \notin M$ and $j_2 = \ell_2\alpha \notin M$ where $\ell_1 \prec k_1$ and $k_2 \prec \ell_2$ in the order (1). If $\overline{[k_1,k_2]} \neq M\alpha^{-1}$ then there exists a number k_3 between ℓ_2 and ℓ_1 such that $i_3 = k_3\alpha \in M$. We see that the sequence $1\alpha, 2\alpha, \ldots, n\alpha$ contains as a subsequence a cyclic shift of i_1, i_2, j_2, i_3, j_1. Since i_1, i_2, i_3 are between i and j whereas j_1, j_2 are not, the latter sequence is easily seen to be neither cyclic nor anti-cyclic. This contradicts α being either an orientation preserving or orientation reversing transformation. $\square$

Now, for each DFA $\mathscr{A} = \langle\{1,2,\ldots,n\},\Sigma,\delta\rangle$ whose transition semigroup is a subsemigroup in $\mathcal{OR}_n$, we construct its *interval automaton* $\mathscr{I} = \langle Q,\Sigma,\delta'\rangle$. The state set Q of the interval automaton consists of all intervals together with the extra state denoted by ∞; thus, $|Q| = n^2+1$. The transition function $\delta' : Q \times \Sigma \to Q$ is defined as follows. First of all, $\delta'(\infty,x) = \infty$ for each letter $x \in \Sigma$. Now consider an arbitrary interval $[i,j] \in Q$ and an arbitrary letter $x \in \Sigma$; let $k = \delta(i,x)$, $\ell = \delta(j,x)$. If $k = \ell$, we set

$$\delta'([i,j],x) = \begin{cases} [k,k] & \text{if } \delta(h,x) = k \text{ for all } h \text{ between } i \text{ and } j, \\ \infty & \text{if } \delta(h,x) \neq k \text{ for some } h \text{ between } i \text{ and } j. \end{cases}$$

If $k \neq \ell$, then we set

$$\delta'([i,j],x) = \begin{cases} [k,\ell] & \text{if the transformation } \delta(\_,x) \text{ is orientation preserving,} \\ [\ell,k] & \text{if the transformation } \delta(\_,x) \text{ is orientation reversing.} \end{cases}$$

Fig. 7 shows the interval automaton built for the weakly orientable automaton presented on Fig. 6. In order to keep the picture readable we have "opened" the arrows going from $[5,4]$ to $[1,5]$, from $[5,3]$ to $[1,4]$, etc. The reader may imagine that these arrows go around the page and only their beginnings and their ends are seen on the front side of the page.

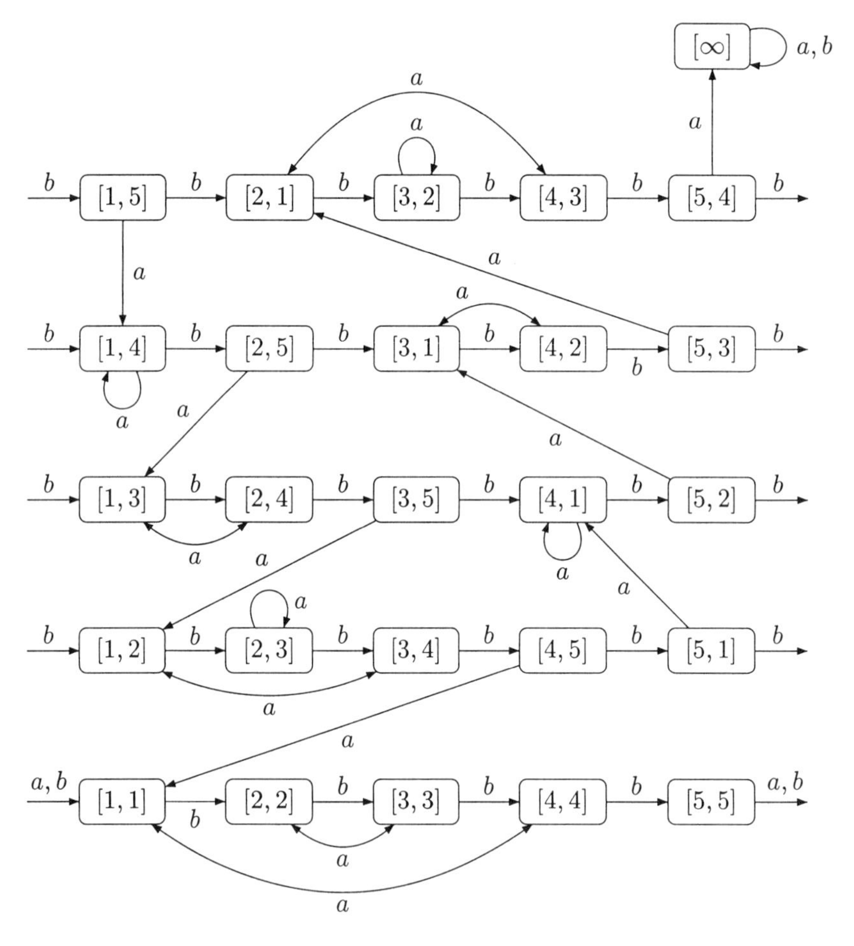

Figure 7. The interval automaton for the automaton from Fig. 6

We shall need the following simple observation:

Lemma 2.2. *If $x \in \Sigma$ and an interval $[i,j] \in Q$ satisfy $\delta'([i,j],x) = \infty$ and $\delta(i,x) = k$, then $\delta'([j,i],x) = [k,k]$.*

For instance, in the automaton shown on Fig. 7 one has $\delta([5,4],a) = \infty$ and, in the accordance with Lemma 2.2, $\delta([4,5],a) = [1,1]$.

Proof. Let α denote the transformation $\delta(\_,x)$. By the definition of the transition function δ', the equality $\delta'([i,j],x) = \infty$ means that $i\alpha = j\alpha = k$ but $h\alpha \neq k$ for some h between i and j. If, besides that, $\delta'([j,i],x) = \infty$, we also have $g\alpha \neq k$ for some g between j and i. But then the sequence $1\alpha, 2\alpha, \ldots, n\alpha$ has as a subsequence the "two-hump" sequence $k, h\alpha, k, g\alpha$ which, as we observed at the beginning of this section, is neither cyclic nor anti-cyclic. This means that the transformation α is neither orientation preserving nor orientation reversing, a contradiction. Thus, $g\alpha = k$ for all g between j and i which means that $\delta'([j,i],x) = [k,k]$. □

As usual, we extend the function δ' to $Q \times \Sigma^*$. Inducting on the length of $w \in \Sigma^*$ one can easily verify the following properties of this extension:

$\delta'(\infty, w) = \infty$;
if $k = \delta(i,w) \neq \ell = \delta(j,w)$, then

$$\delta'([i,j],w) = \begin{cases} [k,\ell] & \text{if the transformation } \delta(\_,w) \text{ is orientation preserving,} \\ [\ell,k] & \text{if the transformation } \delta(\_,w) \text{ is orientation reversing;} \end{cases}$$

if $\delta(i,w) = \delta(j,w) = k$, then either $\delta'([i,j],w) = [k,k]$ or $\delta'([i,j],w) = \infty$.

From this observation we immediately obtain

Lemma 2.3. *If $w \in \Sigma^*$ and $[i,j] \in Q$ satisfy $\delta'([i,j],w) \neq \infty$, then*

$$\delta(\overline{[i,j]},w) \subseteq \overline{\delta'([i,j],w)}.$$

Given two intervals, $[i,j]$ and $[k,\ell]$, we say that $[i,j]$ is a *subinterval* of $[k,\ell]$ if the sequence k,i,j,ℓ is cyclic.

Lemma 2.4. *If $w \in \Sigma^*$ and $[k,\ell] \in Q$ satisfy $\delta'([k,\ell],w) \neq \infty$, then for any subinterval $[i,j]$ of $[k,\ell]$, $\delta'([i,j],w) \neq \infty$ and $\delta'([i,j],w)$ is a subinterval of $\delta'([k,\ell],w)$.*

Proof. We induct on the length of the word w. The induction base (when w is empty) is obvious. Let $w = vx$ for some word $v \in \Sigma^*$ and some letter $x \in \Sigma$. Clearly, $\delta'([k,\ell],w) \neq \infty$ implies $\delta'([k,\ell],v) \neq \infty$, whence the induction assumption applies to the word v. Denoting $I = \delta'([i,j],v)$ and $J = \delta'([k.\ell],v)$, we then conclude that I is a subinterval of J and it remains to check that $\delta'(I,x)$ is a subinterval of $\delta'(J,x)$. In other words, we have shown that it suffices to prove the lemma for the case when w is just a single letter.

Thus, suppose that $w \in \Sigma$. If $\delta(i,w) \neq \delta(j,w)$ or if $\delta(i,w) = \delta(j,w) = \delta(h,w)$ for all h between i and j, then the claim immediately follows from the definition of the function δ'. Consider the situation when $\delta(i,w) = \delta(j,w) = m$ but $\delta(h,w) \neq m$ for some h between i and j. By Lemma 2.2, we then have $\delta(g,w) = m$ for all g between j and i. Since the sequence k,i,j,ℓ is cyclic, so is the sequence j,ℓ,k,i whence we see that both k and ℓ are between j and i. Therefore $\delta(k,w) = \delta(\ell,w) = m$. On the other hand, $\delta(h,w) \neq m$ and h is between k and ℓ. This means that $\delta'([k,\ell],w) = \infty$, a contradiction. □

Lemma 2.5. *Let* $\mathscr{A} = \langle\{1,2,\ldots,n\},\Sigma,\delta\rangle$ *be a weakly orientable automaton,* $w \in \Sigma^*$, $\alpha = \delta(\_,w)$. *Then for each* $k \in \{1,2,\ldots,n\}$ *there is an interval* $[i,j] \in Q$ *representing the set* $k\alpha^{-1}$ *and such that* $\delta'([i,j],w) = [k,k]$.

Proof. Again, we induct on the length of w with the induction base being obvious. Now let $w = xv$ for some letter $x \in \Sigma$ and some word $v \in \Sigma^*$ and denote the transformations $\delta(\_,x)$ and $\delta(\_,v)$ by respectively β and γ. Applying the induction assumption to the word v, we conclude that there exists an interval $[g,h]$ representing the set $k\gamma^{-1}$ and such that $\delta'([g,h],v) = [k,k]$.

First suppose that $\overline{[g,h]}\beta^{-1} \neq \{1,2,\ldots,n\}$. Then by Lemma 2.1 the set $\overline{[g,h]}\beta^{-1}$ can be (uniquely) represented by an interval $[i,j]$. If $\delta'([i,j],x) = \infty$, then by Lemma 2.2 $\delta'([j,i],x) = [\ell,\ell]$ where $\ell = i\beta$. This means that $q\beta = \ell$ for all q between j and i, and therefore, $\overline{[j,i]}\beta = \{\ell\} \subseteq \overline{[g,h]}$. On the other hand, the interval $[i,j]$ has been chosen to satisfy $\overline{[i,j]}\beta \subseteq \overline{[g,h]}$. Since

$$\overline{[i,j]} \cup \overline{[j,i]} = \{1,2,\ldots,n\},$$

we conclude that $\overline{[g,h]}\beta^{-1} = \{1,2,\ldots,n\}$. Thus, we have arrived at a contradiction. This shows that $\delta'([i,j],x) \neq \infty$ whence $\delta'([i,j],x)$ is a subinterval of the interval $[g,h]$. By Lemma 2.4 $\delta'([i,j],xv) = \delta'([i,j],w)$ is then a subinterval in $\delta'([g,h],v) = [k,k]$ whence $\delta'([i,j],w) = [k,k]$.

Now suppose that $\overline{[g,h]}\beta^{-1} = \{1,2,\ldots,n\}$. Then $\{1,2,\ldots,n\}\beta \subseteq \overline{[g,h]}$ and we can choose $[p,q]$ to be the least (with respect to the size of the set $\overline{[p,q]}$) subinterval of $[g,h]$ such that $\{1,2,\ldots,n\}\beta$ is still contained in $\overline{[p,q]}$. We notice that then both p and q are easily seen to belong to $\{1,2,\ldots,n\}\beta$. If $p = q$, then for any representation of the set $\{1,2,\ldots,n\}$ by an interval $[i,j]$, we have $\delta'([i,j],x) = [p,p]$ and hence $\delta'([i,j],w) = \delta'([p,p],v) = [k,k]$.

Finally, consider the case when $p \neq q$. Let first β be an orientation preserving transformation. The set $q\beta^{-1}$ is a proper subset of $\{1,2,\ldots,n\}$ and by Lemma 2.1 $q\beta^{-1}$ has a unique representation by an interval $[m,j]$. Let $j \prec i$ in the order (1). The element $i\beta$ is between p and q (since $\{1,2,\ldots,n\}\beta \subseteq \overline{[p,q]}$ and is not equal to q. If $i\beta \neq p$, then the sequence $q, i\beta, p$ is not cyclic, but some of its cyclic shifts occurs as a subsequence in $1\beta, 2\beta, \ldots, n\beta$. This contradicts our assumption that β is orientation preserving. Thus, $i\beta = p$ and $\delta'([i,j],x) = [p,q]$.

Similarly, if $p \neq q$ and β is an orientation reversing transformation, then consider $p\beta^{-1}$ which is also a proper subset of $\{1,2,\ldots,n\}$. Take an interval $[m,j]$ representing this subset and let $j \prec i$ in the order (1). Then, as above, $i\beta$ is between p and q but this time we have $i\beta \neq p$. If we suppose that $i\beta \neq q$ then the sequence $p, i\beta, q$ is not anti-cyclic but a cyclic shift of it appears as a subsequence in $1\beta, 2\beta, \ldots, n\beta$. This is impossible as β is assumed to be orientation reversing. Thus, $i\beta = q$ and we again get $\delta'([i,j],x) = [p,q]$.

Now by Lemma 2.4 $\delta'([i,j],w) = \delta'([i,j],xv) = \delta'([p,q],v)$ is a subinterval in $\delta'([g,h],v) = [k,k]$ whence $\delta'([i,j],w) = [k,k]$. □

We are ready to prove Theorem 2.2. Take a weakly orientable synchronizing automaton $\mathscr{A} = \langle\{1,2,\ldots,n\},\Sigma,\delta\rangle$. Let w be a reset word for $\mathscr{A}$. Then by the definition there is a number k such that $\delta(i,w) = k$ for each $i \in \{1,2,\ldots,n\}$. By Lemma 2.5 w then labels a path in the interval automaton $\mathscr{I} = \langle Q,\Sigma,\delta'\rangle$ of $\mathscr{A}$ from some interval representing the set $\{1,2,\ldots,n\}$ to the interval $[k,k]$. Conversely, if w is a word labelling a path in $\mathscr{I}$ from an interval representing $\{1,2,\ldots,n\}$ to an interval representing a singleton, then by Lemma 2.3 w should be a reset word for $\mathscr{A}$. Observe that any path with the above property avoids the sink state ∞; if we deal with a path of minimum length then the path visits each state in $Q \setminus \{\infty\}$ at most once. Moreover, the path visits exactly one interval representing $\{1,2,\ldots,n\}$ (of n) and exactly one interval of the form $[k,k]$ (of n). Therefore the path involves at most $|Q|-1-(n-1)-(n-1) = n^2-2n+2$

states whence the length of the corresponding reset word does not exceed $n^2 - 2n + 1 = (n-1)^2$ as required. □

3. Semigroups of order preserving or order reversing transformations

A transformation $\alpha \in \mathcal{T}_n$ is said to be *order preserving* if $i \leq j$ implies $i\alpha \leq j\alpha$ for all $i, j \in \{1, 2, \ldots, n\}$ where $\leq$ denotes the usual non-strict linear order on $\{1, 2, \ldots, n\}$. Similarly, α is called *order reversing* if $i \leq j$ implies $j\alpha \leq i\alpha$ for all $i, j \in \{1, 2, \ldots, n\}$. All order preserving transformations of $\{1, 2, \ldots, n\}$ form a subsemigroup $\mathcal{O}_n$ in the full transformation semigroup $\mathcal{T}_n$; adding to $\mathcal{O}_n$ all order reversing transformations of $\{1, 2, \ldots, n\}$, one obtains a bigger subsemigroup that is denoted by $\mathcal{D}_n$.

A DFA $\mathscr{A}$ with n states is said to be *monotonic* if its states can be indexed by the numbers $1, 2, \ldots, n$ so that the transition semigroup of $\mathscr{A}$ becomes a subsemigroup in $\mathcal{O}_n$. Since $\mathcal{O}_n$ is a subsemigroup in $\mathcal{OP}_n$, each monotonic automaton is orientable in the sense of the previous section whence Theorem 2.1 applies to this class of automata. Moreover, by using the interval automaton construction, one can easily obtain a slightly better upper bound on the minimum length of a reset word for a monotonic synchronizing automaton with n states, namely, $\binom{n}{2}$. (This upper bound follows from the fact that when applying the construction to a monotonic automaton one need considering only intervals $[i, j]$ with $i \leq j$.) The authors [1] have radically improved the latter bound by showing that each monotonic synchronizing automaton with n states can be in fact reset by a word of length $n - 1$ and this is already an exact bound. In the language adopted in the present paper, the result from [1] can be stated as follows:

Theorem 3.1. *The Černý function for the semigroups of all order preserving transformations of* $\{1, 2, \ldots, n\}$ *is equal to* $n - 1$.

In the flavour of our generalization of Theorem 2.1 presented in the previous section, it is natural to introduce *weakly monotonic* automata whose transition semigroups are (under a suitable numbering of states) subsemigroups in $\mathcal{D}_n$. Thus, the relationship between weakly monotonic and monotonic automata is basically the same as the way weakly orientable automata relate to orientable ones, and this analogy may suggest to the reader that the transformation semigroups $\mathcal{D}_n$ and $\mathcal{O}_n$ should have the same Černý function (as $\mathcal{OP}_n$ and $\mathcal{OR}_n$ do). The following series of examples shows that this is not the case. Namely, for each $\ell = 1, 2, \ldots$, we exhibit a weakly monotonic synchronizing automaton $\mathscr{A}_\ell$ with $n = 4\ell + 3$ states such that the minimum length of its reset words is $2n - 3$.

It is convenient to identify the state set Q_ℓ of the automaton $\mathscr{A}_\ell$ with the chain

$$-2\ell - 1 < -2\ell < \cdots < -1 < 0 < 1 < \cdots < 2\ell + 1. \tag{2}$$

The input alphabet Σ of $\mathscr{A}_\ell$ consists of three letters denoted a, b and c. The transition function $\delta : Q_\ell \times \Sigma \to Q_\ell$ is defined as follows:

$$\delta(j, a) = \begin{cases} -2\ell - 1 & \text{if } j < 0, \\ j - 1 & \text{if } j = 2k + 1 > 0, \\ j & \text{if } j = 2k \geq 0; \end{cases} \tag{3}$$

$$\delta(j, b) = \begin{cases} 2\ell + 1 & \text{if } j > 0, \\ 0 & \text{if } j = 0, \\ j & \text{if } j = 2k + 1 < 0, \\ j + 1 & \text{if } j = 2k < 0; \end{cases} \tag{4}$$

$$\delta(j, c) = -j. \tag{5}$$

The following picture shows the action of a and b for $\ell = 2$.

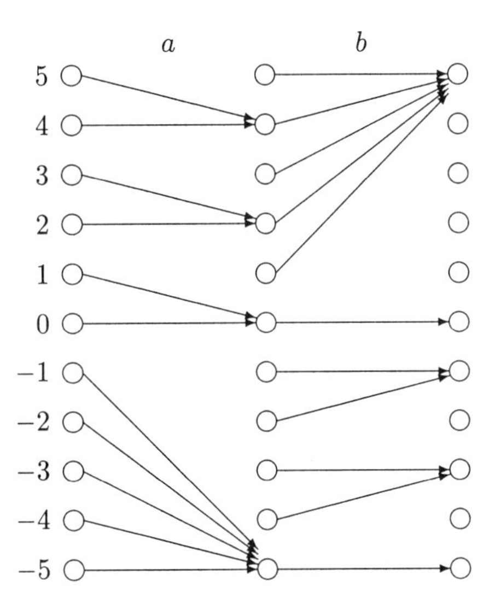

Figure 8. The action of a and b in the automaton $\mathscr{A}_2$

Lemma 3.1. *For each $\ell = 1, 2, \ldots$, the automaton $\mathscr{A}_\ell$ is synchronizing.*

Proof. First we calculate the action of the word cbc on Q_ℓ:

$$\delta(j, cbc) = \begin{cases} -2\ell - 1 & \text{if } j < 0, \\ 0 & \text{if } j = 0, \\ j & \text{if } j = 2k + 1 > 0, \\ j - 1 & \text{if } j = 2k > 0. \end{cases}$$

For $\ell = 2$ this action (together with that of a) is shown on Fig. 9.

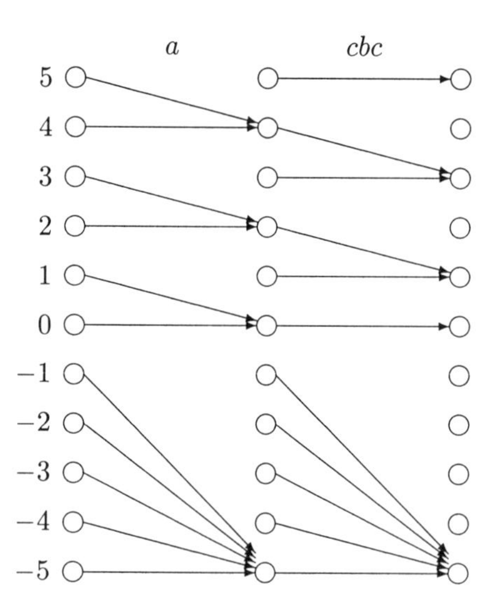

Figure 9. The action of a and cbc in the automaton $\mathscr{A}_2$

Now, inducting on $k \geq 0$, one can easily check that $\delta(2k, (acbc)^k a) = 0$ and $\delta(2k + 1, (acbc)^k a) = 0$. In particular, $\delta(j, (acbc)^\ell a) = 0$ for all $j = 0, 1, \ldots, 2\ell + 1$. Since each negative state is sent to $-2\ell - 1$ by both a and cbc, one has $\delta(Q_\ell, (acbc)^\ell a) = \{-2\ell - 1, 0\}$. Applying the reflection c to the latter set, we obtain the set $\{0, 2\ell + 1\}$ which, as already observed, is compressed to $\{0\}$ by the word $(acbc)^\ell a$. Altogether we get the equality

$$\delta(Q_\ell, (acbc)^\ell ac(acbc)^\ell a) = \{0\},$$

meaning that $(acbc)^\ell ac(acbc)^\ell a$ is a reset word for the automaton $\mathscr{A}_\ell$. $\square$

Observe that the length of the reset word constructed in the above proof is equal to $8\ell + 3 = 2n - 3$.

The proof of the next lemma involves the notion of an interval of the chain (2). As above, such intervals are denoted $[i, j]$ where $i, j \in Q_\ell$ but, in contrast to Section 2, here we will always assume that $i \leq j$. With this assumption every subset of Q_ℓ representable by an interval has only one such representation and hence there is no point anymore in distinguishing between an interval $[i, j]$ and the set it represents. The latter set will therefore be denoted by $[i, j]$ as well. By the *width* of $[i, j]$ we mean the difference $j - i$.

Lemma 3.2. *The length of any reset word for the automaton $\mathscr{A}_\ell$ is at least $2n - 3$.*

Proof. First of all, we observe that since the action of each letter in Σ fixes the state 0, so does the action of every word $v \in \Sigma^*$. Therefore 0 belongs to the image of every transformation $\delta(\_, v)$.

Now let w be a reset word of minimum length for $\mathscr{A}_\ell$. Then it easy to see that w can be decomposed as

$$w = u_1 c u_2 c \cdots c u_m$$

where $u_1, u_2, \ldots, u_m$ are non-empty words over $\{a, b\}$. Indeed, since $\delta(\_, c^2)$ is the identity transformation, c^2 cannot occur as a factor in w and c can be neither the first nor the last letter in w (otherwise one could have reduced w to a shorter reset word).

For $k = 0, 1, \ldots, m$, let $w_k = u_1 c u_2 c \cdots c u_k$ (thus, w_0 is the empty word) and let $[i(k), j(k)]$ denote the interval of minimum width containing the set $\delta(Q_\ell, w_k)$. (Observe that this interval contains 0.) We claim that the width of $[i(k), j(k)]$ decreases in comparison with the width of $[i(k-1), j(k-1)]$ at most by 1. Since the width of $[i(0), j(0)]$ is $n - 1$ while the width of $[i(m), j(m)]$ should be 0, the claim implies that $m \geq n - 1$ whence the length of w is at least $2n - 3$, as required.

It remains to justify the above claim. From (3) and (4) one immediately sees that for every non-singleton interval $[i, j]$ containing 0 the width of each the intervals $[\delta(i, a), \delta(j, a)]$ and $[\delta(i, b), \delta(j, b)]$ is at least $j - i - 1$. But the same inequality holds true for the width of the interval $[\delta(i, u), \delta(j, u)]$ where u is an arbitrary word over $\{a, b\}$. Indeed, one can induct on the length of u using the observation just made as the induction base. The induction step then follows from the equalities

$\delta(\_,a^2)=\delta(\_,a)$, $\delta(\_,b^2)=\delta(\_,b)$ and the inclusions

$$[\delta(i,a),\delta(j,a)]\subseteq[\delta(i,ab),\delta(j,ab)],$$
$$[\delta(i,a),\delta(j,a)]\subseteq[\delta(i,ba),\delta(j,ba)]$$

that one gets from (3) and (4). □

From Lemmas 3.1 and 3.2 we obtain the following lower bound for the Černý function for semigroups $\mathcal{D}_n$:

Proposition 3.1. *The Černý function for the semigroups of all order preserving or order reversing transformations of* $\{1,2,\dots,n\}$ *is no less than* $2n-3$ *for each* $n=4\ell+3$, $\ell=1,2,\dots$.

We do not have a similar lower bound for those values of n whose residues modulo 4 are not equal to 3.

The automata $\mathscr{A}_\ell$ are so easy to handle because they possess a letter acting as a reflection. Let us call a DFA $\mathscr{A}=\langle\{1,2,\dots,n\},\Sigma,\delta\rangle$ *reflexive* if some letter (which we always will denote by c) acts as a reflection, that is, $\delta(i,c)=n-i+1$ for all $i=1,2,\dots,n$. Within the class of reflexive weakly monotonic automata we have an upper bound for the Černý function that matches the lower bound of Proposition 3.1. This upper bound is obtained by a suitable reduction of reflexive weakly monotonic automata to monotonic ones.

Thus, let $\mathscr{A}=\langle\{1,2,\dots,n\},\Sigma,\delta\rangle$ be a reflexive weakly monotonic automaton. For brevity, we call a letter $x\in\Sigma\setminus\{c\}$ order preserving (order reversing) if $\delta(\_,x)\in\mathcal{O}_n$ (respectively if $\delta(\_,x)\in\mathcal{D}_n\setminus\mathcal{O}_n$). We assign each letter $x\in\Sigma\setminus\{c\}$ two new letters x' and x'' and let $\widehat{\Sigma}=\{x',x''\mid x\in\Sigma\setminus\{c\}\}$. Then we consider the automaton $\widehat{\mathscr{A}}=\langle\{1,2,\dots,n\},\widehat{\Sigma},\hat{\delta}\rangle$ whose transition function $\hat{\delta}$ is defined as follows: for each $i=1,2,\dots,n$,

$$\hat{\delta}(i,x')=\begin{cases}\delta(i,x) & \text{if } x \text{ is order preserving,}\\ \delta(i,xc) & \text{if } x \text{ is order reversing;}\end{cases}\tag{6}$$

$$\hat{\delta}(i,x'')=\begin{cases}\delta(i,cxc) & \text{if } x \text{ is order preserving,}\\ \delta(i,cx) & \text{if } x \text{ is order reversing.}\end{cases}\tag{7}$$

Obviously, all transformations $\hat{\delta}(\_,x')$, $\hat{\delta}(\_,x'')$ are order preserving whence the automaton $\widehat{\mathscr{A}}$ is monotonic. The following lemma relates the behavior of $\mathscr{A}$ with that of $\widehat{\mathscr{A}}$:

Lemma 3.3. *For any word* $w\in\Sigma^*$ *there exists a word* $\hat{w}\in\widehat{\Sigma}^*$ *such that either* $\hat{\delta}(\_,\hat{w})=\delta(\_,w)$ *or* $\hat{\delta}(\_,\hat{w})=\delta(\_,wc)$.

Proof. We induct on the length of w. If w is empty then, of course, the empty word can play the role of $\hat{w}$. Suppose that $w = vx$ for some word $v \in \Sigma^*$ and some letter $x \in \Sigma$. By the induction assumption there exists a word $\hat{v} \in \widehat{\Sigma}^*$ such that either $\hat{\delta}(\_, \hat{v}) = \delta(\_, v)$ or $\hat{\delta}(\_, \hat{v}) = \delta(\_, vc)$. If $x \in \Sigma \setminus \{c\}$ then we set $\hat{w} = \hat{v}x'$ in the first case and $\hat{w} = \hat{v}x''$ in the second case, and if $x = c$, then we simply set $\hat{w} = \hat{v}$. Verifying that the word $\hat{w}$ defined this way has the desired property in each case amounts to a straightforward calculation which is left to the reader. □

Proposition 3.2. *For each reflexive weakly monotonic automaton with n states there exists a reset word of length at most $2n - 3$.*

Proof. Lemma 3.3 readily implies that if a reflexive weakly monotonic automaton $\mathscr{A}$ is synchronizing then so is its monotonic counterpart $\widehat{\mathscr{A}}$. Indeed, if w is a reset word for $\mathscr{A}$, then the word wc is also a reset word, and therefore, the word $\hat{w}$ resets the automaton $\widehat{\mathscr{A}}$.

By Theorem 3.1 the automaton $\widehat{\mathscr{A}}$ has a reset word $v \in \widehat{\Sigma}^*$ whose length does not exceed $n - 1$. We convert this word into a word $u \in \Sigma^*$ using the following process:

1) for each order preserving letter $x \in \Sigma \setminus \{c\}$ substitute x for x' and cxc for x'';
2) for each order reversing letter $x \in \Sigma \setminus \{c\}$ substitute xc for x' and cx for x'';

From (6) and (7) we see that $\delta(\_, u) = \hat{\delta}(\_, v)$ whence u is a reset word for the automaton $\mathscr{A}$. Now we can delete from u all occurrences of the factor c^2—this does not change the transformation $\delta(\_, u)$ because $\delta(\_, c^2)$ is the identity transformation. If the first and/or the last letter of the resulting word is c we can delete it as well still having a reset word (say, w) because $\delta(\_, c)$ is a permutation. For each occurrence of a letter from $\widehat{\Sigma}$ in v, the word w contains exactly one occurrence of the corresponding letter from $\Sigma \setminus \{c\}$ so it has at most $n - 1$ such occurrences altogether. Since the letter c is neither the first nor the last letter in w and the factor c^2 does not occur in w, there are at most $n - 2$ occurrences of c in w. Therefore the length of w is at most $2n - 3$. □

We conjecture that the upper bound from Proposition 3.2 holds true for all weakly monotonic automata. To prove this, one probably needs a clever modification of the proof of Theorem 3.1 from [1]. For the time being, the only upper bound we know for the Černý function for semigroups of order reversing transformations is $\binom{n}{2}$. (This can be easily obtained from the interval automaton argument similar to the above proof of Theorem 2.2.) However it seems to be very unlikely that weakly monotonic synchronizing automata without reflections require reset words of quadratic length while the presence of a reflection yields a linear upper bound.

References

1. D. S. Ananichev, M. V. Volkov, *Synchronizing monotonic automata*, Developments in Language Theory. Proc. 7th Internat. Conf. DLT 2003 [Lect. Notes Comp. Sci. **2710**], Springer, 2003, 111–121.
2. Y. Benenson, T. Paz-Elizur, R. Adar, E. Keinan, Z. Livneh, E. Shapiro, *Programmable and autonomous computing machine made of biomolecules*, Nature **414**, no.1 (November 22, 2001) 430–434.
3. Y. Benenson, R. Adar, T. Paz-Elizur, Z. Livneh, E. Shapiro, *DNA molecule provides a computing machine with both data and fuel*, Proc. National Acad. Sci. USA **100** (2003) 2191–2196.
4. J. Černý, *Poznámka k homogénnym eksperimentom s konecnými avtomatami*, Mat.-Fyz. Cas. Slovensk. Akad. Vied. **14** (1964) 208–216 [in Slovak].
5. D. Eppstein, *Reset sequences for monotonic automata*, SIAM J. Comput. **19** (1990) 500–510.
6. P. Frankl, *An extremal problem for two families of sets*, Eur. J. Comb. **3** (1982) 125–127.
7. A. Mateescu and A. Salomaa, *Many-valued truth functions, Černý's conjecture and road coloring*, EATCS Bull. **68** (1999) 134–150.
8. B. K. Natarajan, *An algorithmic approach to the automated design of parts orienters*, Proc. 27th Annual Symp. Foundations Comput. Sci., IEEE, 1986, 132–142.
9. B. K. Natarajan, *Some paradigms for the automated design of parts feeders*, Internat. J. Robotics Research, **8**, no.6 (1989) 89–109.
10. D. Perrin, M.-P. Schützenberger, *Synchronizing prefix codes and automata and the roas coloring problem*, Symbolic dynamics and its applications, Proc. AMS Conf. in honor of R. L. Adler [Contemp. Math. **135**], AMS, 1992, 295–318.
11. J.-E. Pin, *On two combinatorial problems arising from automata theory*, Ann. Discrete Math. **17** (1983) 535–548.

NEAR PERMUTATION SEMIGROUPS*

JORGE M. ANDRÉ[†]
Departamento de Matemática,
Faculdade de Ciências e Tecnologia da Universidade Nova de Lisboa,
Monte da Caparica, 2829–516 Caparica, Portugal
E-mail: jmla@fct.unl.pt

A transformation semigroup over a set X with N elements is said to be a near permutation semigroup if it is generated by a group of permutations on N elements and by a set of transformations of *rank* $N-1$. In this paper we make a summary of the author's thesis, in which are exhibited ways of checking whether those specific transformation semigroups belong to various structural classes of semigroups, just by analysis of the generators and without first having to generate them.

1. Introduction

In this paper we make a summary of the more important results of the author's thesis [6] about near permutation semigroups $S = \langle G, \mathcal{U} \rangle$, that is semigroups generated by a group G of permutations on N elements, for some $N \in \mathbb{N}$, and by a set $\mathcal{U}$ of transformations of *rank* $N - 1$. It is known that the number of elements of the transformation semigroups on a set X with a finite number N of elements, grows exponentially with the cardinality of the set X. We aim to find efficient ways to determine the structure of a given near permutation semigroup, without first having to generate it. The motivation for the study of those semigroups is due to the following. The semigroup $\mathcal{T}_N$ of all full transformations on a set X with N elements, for some $N \in \mathbb{N}$, is generated by the the group of all permutations on the set X and by any idempotent of *rank* $N-1$. Similarly the symmetric inverse semigroup $\mathcal{I}_N$ on X, that is the semigroup of all

*Work partially supported by the Sub-programa Ciência e Tecnologia do 2º Quadro Comunitário de Apoio (grant number BD/18012/98) and project POCTI/32440/MAT/1999 (CAUL) of FCT and FEDER.

†Author's second address: Centro de Álgebra da Universidade de Lisboa, Av. Prof. Gama Pinto, 2, 1649–003 Lisboa, Portugal.

injective partial transformations on a set with N elements, is generated by its group of units and by any idempotent of *rank* $N-1$. For the two previous results refer to [8]. Other well known semigroups like the semigroup $\mathcal{OP}_N$ of all orientation preserving mappings and the semigroup $\mathcal{OPR}_N$ of all orientation preserving and reversing mappings, on a finite chain of order N, are generated by a group of permutations and by an idempotent of *rank* $N-1$ (see [7]). Moreover, McAlister [12] proved that all semigroups generated by a group of permutations on N elements and by an idempotent of *rank* $N-1$ are regular. He also obtained necessary and sufficient conditions for a transformation semigroup generated by a group of permutations on N elements and by an idempotent of *rank* $N-1$ to be inverse. In our work of [3] and [1], we considered a set $\mathcal{U}$ of transformations of *rank* $N-1$ instead of the idempotent e.

To obtain most of the results the author made use of computational tools such as the PC software [9] and [11].

2. Preliminaries

Let $X = \{1, 2, ..., N\}$, where $N \in \mathbb{N}$. As usual, we denote by $\mathcal{T}_N$ the monoid of all full transformations of X (under composition), and by $\mathcal{S}_N$ the symmetric group on X. We call a semigroup generated by a group of permutations on N elements and by a set of transformations of *rank* $N-1$ a *near permutation semigroup*. Denote by $\binom{x}{y}$, where $x, y \in X$, the idempotent of $\mathcal{T}_N$ which maps x to y and i to itself for $i \neq x$. An useful observation is that an element a belongs to a subgroup of $\mathcal{T}_N$ if and only if $rank\ a = rank\ a^2$. Given $z \in \mathcal{T}_N$ denote by $Im\ z$ the image set of the transformation z. It is also useful to note that given an element e of $\mathcal{T}_N$, then e is idempotent if and only if it fixes all elements of $Im\ e$. Given G a group of permutations on X and $x \in X$ define the sets $O_G(x) = \{y \in X : y = xg, \text{for some } g \in G\}$ and $Stab_G(x) = \{g \in G : xg = x\}$, which we also call the *orbit of x under G* and the *stabilizer of x under G*, respectively.

A *directed graph* Γ consists of two sets, a non empty set $V(\Gamma)$ of *vertices* and a set $A(\Gamma)$ of *arrows* of Γ, together with a mapping $\Psi : A(\Gamma) \longrightarrow V(\Gamma) \times V(\Gamma)$.

Given $a \in A(\Gamma)$ and $a\Psi = (v_1, v_2)$, where $v_1, v_2 \in V(\Gamma)$, we call v_1 and v_2 the *initial vertex* and the *terminal vertex*, respectively. We say that a is an arrow from v_1 to v_2.

A *path* between two vertices v_1 and v_2, in a graph Γ, is a sequence of arrows such that the terminal vertex of one arrow is the initial vertex of the next and the first arrow of the sequence has initial vertex v_1 and the last arrow has terminal vertex v_2.

Let S be a near permutation semigroup and u be a transformation in S of *rank* $N-1$. We denote by $\alpha(u)$ and $\beta(u)$ the two unique distinct elements $a, b \in X$ such that $(a, b) \in Ker\ u$. We denote by $\gamma(u)$ the unique element of $X \setminus Im\ u$. If u is being denoted by u_i for some $i \in \{1, ..., m\}$, we abbreviate $\alpha(u)$, $\beta(u)$ and $\gamma(u)$, respectively, by α_i, β_i and γ_i. Given $(a, b) \in Ker\ u$, with $a \neq b$, in order to choose which one a or b is $\alpha(u)$, we set $\alpha(u)$ as the minimum of the set $\{a, b\}$, for each $u \in \mathcal{U}$.

3. Regular near permutation semigroups

Let $S = \langle G, \mathcal{U} \rangle$ be a near permutation semigroup and $\mathcal{U} = \{u_1, ..., u_m\}$. Consider the directed graph $\Gamma(G; \mathcal{U})$, where the vertices are the elements of $\mathcal{U}$ and there is an arrow, $u_i \to u_j$ between two vertices u_i and u_j, if $\gamma_i g \in \{\alpha_j, \beta_j\}$. Notice that given $u_i, u_j \in \mathcal{U}$, the existence of $u_i \to u_j$ is equivalent to saying that *rank* $u_i g u_j = N - 1$, for some $g \in G$. In that case we can also refer that property by writing $u_i \to_g u_j$. In the graph $\Gamma(G; \mathcal{U})$, represent each vertex u_i, where $i = 1, ..., m$, with the triple $(\alpha_i, \beta_i, \gamma_i)$, where $(\alpha_i, \beta_i) \in Ker\ u_i$, $\alpha_i < \beta_i$ and $\gamma_i \in X \setminus Im\ u_i$. We call $\Gamma(G; \mathcal{U})$ a *near permutation graph* of S. For these defined graphs, if we consider more than one vertex representing the same $\mathcal{H}$-class, we obtain equivalent graphs to the ones where there is only one vertex for each $\mathcal{H}$-class in the sense that both graphs give us the same information for our study. Thus, when we consider the graph $\Gamma(G; \mathcal{U})$ of a near permutation semigroup S, we only include in the set of vertices the elements of $\mathcal{U}$ that belong to different $\mathcal{H}$-classes of $\mathcal{T}_N$. Hence we choose exactly one vertex, from the vertices that have the same triples $(\alpha_i, \beta_i, \gamma_i)$.

Note that in the graph $\Gamma(G; \mathcal{U})$ there exists at most one arrow between two vertices.

A special case occurs when $\mathcal{U}$ is a group. In this case, the graph $\Gamma(G; \mathcal{U})$ is complete, that is there exists exactly one arrow between any two distinct vertices.

Let $\Gamma(G; \mathcal{U})$ be the near permutation graph of a near permutation semigroup $S = \langle G, \mathcal{U} \rangle$. Then there exists a binary relation ε on $\mathcal{U}$ induced by the graph $\Gamma(G; \mathcal{U})$, defined by: given $v_0, v_0' \in \mathcal{U}$,

$$v_0\ \varepsilon\ v_0' \quad \textit{if and only if} \text{ there exists a path } v_0 \to v_1 \to v_2 \to ... \to v_k \to v_0'$$

in $\Gamma(G; \mathcal{U})$, for some $v_1, v_2, ..., v_k \in \mathcal{U}$.

Clearly the relation ε is transitive. Note also that the existence of a path $v_1 \rightarrow_{g_1} v_2 \rightarrow_{g_2} ... v_{l-1} \rightarrow_{g_{l-1}} v_l$ in $\Gamma(G;\mathcal{U})$, is equivalent to the equality

$$rank\ v_1 g_1 v_2 ... g_{l-1} v_l = N-1.$$

Observe that, given $u, v \in \mathcal{U}$ and $x \in S$, if $rank\ uxv = N-1$ then exists a path, relative to the graph $\Gamma(G;\mathcal{U})$, from u to v.

The first two results show us that we can use the near permutation graph of a semigroup S in order to determine whether the generators or the elements of S with $rank\ N-1$ are regular.

Proposition 3.1. [3] *Let $S = \langle G, \mathcal{U} \rangle$ be a near permutation semigroup. Then, the elements of $\mathcal{U}$ are regular if and only if the relation ε is reflexive.*

Theorem 3.1. [3] *Let $S = \langle G, \mathcal{U} \rangle$ be a near permutation semigroup. Then the elements of S with rank $N-1$ are regular if and only if the relation ε, induced by the graph $\Gamma(G;\mathcal{U})$, is an equivalence relation.*

Before determining a characterization of the idempotents of $rank\ N-1$, we need some more notation. For each $u_i \in \mathcal{U}$, we fix a u_i' an inverse of u_i in S, and consider $u_i u_i' = \binom{x_i}{y_i}$, where $\{x_i, y_i\} = \{\alpha_i, \beta_i\}$, for $i = 1, ..., m$. We also define $T = \{\binom{x_1}{y_1}, ..., \binom{x_m}{y_m}\}$ and $T' = \{\binom{\beta}{\alpha} \in S : \binom{\alpha}{\beta} \in T\}$.

According to the previous definitions, it is important to know when both idempotents $\binom{\alpha_i}{\beta_i}$ and $\binom{\beta_i}{\alpha_i}$ belong to S. The next proposition gives an answer to this question and in particular tells us how to determine the set T' for each semigroup S.

Proposition 3.2. [3] *Let $S = \langle G, \mathcal{U} \rangle$ be a near permutation semigroup such that the relation ε induced by the graph $\Gamma(G;\mathcal{U})$ is an equivalence relation. Then, for each $i \in \{1, 2, ..., m\}$, the idempotents $\binom{\alpha_i}{\beta_i}$ and $\binom{\beta_i}{\alpha_i}$ belong to S if and only if $\{\alpha_i, \beta_i\} = \{\gamma_p g_1, \gamma_q g_2\}$, for some $g_1, g_2 \in G$ and $u_p, u_q \in \mathcal{U}$.*

Now we can easily determine the $rank\ N-1$ idempotents of S.

Corollary 3.1. [3] *Let $S = \langle G, \mathcal{U} \rangle$ be a near permutation semigroup. Then the set of idempotents of S with rank $N-1$ is $\bigcup_{g \in G} g^{-1}(T \cup T')g$.*

The Proposition 3.2 is very important in order to obtain an easy to check sufficient condition for S to be regular.

Theorem 3.2. [3] *Let $S = \langle G, \mathcal{U} \rangle$ be a near permutation semigroup. If the relation ε induced by the graph $\Gamma(G;\mathcal{U})$ is an equivalence relation and, for each $i \in \{1, ..., m\}$, there exist $g_i, g_i' \in G$ and $u_p, u_q \in \mathcal{U}$, such that $\{\alpha_i, \beta_i\} = \{\gamma_p g_i, \gamma_q g_i'\}$, then S is regular.*

Now we present some more useful and efficient sufficient conditions for S to be regular which do not involve the determination of the idempotents of S with *rank* $N-1$.

Theorem 3.3. [3] *Let* $S = \langle G, \mathcal{U} \rangle$ *be a near permutation semigroup. If for each* $i \in \{1, ..., m\}$, *the elements* α_i, β_i *and* γ_i *belong to the same orbit under* G, *then* S *is regular.*

The next sufficient condition is a particular case of the previous and says that if for any two elements x and y of X exists $g \in G$ such that $xg = y$ then S is regular.

Theorem 3.4. [3] *Let* $S = \langle G, \mathcal{U} \rangle$ *be a near permutation semigroup. If* G *acts transitively on* X, *then* S *is regular.*

Now we have another condition for the study of the regularity of S which is independent of the previous and also very useful in some cases.

Theorem 3.5. [3] *Let* $S = \langle G, \mathcal{U} \rangle$ *be a near permutation semigroup. If the relation* ε *induced by the graph* $\Gamma(G;\mathcal{U})$ *is reflexive and, for each* $i \in \{1, ..., m\}$, *the elements* α_i *and* β_i *belong to the same orbit under* G *then* S *is regular.*

Next we have a theorem which provides a more general sufficient condition for the study of the regularity of S, which turned out to be also necessary in all the examples tested.

Theorem 3.6. [3] *Let* $S = \langle G, \mathcal{U} \rangle$ *be a near permutation semigroup. Suppose that the relation* ε *induced by the graph* $\Gamma(G;\mathcal{U})$ *is an equivalence relation. If for all* $\binom{\alpha_k}{\beta_k} \in T$ *such that* $\binom{\beta_k}{\alpha_k} \notin T \cup T'$ *and for all* $\gamma_s g$ *which are different from* α_k *and* β_k, *with* $s \in \{1, ..., m\}$ *and* $g \in G$, *there exist* $g' \in G$ *and* $\binom{a}{b} \in T \cup T'$ *satisfying:*

i) $\gamma_s g = ag'$;

ii) α_k *belongs to the orbit of* β_k *under the group* $H_{\binom{ag'}{bg'}}$;

iii) $bg' \neq \beta_k$;

then S *is regular.*

The following theorem shows that the study of the regularity of S, when $\mathcal{U}$ is a set of idempotents, depends only on its near permutation graph.

Theorem 3.7. [3] *Let $S = \langle G, E\rangle$ be a near permutation semigroup, where E is a set of idempotents. Then S is regular if and only if the relation ε induced by the graph $\Gamma(G; E)$ is an equivalence relation.*

In [10] was obtained a necessary and sufficient condition for a semigroup generated by a group of permutations on a set with N elements and by a set of idempotents of $rank\ N-1$, to contain all singular transformations of $\mathcal{T}_N$. With the use of the tools and results obtained in [3] we have a different approach to that problem, in which case the semigroup is generated by a group of permutations and by a set of $rank\ N-1$ transformations.

But first we need to define several sets. Let be $\mathcal{U} = \{u_1, ..., u_m\}$ and $I = \{1, ..., m\}$. Let also be $A = \{\{x, y\} : x, y \in X,\ x \neq y\}$, $B = \{\{\alpha_i, \beta_i\} : i \in I\}$, $C = \{a \in X \setminus Im\ u_i : i \in I\}$ and $D = \{\alpha_i : i \in I\} \cup \{\beta_i : i \in I\}$. Now, the group G acts on the sets B and C (on the right), in the following ways:

$$BG = \{\{\alpha_i g, \beta_i g\} : i \in I,\ g \in G\} \text{ and } CG = \{ag : a \in C,\ g \in G\}.$$

The proof of the theorem is in [4].

Theorem 3.8. *Let $S = \langle G, \mathcal{U}\rangle$ be a near permutation semigroup. Then S contains $\mathcal{T}_N \setminus \mathcal{S}_N$ if and only if*

1) The relation ε is an equivalence relation;

2) $BG = A$;

3) $D \subseteq CG$.

4. Near permutation semigroups generated by two groups

An interesting special case of the general case considered previously is when $\mathcal{U}$ is a group. In fact, in the case considered by McAlister in [12], the set $\mathcal{U}$ has only one element, which is an idempotent. Thus it is a group.

Let $S = \langle G, H\rangle$ be a near permutation semigroup, where H is a group. We can rearrange the set X so that the identity of H is $e = \binom{1}{2}$.

From now on, given $S = \langle G, H\rangle$ a near permutation semigroup, where H is a group, let $e = \binom{1}{2}$ be the identity of H.

The assumption that $2 \in O_G(1)$ is crucial in the study of the case $\mathcal{U} = \{e\}$. This was shown by McAlister in [12].

Theorem 4.1. *Let $S = \langle G, e\rangle$ be a near permutation semigroup, where e is an idempotent. If $2 \notin O_G(1)$, then S is an $\mathcal{R}$-unipotent semigroup.*

It was also proved in [12] that $S = \langle G, e\rangle$ is always a regular semigroup.

Theorem 4.2. *Let $S = \langle G, e\rangle$ be a near permutation semigroup, where e is an idempotent. Then S is regular.*

The knowledge of the orbit of 1 under G is also very important in the more general case, where $S = \langle G, \mathcal{U}\rangle$, as the following results show.

Theorem 4.3. [3] *Let $S = \langle G, H\rangle$ be a near permutation semigroup, where H is a group. If $2 \in O_G(1)$, then S is a regular semigroup.*

The next result is obtained by considering a near permutation semigroup generated by a group and by an idempotent and which contains S. But first we need to define the following. Let $S = \langle G, H\rangle$ be a near permutation semigroup, where H is a group with identity $\binom{1}{2}$. Since the elements of H are group elements of $rank\ N-1$, and every element h of the group H satisfies $1 \notin Im\ h$ and $(1,2) \in Ker\ h$, we can associate to each $h \in H$ a permutation $\overline{h}$ such that $1\overline{h} = 1$ and that $x\overline{h} = xh$, for all $x \in X \setminus \{1\}$. From now on, we denote by $\overline{h}$ the corresponding unique permutation of the transformation $h \in H$. The set of all such elements $\overline{h}$ is a group of permutations, that we denote by $\overline{H}$. The semigroup $\langle G, \overline{H}\rangle$ is a group, because all its generators are permutations. We denote $\langle G, \overline{H}\rangle$ by $G(S)$ and the set $\{g \in G :\ 1g \neq 1\}$ by G'. Observe that $S = \langle G, e\overline{H}\rangle$ and S is contained in the semigroup $\langle G(S), e\rangle$. Conversely if G and W are subgroups of $\mathcal{S}_N$ such that the elements of W fix 1, that is $1w = 1$ for all $w \in W$, then $S = \langle G, eW\rangle$ is a near permutation semigroup, where eW is a group of transformations of $rank\ N-1$. Now, we have the semigroup $S = \langle G, H\rangle$ between two semigroups generated by a group and an idempotent, that is $S' = \langle G, e\rangle \subseteq S \subseteq \langle G(S), e\rangle$.

Proposition 4.1. [3] *Let $S = \langle G, H\rangle$ be a near permutation semigroup, where H is a group. If $2 \notin O_{G(S)}(1)$, then S is an $\mathcal{R}$-unipotent semigroup.*

To study a near permutation semigroup $S = \langle G, H\rangle$, it will be very important to know some results about the local submonoid eSe, where e is the identity of H. Consider H_e as being the $\mathcal{H}$-class of e.

Proposition 4.2. [3] *Let $S = \langle G, H\rangle$ be a near permutation semigroup, where H is a group with identity e. If $2 \notin O_G(1)$, then eSe is generated by H_e together with the set $\{ege : g \in G, 1g \neq 1\}$.*

Moreover, the group H_e is generated by H together with the set $\{ege : g \in G, 1g = 1\}$.

In order to use the results of the previous section it is important the following result.

Proposition 4.3. [3] *Let $S = \langle G, H \rangle$ be a near permutation semigroup, with H a group. Suppose that $2 \notin O_G(1)$. Then there exists an isomorphism ψ between the submonoid eSe of S and a submonoid of $\mathcal{T}_{\{2,\ldots,N\}}$, which has as group of units $H_e\psi$. Moreover, eSe is isomorphic to the near permutation semigroup $W_S = \langle H_e\psi, V\psi \rangle$, where H_e is the $\mathcal{H}$-class of e and $V = \{g \in G : 1g \neq 1\}$.*

Using the isomorphic copy of the local submonoid eSe, we can say that eSe is a near permutation semigroup $\langle G_S, \mathcal{U}_S \rangle$, where $G_S = H_e\psi$ and $\mathcal{U}_S = V\psi$. We recall that it is convenient to consider $\mathcal{U}_S = \{u_1, ..., u_m\}$ and $I = \{1, ..., m\}$. Now we can refer some more results about the regularity of S with the help of the technique described before.

Proposition 4.4. [3] *Let $S = \langle G, H \rangle$ be a near permutation semigroup, where H is a group and $W_S = \langle G_S, \mathcal{U}_S \rangle$. If, for each $i \in I$, the elements α_i, β_i and γ_i belong to the same orbit under G_S, then S is regular.*

According to the results of the previous section is easy to conclude that all *rank* $N-1$ elements of S are regular. Next we provide a tool for checking if the elements of *rank* $N-2$ are regular, which is obtained using Theorem 3.1 on the local submonoid eSe of S.

Theorem 4.4. [3] *Let $S = \langle G, H \rangle$ be a near permutation semigroup, where H is a group and $W_S = \langle G_S, \mathcal{U}_S \rangle$, as defined above. Then the elements of S with rank $N-2$ are regular if and only if the relation ε induced by the graph $\Gamma(G_S; \mathcal{U}_S)$, where the vertices are the elements of the set $\mathcal{U}_S = \{u_1, ..., u_m\}$, is an equivalence relation.*

We have one more sufficient condition for S to be regular which uses Theorem 3.5.

Proposition 4.5. [3] *Let $S = \langle G, H \rangle$ be a near permutation semigroup, where H is a group and $W_S = \langle G_S, \mathcal{U}_S \rangle$. If the relation ε induced by the graph $\Gamma(G; \mathcal{U}_S)$ is an equivalence relation and, for each $i \in I$, the elements α_i and β_i belong to the same orbit under G_S, then S is regular.*

It is important to observe that, given a near permutation semigroup $S = \langle G, H \rangle$, we can use the more general result of Theorem 3.6, considering the local submonoid eSe of S and its isomorphic copy $W_S = \langle G_S, \mathcal{U}_S \rangle$.

The results that follow are about inverse near permutation semigroups. The first one has to do with the semigroup $\langle G, e\rangle$ and was proved in [12].

Theorem 4.5. *Let $S = \langle G, H\rangle$ be a near permutation semigroup, where H is a group with identity e. The subsemigroup $S' = \langle G, e\rangle$ of S is inverse if and only if $2 \notin O_G(1)$ and $Stab_G(1) \subseteq Stab_G(2)$.*

With the group $G(S)$ previously defined it is possible to say more about when S is inverse. Next, we obtain a sufficient but not necessary condition for S to be inverse.

Proposition 4.6. [1] *Let $S = \langle G, H\rangle$ be a near permutation semigroup, where H is a group with identity e. If $2 \notin O_{G(S)}(1)$ and $Stab_{G(S)}(1) \subseteq Stab_{G(S)}(2)$, then S is inverse.*

It follows a theorem that provides a necessary and sufficient condition for $S = \langle G, H\rangle$ to be inverse. Recall that $G(S) = \langle G, \overline{H}\rangle$, where $\overline{H}$ is the group of permutations associated with H.

Theorem 4.6. [1] *Let $S = \langle G, H\rangle$ be a near permutation semigroup, where H is a group. Then S is inverse if and only if*

(i) $2 \notin O_{G(S)}(1)$;

(ii) If $1g_1\overline{h}_1...g_{n-1}\overline{h}_{n-1}g_n = 1$ and $1g_1\overline{h}_1...g_{s-1}\overline{h}_{s-1}g_s \neq 1$, for all s such that $1 < s < n$, then $2g_1\overline{h}_1...g_{n-1}\overline{h}_{n-1}g_n = 2$, where $g_1, ..., g_n \in G$ and $\overline{h}_1, ..., \overline{h}_{n-1} \in \overline{H}$.

Notice that Theorem 4.5, proved by McAlister in [12], is an easy consequence of Theorem 4.6. Moreover, it is obvious that if $Stab_{G(S)}(1) \subseteq Stab_{G(S)}(2)$ then $2h = 2$, for all $h \in H$.

Observe also that the conditions of Theorem 4.6 can be simplified when, for $S = \langle G, H\rangle$, the elements of H satisfy the equality $2h = 2$.

Corollary 4.1. [1] *Let $S = \langle G, H\rangle$ be a near permutation semigroup, where H is a group. Let $G(S) = \langle G, \overline{H}\rangle$, where $\overline{H}$ is the group of permutations associated with H. If $2h = 2$, for all $h \in H$, then S is inverse if and only if $2 \notin O_{G(S)}(1)$ and the $Stab_{G(S)}(1) \subseteq Stab_{G(S)}(2)$.*

Theorem 4.6 gives necessary and sufficient conditions for a near permutation semigroup to be inverse. However these are somewhat technical. A consequence of the first, and simplest, of them is that the semigroup $\langle G(S), e\rangle$ is $\mathcal{R}$-unipotent. In particular, its idempotents form an $\mathcal{R}$-unipotent band. By studying the maximal subsemilattices of this band,

and maximal inverse subsemigroups of $\langle G(S), e\rangle$, one can give alternative, and more structural, necessary and sufficient conditions for $\langle G, H\rangle$ to be inverse. This approach is contained in Chapter 4 of the author's thesis [6], and proofs will appear in [2] and [5].

Our next aim is to present a set of necessary and sufficient conditions for a semigroup $S = \langle G, H\rangle$ to be inverse and to have semilattice of idempotents $E(S)$ generated by its idempotents of *rank* greater than or equal to $N-1$.

Proposition 4.7. [1] *Let S be a semigroup with a subgroup of units G, a subgroup H and a subsemilattice E that contains the identities of G and of H. Let also $P = \{x \in G \cup H : x^{-1}vx,\ xvx^{-1} \in E, \text{ for all } v \in E\}$, where x^{-1} denotes the group inverse of x within the respective group. Then $L = \langle P, E\rangle$ is an inverse subsemigroup of S, with semilattice of idempotents E.*

Next we obtain a necessary and sufficient condition for a semigroup S to be inverse and its semilattice is generated by the idempotents of S of *rank* greater or equal to $N-1$.

Proposition 4.8. [1] *Let $S = \langle G, H\rangle$ be a near permutation semigroup, where H is a group. Then S is inverse with semilattice of idempotents $E(S) = \langle\{\left(\begin{smallmatrix} 1g \\ 2g \end{smallmatrix}\right) : g \in G\}\rangle$ if and only if*

i) $1gh \in O_G(1)$, *for all* $g \in G'$ *and* $h \in H$;
ii) $2 \notin O_{G(S)}(1)$;
iii) $Stab_G(1) \subseteq Stab_G(2)$;
iv) $1g_1hg_2 = 1$ *implies* $2g_1hg_2 = 2$, *for all* $g_1 \in G', g_2 \in G, h \in H_e$.

The following proposition describes the idempotents of *rank* greater than or equal to $N-2$ of an inverse near permutation semigroup $S = \langle G, H\rangle$.

Proposition 4.9. [1] *Let $S = \langle G, H\rangle$ be an inverse near permutation semigroup, where H is a group with an identity e. Then the set of idempotents of S with rank greater than or equal to $N-2$ is*

$$A = \{\left(\begin{smallmatrix} 1g \\ 2g \end{smallmatrix}\right) : g \in G\} \cup \{\left(\begin{smallmatrix} 1g \\ 2g \end{smallmatrix}\right)\left(\begin{smallmatrix} 1g_1h_1g_2 \\ 2g_1h_1g_2 \end{smallmatrix}\right) : g, g_2 \in G, g_1 \in G', h_1 \in H_e\} \cup \{1\}.$$

Let $S = \langle G, H\rangle$ be a near permutation semigroup. We determine necessary and sufficient conditions for S to be an inverse semigroup whose semilattice of idempotents is generated by the idempotents of *rank* greater than or equal to $N-2$.

Proposition 4.10. [1] *Let $S = \langle G, H \rangle$ be a near permutation semigroup where H is a group with identity e. Then S is inverse, with semilattice of idempotents $E(S)$ generated by*

$$A = \{\left(\begin{smallmatrix} 1g \\ 2g \end{smallmatrix}\right) : g \in G\} \cup \{\left(\begin{smallmatrix} 1g \\ 2g \end{smallmatrix}\right)\left(\begin{smallmatrix} 1g_1hg_2 \\ 2g_1hg_2 \end{smallmatrix}\right) : g, g_2 \in G, g_1 \in G', h \in H_e\} \cup \{1\}$$

if and only if the following conditions 1), 2), 3) and 4) hold, where

1) $2 \notin O_{G(S)}(1)$;

2) $1g_1hg_2 = 1$ *implies* $2g_1hg_2 = 2$, *for all* $g_1 \in G', g_2 \in G, h \in H_e$;

3) For all $g_1 \in G', g_2 \in G, h_1, h_2 \in H_e$, *there exist* $g_3 \in G', g_4 \in G, h_3 \in H_e$ *such that* $1g_1h_1g_2h_2 = 1g_3h_3g_4$ *and* $2g_1h_1g_2h_2 = 2g_3h_3g_4$;

4) $Stab_G(1) \subseteq Stab_G(2)$.

Acknoledgements

Most of the work on this paper was undertaken during the preparation of my PhD thesis. I am very grateful to the orientation of my supervisors Donald McAlister and Gracinda Gomes.

References

1. J.M. André, Inverse near permutation semigroups, *Semigroup Forum*, to appear.
2. J.M. André, On subsemilattices of near permutation semigroups, in preparation.
3. J.M. André, Regularity on near permutation semigroups, *Comm. Algebra,* to appear.
4. J.M. André, Semigroups that contain all singular transformations, *Semigroup Forum*, to appear.
5. J.M. André, The maximum inverse subsemigroup of a near permutation semigroup, in preparation.
6. J.M. André, Ph. D. Thesis, University of Lisbon, 2002.
7. Robert E. Arthur, and N. Ruskuc, Presentations for two extensions of the monoid of order-preserving mappings on a finite chain, *Southeast Asian Bull. Math.* **24** (2000) 1–7.
8. J. M. Howie, *Fundamentals of Semigroup Theory,* Clarendon Press, Oxford, 1995.
9. S. G. Linton, G. Pfeiffer, E. Robertson and N. Ruskuc, Gap package *PC software* (1997).
10. K. Kearnes, A. Szendrei and J. Wood, Generating singular transformations, *Semigroup Forum* **63**(3) (2001) 441–448.
11. D. B. McAlister, Semigroup for Windows *PC software* (1997).
12. D. B. McAlister, Semigroups generated by a group and an idempotent, *Comm. Algebra* **26** (1998) 515–547.

THE ORIGINS OF INDEPENDENCE ALGEBRAS

JOÃO ARAÚJO

Centro de Álgebra da Universidade de Lisboa,
Av. Prof. Gama Pinto, 2,
1649–003 Lisboa, Portugal
E-mail: mjoao@ptmat.lmc.fc.ul.pt

JOHN FOUNTAIN

Department of Mathematics,
University of York,
Heslington, York YO10 5DD, U.K.
E-mail: jbf1@york.ac.uk

In this paper we describe how independence algebras could have been discovered and how v^*-algebras *probably* were discovered. We also provide a set of equivalent definitions for independence algebras (and hence for v^*-algebras) and mention some of the uses of these algebras in semigroup theory.

1. A Heuristic Approach to Independence Algebras

Let S be a monoid (that is, a semigroup with identity). We say that two elements $a, b \in S$ are $\mathcal{J}$-related if $SaS = SbS$. It is well known (see [26]) that $\mathcal{J}$ is an equivalence relation and that we can introduce a partial order in the set $S/\mathcal{J}$ of $\mathcal{J}$-classes as follows: for all $x, y \in S$,

$$J_x \leqslant_{\mathcal{J}} J_y \Leftrightarrow x \in SyS.$$

Now suppose that we have proved an interesting theorem about semigroups in which $(S/\mathcal{J}, \leqslant_{\mathcal{J}})$ is a well ordered chain. Clearly our next aim would be to provide natural examples of semigroups with this property.

When $S = T(X)$, the transformation monoid on a set X, then

$$(\forall_{x,y \in S})\ x \in SyS \Leftrightarrow |\operatorname{im}(x)| \leqslant |\operatorname{im}(y)|.$$

Similarly, when $S = \operatorname{End}(V)$, the endomorphism monoid of a vector space V, we have

$$(\forall_{x,y \in S})\ x \in SyS \Leftrightarrow \operatorname{rank}(x) \leqslant \operatorname{rank}(y),$$

where $\text{rank}(x)$ is the dimension of $\text{im}(x)$.

For both $S = T(X)$ and $S = \text{End}(V)$ one of $J_x \leqslant_{\mathcal{J}} J_y$ or $J_y \leqslant_{\mathcal{J}} J_x$ holds for all $x, y \in S$. Thus these are two examples of monoids in which $(S/\mathcal{J}, \leqslant_{\mathcal{J}})$ is a well ordered chain.

Following this path, one may observe, considering the two examples above, that V is a universal algebra and $\text{End}(V)$ is the monoid of all mappings $f : V \to V$ which preserve the operations of the algebra V. In the same way, the set X is a universal algebra (with no operations) and $T(X)$ is the monoid of all mappings $f : X \to X$ that preserve the operations of the algebra X. Thus a natural monoid to be considered is $\text{End}(\mathcal{A})$, where $\mathcal{A}$ is a universal algebra. This is not only a natural example of a monoid, but, in fact, every monoid is isomorphic to $\text{End}(\mathcal{A})$, for some algebra $\mathcal{A}$. (See, for example, [20] or [22].) Clearly then, we need to impose some conditions on our algebra if its endomorphism monoid is to have the property we desire.

For an element $a \in \text{End}(\mathcal{A})$, we denote the image of a by $\text{im}(a)$. Now let $a, b \in S = \text{End}(\mathcal{A})$, and suppose, by analogy with the two examples above, that we could prove that $a \in SbS$ if and only the dimension of the image of a (in short, $\dim(\text{im}(a))$) is less than or equal to the dimension of the image of b, that is, suppose we were able to prove something of the form

$$a \in SbS \Leftrightarrow \dim(\text{im}(a)) \leqslant \dim(\text{im}(b)). \tag{1}$$

If such a proposition were true, then S would be an example of a monoid in which $(S/\mathcal{J}, \leqslant_{\mathcal{J}})$ is a chain. The problem here is that $\dim(\text{im}(a))$ might not be defined. Next we consider how we might obtain a well defined notion of dimension.

We say that B is a basis for $\text{im}(a)$ if B is a minimal generating set for $\text{im}(a)$, that is,

$$\begin{aligned}
&(1)\ B \subseteq \text{im}(a);\\
&(2)\ \langle B \rangle = \text{im}(a);\\
&(3)\ (\forall_{b \in B})\ \langle B \setminus b \rangle \subsetneqq \text{im}(a).
\end{aligned}$$

A finite algebra always has a basis. However, two bases of an algebra might have different cardinalities and hence for that algebra we cannot define the dimension (i.e., the number of elements of any basis).

Thus, in order to be able to speak about *dimension* of an algebra, we must restrict our study to those algebras which have a basis and in which all the bases have the same cardinality. It is well known (in matroid theory) that any algebra satisfying the *exchange property* has bases and all the bases have the same cardinality. Thus the obvious thing to do now, in order to be able to speak about the *dimension* of an algebra, is to restrict our study to those algebras which satisfy the exchange property.

Matroids are usually taken to be finite, but for considering the notion of independence, finiteness is irrelevant. Infinite matroids are treated by Oxley in [33]; the following definition, which is equivalent to his, is taken from [11]. A *matroid* is a pair $(X, \langle\cdot\rangle)$ where X is a set and $\langle\cdot\rangle : P(X) \to P(X)$ is a mapping defined on $P(X)$, the power set of X, and satisfying the following conditions [C], [EP] and [Fin]:

[C] The mapping $\langle\cdot\rangle : P(X) \to P(X)$ is a *closure operator*, that is,

$(C1)$ $Y \subseteq \langle Y\rangle$, for all $Y \in P(X)$;

$(C2)$ $Z \subseteq Y \Rightarrow \langle Z\rangle \subseteq \langle Y\rangle$, for all $Y, Z \in P(X)$;

$(C3)$ $\langle Y\rangle = \langle\langle Y\rangle\rangle$, for all $Y \in P(X)$;

[EP] for all $x, y \in X$ and $Y \in P(X)$,

if $x \in \langle Y \cup \{y\}\rangle$ and $x \notin \langle Y\rangle$, then $y \in \langle Y \cup \{x\}\rangle$;

[Fin] for all $Y \in P(X)$, if $x \in \langle Y\rangle$, then $x \in \langle Y'\rangle$ for some finite $Y' \subseteq Y$.

A closure operator which satisfies [Fin] is called an *algebraic closure operator*. The general techniques of matroid theory guarantee, as mentioned above, that every matroid has a basis (i.e., a minimal generating set) and all the bases have the same cardinality.

Let $\mathcal{A}$ be an algebra with universe A. Given a subset X of A, we denote by $\langle X\rangle$ the subalgebra of $\mathcal{A}$ generated by X. It is obvious that $\langle\cdot\rangle$ is an algebraic closure operator in any algebra. We say that $\mathcal{A}$ is a *weak independence algebra* if $(A, \langle\cdot\rangle)$ is a matroid, that is, $\langle\cdot\rangle$ satisfies [EP] (we also say that $\mathcal{A}$ satisfies [EP]). Thus we can define the *rank* or *dimension* of a weak independence algebra to be the cardinality of any basis for the algebra.

We remark that any subalgebra of a weak independence algebra is clearly also a weak independence algebra, and hence we can also speak of the rank or dimension of a subalgebra. In particular, if $\mathcal{A}$ is a weak independence algebra and $a \in \mathrm{End}(\mathcal{A})$, then the dimension of $\mathrm{im}(a)$ is well defined. We refer to this cardinal as the *rank* of a, and denote it by $\mathrm{rank}(a)$.

Recall that our aim is to find a class of weak independence algebras such that for any algebra $\mathcal{A}$ in the class and any endomorphisms a, b of $\mathcal{A}$, we have

$$a \in SbS \Leftrightarrow \mathrm{rank}(a) \leqslant \mathrm{rank}(b) \tag{2}$$

where $S = \mathrm{End}(\mathcal{A})$. Whether or not (2) holds for a particular $\mathcal{A}$, at least $\mathrm{rank}(a)$ now has a meaning, in contrast to the situation for general algebras in (1). Thus we are going to look for conditions under which (2) holds.

To do this we need to investigate bases of subalgebras, that is, subsets of an algebra which are minimal generating sets for the subalgebras which they generate. There are several conditions involving such subsets which are equivalent to [EP] and which are very useful when applying [EP].

Before stating these conditions we mention the following easy lemma.

Lemma 1.1. *Let Y be a subset of an algebra $\mathcal{A}$. Then the following are equivalent:*

(1) *Y is a basis for the subalgebra it generates;*
(2) *for every y in Y, we have $y \notin \langle Y \setminus \{y\} \rangle$.*

A subset of an algebra $\mathcal{A}$ satisfying the equivalent conditions of Lemma 1.1 is said to be *independent*. From Exercise 6 on page 50 of [30] we have the following lemma.

Lemma 1.2. *For an algebra $\mathcal{A}$, the following conditions are equivalent:*

(1) *$\mathcal{A}$ satisfies [EP].*
(2) *For every subset X of A and every element $u \in A$, if X is independent and $u \notin \langle X \rangle$, then $X \cup \{u\}$ is independent.*
(3) *For every subset X of A, if Y is a maximal independent subset of X, then $\langle X \rangle = \langle Y \rangle$.*
(4) *For subsets X, Y of A with $Y \subseteq X$, if Y is independent, then there is an independent set Z with $Y \subseteq Z \subseteq X$ and $\langle Z \rangle = \langle X \rangle$.*

A straightforward consequence of Lemma 1.2 is that, in a weak independence algebra, a subset is a basis if and only if it is a maximal independent subset.

We now consider equation (2) again. Let $\mathcal{A}$ be a weak independence algebra and $a, b \in S = \mathrm{End}(\mathcal{A})$. Using Lemma 1.2, it is straightforward to show that, for $a, b \in S$, if $a \in SbS$, then $\mathrm{rank}(a) \leqslant \mathrm{rank}(b)$. For the converse, we want to prove that if $\mathrm{rank}(a) \leqslant \mathrm{rank}(b)$, then there exist $u, v \in S$ such that $a = ubv$. It is easy to show that for some mappings $u', v' \in T(A)$, we have $a = u'bv'$. However we have no guarantee that these mappings belong to S, that is, we do not know if we can choose these two mappings to be endomorphisms of $\mathcal{A}$.

To ensure that we can choose u', v' to be endomorphisms, we use the following existence condition.

[F] For any basis X of $\mathcal{A}$ and a function $\alpha : X \to A$, there is an endomorphism $\overline{\alpha}$ of $\mathcal{A}$ such that $\overline{\alpha}_{|X} = \alpha$.

An algebra satisfying $[EP]$ and $[F]$ is called an *independence algebra.* Now, again using Lemma 1.2, it is possible to prove that for an independence algebra $\mathcal{A}$ and $S = \text{End}(\mathcal{A})$ we have

$$a \in SbS \Leftrightarrow \text{rank}(a) \leqslant \text{rank}(b), \quad \text{for all } a, b \in S.$$

The crucial observation is that because of (4) of Lemma 1.2, condition [F] implies that any function from an independent subset of $\mathcal{A}$ can be extended to an endomorphism of $\mathcal{A}$. The details of the argument can be found in the original paper of Gould [18] where independence algebras are introduced.

We remark that there are some weak independence algebras other than independence algebras which satisfy equation (2). For example, any chain C regarded as a semigroup with multiplication given by $xy = \min\{x, y\}$ is a weak independence algebra and it is known that equation (2) holds for C. However, if C has more than one element, it is not an independence algebra.

Finally we should observe that when Gould defined independence algebras, a lot was already known about them. In contrast to the path described in this section which is based on properties of the endomorphism monoid of the algebra, the original discovery of independence algebras came through an investigation of notions of independence, as we will see in the next section. In order to avoid confusion we will use the term *C-independence* (because it is linked to a *Closure Operator*) for the notion of independence defined above.

2. Thoughts of Independence in Poland

In [28] Marczewski introduced a notion of independence (for universal algebras) and proved that many different notions of independence from numerous branches of mathematics are, in fact, particular cases of his notion.

Let $\mathcal{A} = (A, F)$ be a universal algebra (A is its universe and F is the set of fundamental operations). Marczewski denoted by $A^{(n)}$ the set of all n-ary algebraic operations (or, following [6], the set of all n-ary terms). Now we say that a set $N \subseteq A$ is *independent* if and only if it satisfies the following condition:

$[M_1]$ every map $f : N \to A$ can be extended to a morphism $\phi : \langle N \rangle \to \mathcal{A}$.

(Recall that $\langle N \rangle$ denotes the subalgebra generated by N). We shall call this notion of independence *M-independence* (for Marczewski) in order to distinguish it from the C-independence defined in the previous section. We say that a set is *M-dependent* (*C-dependent*) if it is not M-independent (C-independent). Investigating the interplay between these two concepts of independence led Marczewski and his colleagues to the idea of v^*-algebras [31] several decades before the advent of independence algebras [18], but in fact, the two types of algebras are precisely the same.

After introducing M-independence, Marczewski provides two more characterizations of it. One is the following. A set $N \subseteq A$ is M-independent if it satisfies

> $[M_2]$ for each sequence of different elements $a_1, \ldots, a_n \in N$ and for each pair of $g, h \in A^{(n)}$, if $g(a_1, \ldots, a_n) = h(a_1, \ldots, a_n)$, then g and h are identical in $\mathcal{A}$.

Then Marczewski closes this introductory section saying the following: *If N is a set of independent elements, then no $a \in N$ belongs to the subalgebra generated by the set $N \setminus \{a\}$. The converse implication is not true.* With our definitions this means that in any universal algebra M-independence implies C-independence, but not conversely. Therefore we can ask what is needed so that the algebra satisfies the converse? In a vector space M-independence coincides with C-independence. Thus we can check what makes this work in a vector space to see what properties we should require a universal algebra to satisfy in order to have the two notions of independence coinciding.

Suppose that a finite set of vectors $N = \{x_1, \ldots, x_n\}$ is M-dependent. Then, by $[M_2]$, there exist two linear combinations $s(x_1, \ldots, x_n) = \sum s_i x_i$ and $t(x_1, \ldots, x_n) = \sum t_i x_i$, such that $s(x_1, \ldots, x_n) = t(x_1, \ldots, x_n)$, but $s \neq t$. Moreover, it is well known that $s \neq t$ if and only if there exists an index $i \in \{1, \ldots, n\}$ such that $s_i \neq t_i$. (Since $+$ is commutative we can assume without loss of generality that $s_1 \neq t_1$). Therefore $\sum s_i x_i = \sum t_i x_i$ is equivalent to $x_1 = (s_1 - t_1)^{-1} \sum_{i=2}^{n} (t_i - s_i) x_i$. Thus $x_1 \in \langle x_2, \ldots, x_n \rangle$ and hence N is C-dependent. In general M-independence implies C-independence and now we know what is used to prove the converse in a vector space. Schematically, in a vector space, M-dependence implies C-dependence because of the following:

if $s(x_1, \ldots, x_n) = \sum s_i x_i = \sum t_i x_i = t(x_1, \ldots, x_n)$ and $s \neq t$, then

(1) there exists at least one index (say $i = 1$) such that $s_1 \neq t_1$, and

(2) there exists an $(n-1)$-term h such that $s(x_1,\ldots,x_n) = t(x_1,\ldots,x_n)$ is equivalent to $x_1 = h(x_2,\ldots,x_n)$.

Clearly the hypothesis and the second conclusion can be formulated for any universal algebra (using terms). However, the first conclusion cannot be directly expressed in a general setting. Therefore we need an equivalent way of expressing it. In fact, it is easy to see that, in a vector space V, if we have $s(x_1,\ldots,x_n) = \sum s_i x_i$ and $t(x_1,\ldots,x_n) = \sum t_i x_i$, then $s_1 \neq t_1$ if and only if there exist $a_1', a_1, \ldots, a_n \in V$ such that

$$s(a_1, a_2, \ldots, a_n) = t(a_1, a_2, \ldots, a_n) \text{ but } s(a_1', a_2, \ldots, a_n) \neq t(a_1', a_2, \ldots, a_n).$$

This fact leads to the following definition. Let $\mathcal{A} = (A, F)$ be a universal algebra and let s, t be two n-ary terms. We say that s and t are *distinguishable by the first variable* if there exist $a_1', a_1, \ldots, a_n \in A$ such that $s(a_1,\ldots,a_n) = t(a_1,\ldots,a_n)$, but $s(a_1', a_2, \ldots, a_n) \neq t(a_1', a_2, \ldots, a_n)$. It is clear how one defines *distinguishable by the ith variable*

We mention that in [36], instead of saying that two terms are distinguishable by the ith variable, Urbanik says that the terms *depend on the ith variable.*

After defining distinguishable terms (which corresponds in our scheme above to conclusion (1)), Marczewski goes on to define what he calls v-algebras. We say that an algebra $\mathcal{A} = (A, F)$ is a *v-algebra*, when for every pair of terms $s, t \in A^{(n)}$ $(n = 1, \ldots)$ distinguishable by the first variable there exists a term $h \in A^{(n-1)}$ such that $s(x_1,\ldots,x_n) = t(x_1,\ldots,x_n)$ is equivalent to $x_1 = h(x_2,\ldots,x_n)$.

It is now very easy to prove that, in v-algebras, M-independence coincides with C-independence. After introducing v-algebras and proving some direct consequences of the definition, Marczewski proves the following two (easy) results.

Lemma 2.1. *Let $\mathcal{A}$ be a v-algebra. Then a singleton subset of $\mathcal{A}$ whose sole element is not a constant is M-independent.*

Lemma 2.2. *If $\{a_1,\ldots,a_{n+1}\} \subseteq A$ is M-dependent, but $\{a_1,\ldots,a_n\}$ is M-independent, then there exists an n-ary term h such that $a_{n+1} = h(a_1,\ldots a_n)$.*

After these two lemmas Marczewski proves three more auxiliary lemmas and then proves the main results of the paper. In these theorems, attention is restricted to algebras that contain some elements which are not constants.

Theorem 2.1. *In a v-algebra $\mathcal{A}$ the following are equivalent:*

(1) *B is a basis (i.e., an M-independent set of generators) of $\mathcal{A}$;*
(2) *B is a minimal set of generators;*
(3) *B is a maximal M-independent set.*

Observe that in any algebra every M-independent set of generators is also a minimal set of generators and a maximal M-independent set.

Theorem 2.2. *If $\mathcal{A}$ is a v-algebra, then $\mathcal{A}$ admits a basis.*

Theorem 2.3. *If $\mathcal{A}$ is a v-algebra, then any two bases have the same number of elements.*

In all the results proved after Lemma 2.2, Marczewski uses only general facts about universal algebras and Lemmas 2.1 and 2.2. Therefore it is natural to think that Marczewski was aware of this and so knew that Theorems 2.1, 2.2 and 2.3 are true for any universal algebra which satisfies Lemmas 2.1 and 2.2. Eventually in [31] Narkiewicz introduced v^*-algebras (saying that *the investigation of v^*-algebras was suggested to me by Professor E. Marczewski*) using the following definition.

An algebra $\mathcal{A}$ is a *v^*-algebra* if it satisfies the following conditions:

(I) If $a \in A$ is not a constant, then the set $\{a\}$ is M-independent;
(II) If $\{a_1, \dots, a_n\}$ is an M-independent subset of $\mathcal{A}$, but $\{a_1, \dots, a_{n+1}\}$ is not M-independent, then $a_{n+1} \in \langle a_1, \dots, a_n \rangle$.

Since (I) and (II) come from Lemma 2.1 and Lemma 2.2, it follows that v-algebras are v^*-algebras (but not conversely). Then Narkiewicz provides an equivalent definition for v^*-algebras (suggested to him by Świerezkowski) by proving the following result.

Proposition 2.1. *An algebra $\mathcal{A}$ is a v^*-algebra if and only if it satisfies the following conditions:*

(III) *in $\mathcal{A}$, M-independence coincides with C-independence;*
(IV) *in every subalgebra with a finite basis (i.e., M-independent generating set) consisting of k elements, every M-independent set of k elements forms a basis for the subalgebra.*

In [32] Narkiewicz provided yet another set of defining conditions for v^*-algebras as recorded in the next result.

Proposition 2.2. *An algebra is a v^*-algebra if and only if it satisfies* (III) *of Proposition* 2.1 *and* [EP].

It is very easy to prove that (III) and [EP] are equivalent to [F] and [EP]. Thus the class of v^*-algebras coincides with the class of independence algebras.

It is worth observing that the research carried out in Poland during the sixties led to many other classes of algebras (v_*-algebras, v_*^*-algebras, v'-algebras, separable variables algebras, etc.) and to more general notions of independence. A comprehensive survey paper (containing more than eight hundred references) was written by Głazek [16].

3. The Description of Independence Algebras

As with any algebraic structure, the question arises of whether independence algebras can be classified in some way. Marczewski's paper [29] about v-algebras is followed in the same issue of the same journal by an announcement of Urbanik [36] in which he describes all v-algebras (the proofs appeared in [37]). The characterisation is up to term equivalence where, we recall, that two algebras on the same underlying set are *term equivalent* if their sets of n-ary term operations are the same for each positive integer n.

Characterising the v^*-algebras turned out to be much more complicated but it was eventually accomplished. For a survey paper (and proof of the final cases) see [38] to which we refer the reader for a description of all independence algebras.

It was more than 30 years before any new results about independence algebras (from the perspective of universal algebra) appeared. In [7] Cameron and Szabó prove the following result.

Theorem 3.1. *The subalgebra lattice of an independence algebra of finite dimension is a Boolean lattice or a projective or affine geometry.*

They also reproved the description of finite independence algebras, providing for this case an extraordinarily short proof.

4. Independence Algebras and Semigroups

The monoids $T(X)$ and $\mathrm{End}(V)$ (where X is a set and V is a vector space) have much more in common than the fact that their principal ideals form a well ordered chain. For example, if $T_f(X)$ denotes the set of all members of $T(X)$ with finite image, and $\mathrm{End}_f(V)$ denotes the set of all endomorphisms of V of finite rank, then $T_f(X)$ and $\mathrm{End}_f(V)$ are both completely semisimple semigroups, that is, they are regular and all their principal factors are completely 0-simple or completely simple. The search for algebras whose endomorphism monoids enjoy properties similar to those of $T(X)$ and $\mathrm{End}(V)$ led Gould [18] to rediscover independence algebras. In [18], she described the structure of the endomorphism monoid of an independence algebra, characterising Green's relations, and showing that the set of endomorphisms of finite rank is a completely semisimple subsemigroup. The original motivation for studying these endomorphism monoids was the hope of finding semigroup analogues of her work with Petrich [19] on rings of quotients. This aim has not yet been realised, but in a recent paper [13] a small step in this direction has been made.

However, since the early 1990s, endomorphism monoids of independence algebras, and related semigroups, have been extensively studied and the topic continues to receive a great deal of attention. It is not our purpose to give a comprehensive survey of this rapidly growing body of work, but it seems appropriate to give an account of some of the early results which influenced the development of the topic.

In 1966 Howie described the subsemigroup $E(X)$ of $T(X)$ generated by all the non-identity idempotents [23]. The corresponding result for $\mathrm{End}(V)$ where V is a finite dimensional vector space quickly followed [10], but it was nearly twenty years later before Reynolds and Sullivan [34] found the appropriate analogue in the infinite dimensional case. Their work also uncovered a significant difference between the semigroups $E(X)$ and $E(V)$ where X is an infinite set, V is an infinite dimensional vector space, and where for any algebra $\mathcal{A}$ we denote the subsemigroup of $\mathrm{End}(\mathcal{A})$ generated by the non-identity idempotents by $E(\mathcal{A})$. Sullivan surveyed the parallels and distinctions between $T(X)$ and $\mathrm{End}(V)$ in an influential conference talk in 1990 which was published in [35].

Fountain and Lewin, having seen a preliminary version of Gould's paper [18], realised that independence algebras provided a suitable conceptual framework for unifying those results on products of idempotents which hold for both $T(X)$ and $\mathrm{End}(V)$. They described $E(\mathcal{A})$ for an independence algebra of finite rank in [14], proving the following result. (For a direct proof see [1].)

Theorem 4.1. *If $\mathcal{A}$ is an independence algebra of finite rank n, then*

$$E(\mathcal{A}) = \langle E_1 \rangle = \mathrm{End}(\mathcal{A}) \setminus \mathrm{Aut}(\mathcal{A})$$

where E_1 is the set of idempotents of rank $n-1$ in $\mathrm{End}(\mathcal{A})$.

The results of Howie [23] for finite sets and Erdos [10] are simply special cases.

Let S be a semigroup generated by its set of idempotents E. The *depth* of S is the smallest integer k such that $E^k = S$. Using this notion, Howie [24] refined the result about idempotent generation of $E(X)$ when X is finite by showing that $E(X)$ has depth $[\frac{3}{2}(n-1)]$. In contrast, Ballantine [5], Dawlings [9] and Laffey [27] independently obtained results showing that, for a vector space V of finite dimension n, the semigroup $E(V)$ has depth n. Ballantine's results were extended to a special class of finite rank independence algebras in [17].

In the infinite rank case attention was restricted to strong independence algebras. An independence algebra $\mathcal{A}$ is *strong* when for any independent subsets X and Y of $\mathcal{A}$, if $\langle X\rangle \cap \langle Y\rangle = \langle\emptyset\rangle$, then $X \cup Y$ is independent. For such an algebra $\mathcal{A}$, a characterisation of $E(\mathcal{A})$ generalising the results of [23] and [34] was given in [15]. To describe this result we need the notions of shift, defect and collapse for an endomorphism of a strong independence algebra. First, we remark that the exchange property allows us to define the co-rank of a subalgebra $\mathcal{B}$ of an independence algebra $\mathcal{A}$ as follows. Let X be a basis for $\mathcal{B}$ and extend X to obtain a basis $X \cup Y$ for $\mathcal{A}$. Then the *co-rank* of $\mathcal{B}$ (in $\mathcal{A}$) is defined to be the cardinal $|Y|$.

We also observe that if α is an endomorphism of an independence algebra $\mathcal{A}$, then $\mathrm{im}(\alpha)$ and $\mathrm{fix}(\alpha)$ are subalgebras where

$$\mathrm{fix}(\alpha) = \{a \in A : a\alpha = a\}.$$

We can now define the *defect*, $d(\alpha)$, and the *shift*, $s(\alpha)$ as follows:

$$d(\alpha) = \text{co-rank}\,\mathrm{im}(\alpha),$$
$$s(\alpha) = \text{co-rank}\,\mathrm{fix}(\alpha).$$

It is a little more complicated to define the collapse of α; we start by putting $K = \langle\emptyset\rangle\alpha^{-1}$, the inverse image under α of the subalgebra of constants. Now let T be any subset of A such that $T\alpha$ is a basis for $\mathrm{im}\,\alpha$ and α restricted to T is one-one. Let $M = \langle K \cup T\rangle$ and define the *collapse*, $c(\alpha)$, of α by

$$c(\alpha) = \mathrm{rank}\,K + \text{co-rank}\,M.$$

This definition is independent of the choice of T, so we do have a well defined notion.

Now let $\mathcal{A}$ be a strong independence algebra of infinite rank. Define subsets F and Q of $\mathrm{End}(\mathcal{A})$ as follows:

$$F = \{\alpha \in \mathrm{End}\,\mathcal{A} : 0 < d(\alpha), s(\alpha) < \aleph_0\},$$
$$Q = \{\alpha \in \mathrm{End}\,\mathcal{A} : d(\alpha) = s(\alpha) = c(\alpha) \geqslant \aleph_0\}.$$

In fact, it can be shown that

$$F = \{\alpha \in \operatorname{End} \mathcal{A} : 0 < c(\alpha) = d(\alpha) \leqslant s(\alpha) < \aleph_0\}.$$

Recall that an element s of a monoid S is *unit regular* if $sus = s$ for some unit u of S, and let U be the set of unit regular members of $\operatorname{End}(\mathcal{A})$. Then U is a submonoid of $\operatorname{End}(\mathcal{A})$ and

$$Q = \{\alpha \in U : d(\alpha) = s(\alpha)\}.$$

We can now characterise the subsemigroup $E(\mathcal{A})$ generated by the non-identity idempotents of $\operatorname{End}(\mathcal{A})$.

Theorem 4.2. *If $\mathcal{A}$ is a strong independence algebra of infinite rank, then F and Q are regular idempotent generated subsemigroups of* $\operatorname{End}(\mathcal{A})$ *and*

$$E(\mathcal{A}) = F \cup Q.$$

If α is an endomorphism of a vector space, then it is not difficult to see that the collapse of α is the nullity of α, and this observation shows that the result of Reynolds and Sullivan [34] describing the idempotent generated subsemigroup of $\operatorname{End}(V)$ where V is an infinite dimensional vector space is a consequence of Theorem 4.2. When specialised to sets, the notion of collapse we have defined is not exactly the same as the original definition in [23]. However, the two notions give the same value in the case of infinite collapse, and so the characterisation of $E(X)$ for an infinite set X given in [23] also follows from Theorem 4.2.

The members of Q are called *balanced* endomorphisms, and it is here that there is a difference betwen the set case and the vector space case. In the case of an infinite set, it was shown in [25] that the semigroup of balanced endomorphisms has depth 4 whereas in the case of an infinite dimensional vector space Reynolds and Sullivan [34] show that the depth is 3. In [12] Fountain determined a property of some strong independence algebras which distinguishes between those for which the semigroup of balanced endomorphisms has depth 4 and those for which it has depth 3.

We conclude by emphasising that independence algebras have proved to be very useful in providing unified proofs for analogous results for $T(X)$ and $\operatorname{End}(V)$; in explaining differences between $T(X)$ and $\operatorname{End}(V)$; and also in exporting results from semigroups to linear algebra and to some other universal algebras. (See, for example, [4], [2] and [3]).

Acknowledgements

The first author acknowledges with thanks the support of FCT, within the POCTI project "Fundamental and Applied Algebra" of CAUL, and Fundação Calouste Gulbenkian.

References

1. J. Araújo, Idempotent generated endomorphisms of an independence algebra, *Semigroup Forum*, to appear.
2. J. Araújo, Generators for the semigroup of endomorphisms of an independence algebra, *Algebra Colloq.* **9** (2002), 375–382.
3. J. Araújo, J.D. Mitchell and N. Silva, On embedding countable sets of endomorphisms, *Algebra Universalis*, to appear.
4. J. Araújo and F. C. Silva, Semigroups of linear endomorphisms closed under conjugation, *Comm. Algebra* **28** (2000), 3679–3689.
5. C. S. Ballantine, Products of idempotent matrices, *Linear Algebra and Appl.* **19** (1978), 81–86.
6. S. Burris and H. P. Sankappanavar, *A course in universal algebra*, Springer-Verlag, New York–Heidelberg–Berlin, 1981.
7. P. J. Cameron and C. Szabó, Independence algebras, *J. London Math. Soc.* **61** (2000), 321–334.
8. A. H. Clifford and G. B. Preston, *The algebraic theory of semigroups* Vol. 1, Amer. Math. Soc., Providence, R.I., 1961.
9. R. J. H. Dawlings, Products of idempotents in the semigroup of singular endomorphisms of finite-dimensional vector space, *Proc. Roy. Soc. Edinburgh Sect. A* **91** (1981/82), 123–133.
10. J.A. Erdos, On products of idempotent matrices, *Glasgow Math. J.* **8** (1967), 118–122.
11. C.-A. Faure and A. Frölicher, *Modern projective geometry*, Kluwer, Dordrecht, 2000.
12. J. Fountain, The depth of the semigroup of balanced endomorphisms, *Mathematika* **41** (1994), 199–208.
13. J. Fountain and V. Gould, Endomorphisms of relatively free algebras with weak exchange properties, *Algebra Universalis*, to appear.
14. J. Fountain and A. Lewin, Products of idempotent endomorphisms of an independence algebra of finite rank, *Proc. Edinburgh Math. Soc.* **35** (1992), 493–500.
15. J. Fountain and A. Lewin, Products of idempotent endomorphisms of an independence algebra of infinite rank, *Math. Proc. Camb. Phil. Soc.* **114** (1993), 303–319.
16. K. Głazek, Some old and new problems in the independence theory, *Colloq. Math.* **42** (1979), 127–189.
17. G. M. S. Gomes and J. M. Howie, Idempotent endomorphisms of an independence algebra of finite rank, *Proc. Edinburgh Math. Soc.* **38** (1995), 107–116.
18. V. Gould, Independence algebras, *Algebra Universalis* **33** (1995), 294–318.

19. V. Gould and M. Petrich, A new approach to orders in simple rings in one-sided ideals, *Semigroup Forum* **41** (1990), 267–290.
20. G. Grätzer and J. Sichler, On the endomorphism semigroup (and category) of bounded lattices, *Pacific J. Math.* **35** (1970), 639–647.
21. G. Grätzer, *Universal Algebra*, Springer-Verlag, New York, 1979.
22. Z. Hedrlín and J. Lambek, How comprehensive is the category of semigroups?, *J. Algebra* **11** (1969), 195–212.
23. J. M. Howie, The subsemigroup generated by the idempotents of a full transformation semigroup, *J. London Math Soc.* **41** (1966), 707–716.
24. J. M. Howie, Products of idempotents in finite full transformation semigroups, *Proc. Roy. Soc. Edinburgh Sect. A* **86** (1980), 243–254.
25. J. M. Howie, Some subsemigroups of infinite full transformation semigroups, *Proc. Roy. Soc. Edinburgh Sect. A* **88** (1981), 159–167.
26. J. M. Howie, *Fundamentals of Semigroup Theory*, Oxford University Press, Oxford, 1995.
27. T. J. Laffey, Products of idempotent matrices, *Linear and Multilinear Algebra* **14** (1983), 309–314.
28. E. Marczewski, A general scheme of the notions of independence in mathematics, *Bull. Acad. Pol. Sci.* **6** (1958), 731–736.
29. E. Marczewski, Independence in some abstract algebras, *Bull. Acad. Pol. Sci.* **7** (1959), 611–616.
30. R.N. McKenzie, G. F. McNulty and W. F. Taylor, *Algebra, lattices, varieties*, Vol. I (Wadsworth, Monterey, 1983).
31. W. Narkiewicz, Independence in a certain class of abstract algebras, *Fund. Math.* **50** (1961/62), 333–340.
32. W. Narkiewicz, On a certain class of abstract algebras, *Fund. Math.* **54** (1964), 115–124.
33. J. G. Oxley, Infinite matroids, in: N. White (ed.) *Matroid Applications*, Cambridge University Press, 1992, pp. 73–90.
34. M. A. Reynolds and R. P. Sullivan, Products of idempotent linear transformations, *Proc. Roy. Soc. Edinburgh* **100A** (1985), 123–138.
35. R. P. Sullivan, Transformation semigroups and linear algebra, in: T. E. Hall, P. R. Jones and J. C. Meakin (eds.), *Proc. Monash Conference on Semigroup Theory*, World Scientific, 1991, pp. 290–295.
36. K. Urbanik, Representation theorem for Marczewski's algebras, *Bull. Acad. Pol. Sc.* **7** (1959), 617–619.
37. K. Urbanik, A representation theorem for Marczewski's algebras, *Fund. Math.* **48** (1959/60), 147–167.
38. K. Urbanik, Linear independence in abstract algebras, *Colloq. Math.* **14** (1966), 233–255.

ABELIAN KERNELS, SOLVABLE MONOIDS AND THE ABELIAN KERNEL LENGTH OF A FINITE MONOID

MANUEL DELGADO

Centro de Matemática da Universidade do Porto,
Rua do Campo Alegre, 687,
4169-007 Porto, Portugal
E-mail: mdelgado@fc.up.pt

VíTOR H. FERNANDES*

Departamento de Matemática,
Faculdade de Ciências e Tecnologia da Universidade Nova de Lisboa,
Monte da Caparica, 2829-516 Caparica, Portugal
E-mail: vhf@fct.unl.pt

The notion of abelian kernel of a finite monoid extends the notion of derived subgroup of a finite group. Extensions to finite monoids of the notions of solvable group and derived length of a solvable group appear then naturally. In this paper we study these notions for some classes of finite monoids.

1. Introduction and Preliminaries

In this paper we do not make a clear distinction between what is introduction and preliminaries. In fact, we have decided to put it all in a single section which is divided into two subsections. The first, concerning kernels of finite monoids and related properties, should mainly serve as a general motivation for the subsequent study of some particular classes of monoids. In the second subsection these classes of monoids are introduced. A little history on studies made involving these monoids should make clear their importance.

All monoids considered in this paper are assumed to be finite.

*Author's second address: Centro de Álgebra da Universidade de Lisboa, Av. Prof. Gama Pinto, 2, 1649–003 Lisboa, Portugal.

1.1. *Kernels and related properties*

A conjecture of J. Rhodes, which became known as the *Type II Conjecture*, attracted the attention of many semigroup theorists during about two decades before being solved. Its first solution, given by Ash [10], appeared in the early nineties. Almost at the same time an independent solution was given by Ribes and Zalesskiĭ [48]. For motivation, history and some consequences of the type II conjecture we refer the reader to [38]. A recent proof of the type II conjecture was given by Auinger [12]. The techniques used by Ash and by Ribes and Zalesskiĭ in their solutions are seemingly very different (although some connections between them have been found [19, 7]): algebraic-combinatorial methods are used in Ash's solution, while profinite methods play a crucial role in Ribes and Zalesskiĭ's solution.

Ash's and Ribes and Zalesskiĭ's solutions involve (and are themselves) deep results and brought many new ideas into semigroup theory. But they also attracted also the attention of researchers of other areas of Mathematics, such as Model Theory, as may be inferred by works of Herwig and Lascar [39] (see also [4, 5]) and Coulbois [15, 16].

The type II conjecture proposed an algorithm to compute the *kernel* of a finite monoid, where the notion of kernel was taken relative to the pseudovariety G of all finite groups. This notion, to be defined below, can be given relative to any other pseudovariety of groups in exactly the same way.

The problem of computing the kernel of a finite monoid relative to a pseudovariety H of groups happens to be a particular instance of the widely studied hyperdecidability [2] or tameness [8] properties of H: a tame pseudovariety H of groups is hyperdecidable and the kernel relative to a hyperdecidable pseudovariety of groups is computable. Although kernels have to be relative to pseudovarieties of groups, tameness and hyperdecidability are defined more generally, for pseudovarieties of semigroups. Among the pseudovarieties of groups that have been considered in connection with these properties, we refer the pseudovariety G (although using other terminology, Ash's paper [10] essentially proves its tameness (see [7])), the pseudovariety $\mathsf{G_p}$ of all finite p-groups where p is a prime (studied by Steinberg [49] and Almeida [3] who proved its tameness) and the pseudovariety Ab of all finite abelian groups (see [18, 19, 6]; the joint work with Almeida proves the tameness of Ab).

The kernel of a monoid relative to the pseudovariety Ab will be called an *abelian kernel.*

The feasibility in practice of an algorithm to compute the abelian kernel of a monoid has also interested the first author [20, 26] and an implementation in GAP [50] of an algorithm with the purpose of computing abelian kernels of finite monoids is part of a package that is currently being prepared [21]. Computations achieved using this software helped us to formulate and prove some of the results used in this paper (most of them proved elsewhere).

We assume some knowledge of semigroups, mainly on Green's relations and inverse semigroups (references include [43, 44]). For the basics on kernels in general and abelian kernels in particular, we refer the reader to [38] and [18] respectively.

Given monoids M and N, a *relational morphism of monoids* $\tau : M \twoheadrightarrow N$ is a function from M into the power set of N such that $1 \in \tau(1)$, $\tau(s_1) \neq \emptyset$ and $\tau(s_1)\tau(s_2) \subseteq \tau(s_1 s_2)$, for all $s_1, s_2 \in M$.

Clearly, τ can be viewed as a subset of $M \times N$ (i.e., a relation from M to N) satisfying certain conditions. Homomorphisms, viewed as relations, and inverses of onto homomorphisms are examples of relational morphisms.

The *abelian kernel of a monoid* M is the submonoid $\mathsf{K}_{\mathsf{Ab}}(M) = \bigcap \tau^{-1}(1)$, with the intersection being taken over all relational morphisms $\tau : M \twoheadrightarrow G$, with $G \in \mathsf{Ab}$. Replacing Ab by any other pseudovariety H of groups, one obtains the definition of *kernel relative to* H.

As already mentioned, the abelian kernel of a finite group is a well known subgroup:

Proposition 1.1. [18] *The abelian kernel of a finite group G is precisely its derived subgroup G'.*

Thus, as one example, one can easily check [27] that the abelian kernel of the dihedral group $G = \langle g, h \mid h^2 = g^n = 1, gh = hg^{-1} \rangle$ of order $2n$ is the subgroup $\langle g^2 \rangle$ generated by g^2.

A finite group G is said to be *solvable* if there exists a non negative integer n such that the nth derived subgroup reduces to the neutral element of G. When such an n exists, the least one is said to be the *derived length* of G.

It is easy to see (and is stated in several of our references) that the kernel (relative to any pseudovariety of groups) of a finite monoid is a submonoid containing the idempotents. Thus we can define recursively $\mathsf{K}_{\mathsf{Ab}}^{(n)}(M)$ as follows:

- $\mathsf{K}_{\mathsf{Ab}}^{(0)}(M) = M$;

- $\mathsf{K}_{\mathsf{Ab}}^{(n+1)}(M) = \mathsf{K}_{\mathsf{Ab}}(\mathsf{K}_{\mathsf{Ab}}^{(n)}(M))$, for $n \geq 1$.

Denoting by $E(M)$ the set of idempotents of M, we have that, for any positive integer n, $\langle E(M) \rangle \subseteq \mathsf{K}_{\mathsf{Ab}}^{(n)}(M)$.

In analogy with the group case, we say that M is *solvable* if $\mathsf{K}_{\mathsf{Ab}}^{(n)}(M) = \langle E(M) \rangle$, for some non-negative integer n. In analogy with the group case again, when such an n exists, the least one is said to be the *abelian kernel length* of M. We denote this non-negative integer by $\ell_{\mathsf{Ab}}(M)$.

Clearly, for a solvable group G, $\ell_{\mathsf{Ab}}(G)$ is the derived length of G. In particular, if G is a non-trivial abelian group then $\ell_{\mathsf{Ab}}(G) = 1$. On the other hand, bands and, more generally, idempotent generated monoids have abelian kernel length equal to zero.

Next we recall some results, proved by the authors in earlier papers, which will be used in the sequel.

Theorem 1.1. [24] *Let T be a monoid, let $x_1, x_2, \ldots, x_k, y$ be a set of generators of T such that $y^2 = 1$ and let S be the submonoid of T generated by $x_1, x_2, \ldots, x_k$. If for each $i \in \{1, \ldots, k\}$ there exists $u_i \in S$ such that $x_i y = y u_i$, then $\mathsf{K}_{\mathsf{Ab}}(T) \subseteq S$.* □

Another result that we will need later on is:

Proposition 1.2. [22] *Let M be a monoid that is a disjoint union of a submonoid N and an ideal. Then $\mathsf{K}_{\mathsf{Ab}}(M) \cap N = \mathsf{K}_{\mathsf{Ab}}(N)$.* □

In fact, we will use the following particular case of this result:

Corollary 1.1. *Let M be a monoid with group of units G. Then $\mathsf{K}_{\mathsf{Ab}}(M) \cap G = G'$.* □

For an inverse monoid, the following stronger result holds:

Proposition 1.3. [23] *Let M be an inverse monoid. Let J be a non-trivial $\mathcal{J}$-class of M. If J is a $\leq_{\mathcal{J}}$-maximal $\mathcal{J}$-class among the non-trivial $\mathcal{J}$-classes of M, then an element of J belongs to the Abelian kernel of M if and only if it belongs to the derived subgroup of a maximal subgroup of M.* □

Proposition 1.4. [23] *Let M be an inverse monoid. Then M is solvable if and only if all subgroups of M are solvable.* □

The previous proposition was the basis for the proof of the already mentioned more general analogue for semigroups whose idempotents commute [23]. A different proof for an even more general result (which holds for semigroups whose idempotents generate an aperiodic semigroup) was given in [25].

Natural problems on solvable groups have now their counterpart on solvable monoids. For instance, the determination of the derived length of a solvable group has the "determination of the *abelian kernel length*" as its monoid counterpart. In this paper we treat this problem for the transformation monoids that are the subject of next subsection.

1.2. *On some transformation monoids*

Let $n \in \mathbb{N}$. Let X_n be a chain with n elements. By default we take $X_n = \{1 < 2 < \cdots < n\}$. As usual, we denote by $\mathcal{PT}_n$ the monoid of all (partial) transformations of X_n (under composition), by $\mathcal{T}_n$ the submonoid of $\mathcal{PT}_n$ of all full transformations of X_n, by $\mathcal{I}_n$ the symmetric inverse semigroup on X_n, i.e. the submonoid of $\mathcal{PT}_n$ of all injective transformations of X_n, and by $\mathcal{S}_n$ the symmetric group on X_n, i.e. the subgroup of $\mathcal{PT}_n$ of all injective full transformations (permutations) of X_n.

We say that a transformation s in $\mathcal{PT}_n$ is *order-preserving* if, for all $x, y \in \mathrm{Dom}(s)$, $x \leq y$ implies $xs \leq ys$, and *order-reversing* if, for all $x, y \in \mathrm{Dom}(s)$, $x \leq y$ implies $xs \geq ys$. An immediate but important property is that the product of two order-preserving transformations or two order-reversing transformations is an order-preserving transformation and the product of an order-preserving transformation with an order-reversing transformation, or vice-versa, is an order-reversing transformation. We denote by $\mathcal{PO}_n$ the submonoid of $\mathcal{PT}_n$ of all order-preserving transformations and by $\mathcal{POD}_n$ the submonoid of $\mathcal{PT}_n$ of all order-preserving transformations together with all order-reversing transformations. The full transformation counterparts are $\mathcal{O}_n$, the submonoid of $\mathcal{PO}_n$ of all (order-preserving) full transformations, and $\mathcal{OD}_n$, the submonoid of $\mathcal{POD}_n$ of all (order-preserving or order-reversing) full transformations. We have also the injective versions: $\mathcal{POI}_n$, the inverse submonoid of $\mathcal{I}_n$ of all order-preserving transformations; and $\mathcal{PODI}_n$, the inverse submonoid of $\mathcal{I}_n$ whose elements belong to $\mathcal{POD}_n$.

The order preserving or reversing notions can be generalized in the following way: let $c = (c_1, c_2, \ldots, c_t)$ be a sequence of t ($t \geq 0$) elements from the chain X_n and say that c is *cyclic* (respectively, *anti-cyclic*) if there exists no more than one index $i \in \{1, \dots, t\}$ such that $c_i > c_{i+1}$ (respectively, $c_i < c_{i+1}$), where $c_{t+1} = c_1$. Then, given $s \in \mathcal{PT}_n$ such that $\mathrm{Dom}(s) = \{a_1, \dots, a_t\}$, with $t \geq 0$ and $a_1 < \cdots < a_t$, we say that s is an *orientation-preserving* (respectively, *orientation-reversing*) transformation if the sequence of its images $(a_1 s, \dots, a_t s)$ is cyclic (respectively, anti-cyclic). As in the order case, the product of two orientation-preserving transformations or of two orientation-reversing transformations is an orientation-preserving transformation and the product of an orientation-preserving transformation by an orientation-reversing transformation, or vice-versa, is an orientation-reversing transformation. We denote by $\mathcal{POP}_n$ the submonoid of $\mathcal{PT}_n$ of all orientation-preserving transformations and by $\mathcal{POR}_n$ the submonoid of $\mathcal{PT}_n$ of all orientation-preserving transformations together with all orientation-reversing transformations. The full transformation and injective transformation counterparts are: $\mathcal{OP}_n$, the submonoid of $\mathcal{POP}_n$ of all (orientation-preserving) full transformations; $\mathcal{OR}_n$, the submonoid of $\mathcal{POR}_n$ of all (orientation-preserving or orientation-reversing) full transformations; $\mathcal{POPI}_n$, the inverse submonoid of $\mathcal{I}_n$ whose elements are all orientation-preserving transformations; and $\mathcal{PORI}_n$, the inverse submonoid of $\mathcal{I}_n$ whose elements belong to $\mathcal{POR}_n$.

We have the following diagram with respect to the inclusion (submonoid) relation:

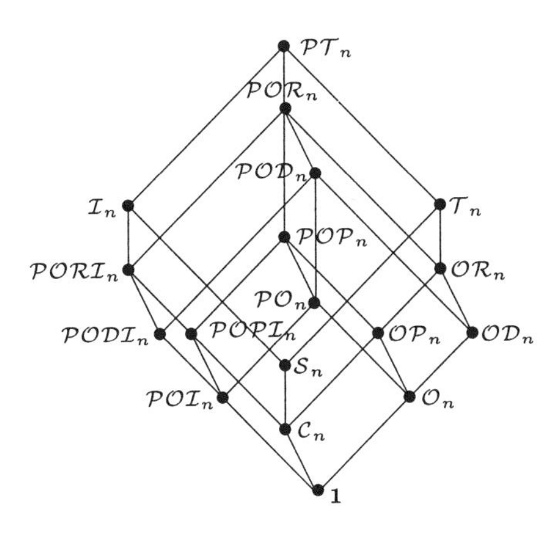

Diagram 1

(denoting by $\mathbf{1}$ the trivial monoid and by $\mathcal{C}_n$ the cyclic group of order n).

Some of these monoids of transformations have been studied since the sixties. In fact, presentations for $\mathcal{O}_n$ and $\mathcal{PO}_n$ were established respectively by Aĭzenštat [1] in 1962 and by Popova [46] in the same year. Some years later (1971) Howie [42] studied some combinatorial and algebraic properties of $\mathcal{O}_n$ and, in 1992, Gomes and Howie [37] established some more properties of $\mathcal{O}_n$, namely its rank and idempotent rank. Also in [37] the monoid $\mathcal{PO}_n$ was studied. The monoid $\mathcal{O}_n$ played also a main role in several other papers [40, 51, 9, 28, 47, 34] where the central topic concerns the problem of the decidability of the pseudovariety generated by the family $\{\mathcal{O}_n \mid n \in \mathbb{N}\}$. This question was posed by J.-E. Pin in 1987 in the "Szeged International Semigroup Colloquium" and, as far as we know, is still unanswered.

The monoid $\mathcal{POI}_n$ has been studied since 1997 by the second author in various papers [28, 29, 31, 32, 34] and also by Cowan and Reilly in [17].

The notion of an orientation-preserving transformation was introduced by McAlister in [45] and also by Catarino and Higgins [14], who have studied several properties of the monoid $\mathcal{OP}_n$. The monoid $\mathcal{OP}_n$ was also considered in [13, 11, 33]. The injective counterpart of $\mathcal{OP}_n$, the monoid $\mathcal{POPI}_n$, was studied by the second author in [30, 33].

Recently, some properties of the monoids $\mathcal{PODI}_n$, $\mathcal{POPI}_n$ and $\mathcal{PORI}_n$ were studied by the second author, Gomes and Jesus in [35].

Finally, in [36] the monoids $\mathcal{OD}_n$, $\mathcal{POD}_n$, $\mathcal{POP}_n$ and $\mathcal{POR}_n$ were the objects in study. In particular, presentations for them all were given.

It remains to remark that the papers mentioned above are not all concerning the monoids in Diagram 1. Many other papers were written in the past four decades about order-preserving transformation monoids and some of their extensions.

The authors themselves have already considered the inverse monoids of Diagram 1. In particular, the abelian kernels of $\mathcal{POI}_n$ and of $\mathcal{POPI}_n$ were determined in [22] and the abelian kernels of $\mathcal{PODI}_n$, $\mathcal{PORI}_n$ and $\mathcal{I}_n$ in [24]. In [23] it is shown, in particular, that a monoid M whose idempotents commute is solvable exactly when all subgroups of M are solvable. This result was generalized in [25]: it is valid for monoids whose idempotents generate an aperiodic semigroup. Since all subgroups of the monoids in Diagram 1 not containing $\mathcal{S}_n$ are solvable, one gets immediately that the monoids $\mathcal{POI}_n, \mathcal{POPI}_n, \mathcal{PODI}_n$ and $\mathcal{PORI}_n$ are solvable as they are inverse and so their idempotents commute.

The following section is devoted to a general result that gives a bound for the abelian kernel length of a finite solvable monoid.

The paper has then two more sections, the first of which is devoted to the inverse monoids of the above diagram; the second is devoted to the non-inverse ones.

2. The abelian kernel length of a solvable monoid

In this section we present an upper bound for the abelian kernel length of a solvable inverse monoid M. It is not difficult to see (it is the easy part of Proposition 1.4) that all subgroups of M must be solvable: moreover, all submonoids of M must be solvable.

For a non-trivial $\mathcal{J}$-class J of M denote by $\ell(J)$ the maximum size of a $\leq_{\mathcal{J}}$-chain of non-trivial $\mathcal{J}$-classes of M having J as the $\leq_{\mathcal{J}}$-minimum. Denote by $\Omega(J)$ the subset of $M/\mathcal{J}$ of all non-trivial $\mathcal{J}$-classes J' such that $\ell(J') = \ell(J)$ and by $\lambda(J)$ the maximum of the derived lengths of the maximal subgroups of M contained in the members of $\Omega(J)$. Notice that two distinct elements of $\Omega(J)$ are not $\leq_{\mathcal{J}}$-comparable.

We are now prepared to state the following result:

Proposition 2.1. *Let M be a solvable inverse monoid and let $\{J_1 <_{\mathcal{J}} J_2 <_{\mathcal{J}} \cdots <_{\mathcal{J}} J_k\}$ be a $\leq_{\mathcal{J}}$-chain of maximum size of non-trivial $\mathcal{J}$-classes of M. Then $\ell_{\mathsf{Ab}}(M) \leq \sum_{i=1}^{k}(1+\max\{0, \lambda(J_i)-1\})$.*

Proof. We proceed by induction on the maximum size k of a $\leq_{\mathcal{J}}$-chain of non-trivial $\mathcal{J}$-classes of M.

First, notice that if M has only trivial $\mathcal{J}$-classes, i.e. if M is a semilattice, then the inequality reduces to $\ell_{\mathsf{Ab}}(M) \leq 0$, which is trivially valid.

So we may suppose that M has non-trivial $\mathcal{J}$-classes, i.e. $k \geq 1$, and assume, by induction hypothesis, the validity of the inequality for any (solvable inverse) monoid which has $k-1$ as the maximum size of a $\leq_{\mathcal{J}}$-chain of non-trivial $\mathcal{J}$-classes.

Let $N = \{x \in M \mid J_x <_{\mathcal{J}} J, \text{ for some } J \in \Omega(J_k)\} \cup E(M)$. Then N is an inverse submonoid of M whose Green relation $\mathcal{J}$ coincides with the restriction to N of the Green relation $\mathcal{J}$ on M and $\{J_1 <_{\mathcal{J}} J_2 <_{\mathcal{J}} \cdots <_{\mathcal{J}} J_{k-1}\}$ is a $\leq_{\mathcal{J}}$-chain of maximum size of non-trivial $\mathcal{J}$-classes of N. Also, given a non-trivial $\mathcal{J}$-class J of N, the value $\lambda(J)$ calculated in N coincides with the value $\lambda(J)$ considered in M.

Since N is solvable, by induction hypothesis we have $\ell_{\mathsf{Ab}}(N) \leq \sum_{i=1}^{k-1}(1+\max\{0, \lambda(J_i)-1\})$.

Now, let $J \in \Omega(J_k)$. Then J is $\leq_{\mathcal{J}}$-maximal among the non-trivial $\mathcal{J}$-classes of M and so, by Proposition 1.3, an element of J belongs to $\mathsf{K}_{\mathsf{Ab}}(M)$ if and only if it belongs to the derived subgroup of a maximal subgroup of J. Observe that, if $\lambda(J_k) = 0$ (respectively, $\lambda(J_k) = 1$), then the subgroups of M contained in J are trivial (respectively, trivial or abelian) and so $J \cap \mathsf{K}_{\mathsf{Ab}}(M)$ consists entirely of idempotents. Since a non-trivial derived subgroup of a maximal subgroup of M contained in J is a maximal subgroup of a $\leq_{\mathcal{J}}$-maximal $\mathcal{J}$-class among the non-trivial $\mathcal{J}$-classes of $\mathsf{K}_{\mathsf{Ab}}(M)$, and the same occurs if we iterate successively the operator K_{Ab}, then it suffices to iterate the operator K_{Ab} not more than $\max\{0, \lambda(J_k) - 1\}$ times to obtain $J \cap \mathsf{K}_{\mathsf{Ab}}^{(1+\max\{0,\lambda(J_k)-1\})}(M) = E(M)$. The result follows. □

We have then an immediate, but useful, consequence:

Corollary 2.1. *Let M be an inverse monoid all of whose subgroups are abelian and let k be the maximum size of a $\leq_{\mathcal{J}}$-chain of non-trivial $\mathcal{J}$-classes of M. Then M is solvable and $\ell_{\mathsf{Ab}}(M) \leq k$.* □

3. On some inverse monoids

In this section, we consider the inverse members of the family of transformation monoids given by Diagram 1, i.e.

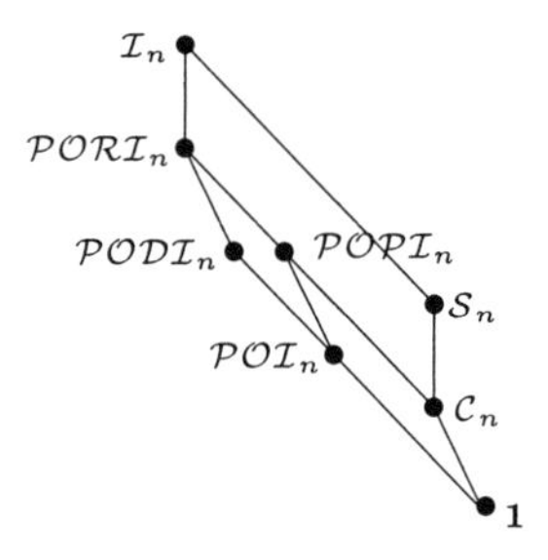

Diagram 2

the monoids of injective transformations. Notice that $\mathcal{C}_n$ can be defined as the (cyclic) group generated by the permutation (n-cycle)

$$g = \begin{pmatrix} 1 & 2 & \cdots & n-1 & n \\ 2 & 3 & \cdots & n & 1 \end{pmatrix}.$$

Our aim is to investigate the abelian kernel lengths of the monoids of Diagram 2. As $\mathcal{C}_n$ is an abelian group we have $\ell_{\mathsf{Ab}}(\mathcal{C}_n) = 1$. As $\mathbf{1}$ denotes the trivial group we have $\ell_{\mathsf{Ab}}(\mathbf{1}) = 0$. Computations show that $\ell_{\mathsf{Ab}}(\mathcal{I}_3) = 2$ and $\ell_{\mathsf{Ab}}(\mathcal{I}_4) = 3$. Notice that $\ell_{\mathsf{Ab}}(\mathcal{I}_n)$ is not defined for $n \geq 5$ since for such n's these monoids are not solvable. It remains to consider the monoids $\mathcal{POI}_n$, $\mathcal{PODI}_n$, $\mathcal{POPI}_n$ and $\mathcal{PORI}_n$.

We start by observing that all these inverse monoids have a similar $\mathcal{J}$-class structure (indeed the same is valid for all monoids of Diagram 1): let

M be any of the monoids $\mathcal{POI}_n$, $\mathcal{PODI}_n$, $\mathcal{POPI}_n$ or $\mathcal{PORI}_n$. Then

$$M/\mathcal{J} = \{J_0 <_\mathcal{J} J_1 <_\mathcal{J} \cdots <_\mathcal{J} J_n\},$$

where J_k, for $k \in \{0, 1, \ldots, n\}$, consists of all elements of M of rank k, i.e. $J_k = \{s \in M \mid |\operatorname{Im}(s)| = k\}$. Notice that the $\mathcal{J}$-classes $J_1, \ldots, J_{n-1}$ are non-trivial. Furthermore, the maximal subgroups of M are abelian (cyclic), except for $M = \mathcal{PORI}_n$, which contains also dihedral groups (see [28, 30, 31, 35]). Thus, all subgroups of M are solvable and, since M is an inverse monoid, by Proposition 1.4, the monoid M is solvable. Moreover, $\mathcal{POI}_n$ is aperiodic [28, 31] (and so, in this case, its group of units J_n is trivial); the group of units J_n of $\mathcal{PODI}_n$ is the cyclic group $\langle h \rangle$ of order two generated by the permutation

$$h = \begin{pmatrix} 1 & 2 & \cdots & n-1 & n \\ n & n-1 & \cdots & 2 & 1 \end{pmatrix}$$

and $\mathcal{PODI}_n$ is generated by $\mathcal{POI}_n \cup \{h\}$ [35]; the group of units J_n of $\mathcal{POPI}_n$ is the order n cyclic group $\mathcal{C}_n$ generated by the n-cycle permutation g and $\mathcal{POPI}_n$ is generated by $\mathcal{POI}_n \cup \{g\}$ [30]; and, at last, for $n \geq 3$, the group of units J_n of $\mathcal{PORI}_n$ is the order $2n$ dihedral group generated by the permutations g and h and $\mathcal{PORI}_n$ is generated by $\mathcal{POI}_n \cup \{g, h\}$ [35].

In [22] the authors gave the following description of the abelian kernels of $\mathcal{POI}_n$ and $\mathcal{POPI}_n$:

Theorem 3.1. *Let M be either of the monoids $\mathcal{POI}_n$ or $\mathcal{POPI}_n$. Then the abelian kernel of M consists of all idempotents and all elements of rank less than $n-1$.* □

Therefore, $\mathsf{K}_{\mathsf{Ab}}(\mathcal{POPI}_n)$ is an inverse monoid with $n-2$ non-trivial $\mathcal{J}$-classes, all $\leq_\mathcal{J}$-comparable. Hence the maximum size of a $\leq_\mathcal{J}$-chain of non-trivial $\mathcal{J}$-classes of $\mathsf{K}_{\mathsf{Ab}}(\mathcal{POPI}_n)$ is exactly $n-2$. Since all subgroups of $\mathsf{K}_{\mathsf{Ab}}(\mathcal{POPI}_n)$ are abelian, by Corollary 2.1, we have $\ell_{\mathsf{Ab}}(\mathsf{K}_{\mathsf{Ab}}(\mathcal{POPI}_n)) \leq n-2$ and so $\ell_{\mathsf{Ab}}(\mathcal{POPI}_n) \leq n-1$. Thus, as $\mathcal{POI}_n$ is a submonoid of $\mathcal{POPI}_n$, we also have $\ell_{\mathsf{Ab}}(\mathcal{POI}_n) \leq n-1$.

On the other hand, $\mathcal{POI}_{n-1}$ may be viewed as a submonoid of $\mathsf{K}_{\mathsf{Ab}}(\mathcal{POI}_n)$ and so, by induction on n, it follows immediately that $\mathsf{K}_{\mathsf{Ab}}^{(n-2)}(\mathcal{POI}_n)$ contains non-idempotent elements, for $n \geq 3$. Hence, $\ell_{\mathsf{Ab}}(\mathcal{POI}_n) \geq n-1$. Thus, we also have $\ell_{\mathsf{Ab}}(\mathcal{POPI}_n) \geq n-1$, by considering again $\mathcal{POI}_n$ as a submonoid of $\mathcal{POPI}_n$.

We have proved:

Theorem 3.2. $\ell_{\mathsf{Ab}}(\mathcal{POI}_n) = n-1 = \ell_{\mathsf{Ab}}(\mathcal{POPI}_n)$. □

The abelian kernels of the monoids $\mathcal{PODI}_n$ and $\mathcal{PORI}_n$ have been recently determined by the authors in [24]. We do not yet have a complete solution for their abelian kernel lengths.

Since $\mathsf{K}_{\mathsf{Ab}}(\mathcal{PODI}_n) \subseteq \mathcal{POI}_n$ and $\mathsf{K}_{\mathsf{Ab}}(\mathcal{PORI}_n) \subseteq \mathcal{POPI}_n$ [24] (note that these inclusions were deduced by applying Theorem 1.1) and considering the equalities of Theorem 3.2 (observe that $\mathcal{POI}_n$ is a submonoid of $\mathcal{PODI}_n$ and $\mathcal{PORI}_n$), we have the following bounds for the abelian kernel lengths of $\mathcal{PODI}_n$ and $\mathcal{PORI}_n$: $n-1 \leq \ell_{\mathsf{Ab}}(\mathcal{PODI}_n), \ell_{\mathsf{Ab}}(\mathcal{PORI}_n) \leq n$.

Furthermore, when n is an even integer, since we have precisely that $\mathsf{K}_{\mathsf{Ab}}(\mathcal{PODI}_n) = \mathsf{K}_{\mathsf{Ab}}(\mathcal{POI}_n)$ [24], as a corollary of Theorem 3.2 we can also state:

Corollary 3.1. *If n is an even integer then $\ell_{\mathsf{Ab}}(\mathcal{PODI}_n) = n-1$.* □

We have reasons to believe that the previous result also holds when n is an odd integer. We conjecture the following stronger result: $\mathsf{K}_{\mathsf{Ab}}^{(2)}(\mathcal{PODI}_n) = \mathsf{K}_{\mathsf{Ab}}^{(2)}(\mathcal{POI}_n)$. In fact, computations achieved with the already mentioned software [21] led to the following results: $\mathsf{K}_{\mathsf{Ab}}^{(2)}(\mathcal{PODI}_3) = \mathsf{K}_{\mathsf{Ab}}^{(2)}(\mathcal{POI}_3)$, $\mathsf{K}_{\mathsf{Ab}}^{(2)}(\mathcal{PODI}_5) = \mathsf{K}_{\mathsf{Ab}}^{(2)}(\mathcal{POI}_5)$ and $\mathsf{K}_{\mathsf{Ab}}^{(2)}(\mathcal{PODI}_7) = \mathsf{K}_{\mathsf{Ab}}^{(2)}(\mathcal{POI}_7)$.

For the monoids $\mathcal{PORI}_n$ the problem seems to be more complex and the precise value for $\ell_{\mathsf{Ab}}(\mathcal{PORI}_n)$ is an open question: our computations showed that $\ell_{\mathsf{Ab}}(\mathcal{PORI}_3) = 2$ and $\ell_{\mathsf{Ab}}(\mathcal{PORI}_4) = 3$, but $\ell_{\mathsf{Ab}}(\mathcal{PORI}_5) = 5$.

4. On some non-inverse monoids

We now consider the non-inverse monoids of Diagram 1:

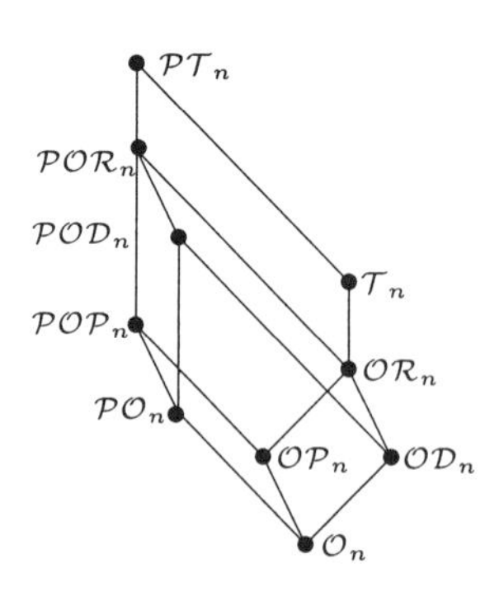

Diagram 3

We will determine their abelian kernels, show that those not containing $\mathcal{S}_n$ are solvable and compute their abelian kernel lengths.

We begin by considering the monoids of order-preserving transformations, i.e. the monoids $\mathcal{O}_n$ (of full transformations) and $\mathcal{PO}_n$ (of partial transformations). Both are generated by idempotents [42] and so we immediately have:

Theorem 4.1. $\mathsf{K}_{\mathsf{Ab}}(\mathcal{O}_n) = \mathcal{O}_n$ *and* $\mathsf{K}_{\mathsf{Ab}}(\mathcal{PO}_n) = \mathcal{PO}_n$. *Moreover,* $\mathcal{O}_n$ *and* $\mathcal{PO}_n$ *are solvable and* $\ell_{\mathsf{Ab}}(\mathcal{O}_n) = 0 = \ell_{\mathsf{Ab}}(\mathcal{PO}_n)$. □

Next, we consider the monoids containing also order-reversing transformations, namely the monoid $\mathcal{OD}_n$ containing only full transformations and the monoid $\mathcal{POD}_n$ containing all partial order-preserving or order-reversing transformations. As the product of two order-preserving transformations or two order-reversing transformations is an order-preserving transformation and the product of an order-preserving transformation with an order-reversing transformation, or vice-versa, is an order-reversing transformation, the monoid $\mathcal{OD}_n$ is generated by $\mathcal{O}_n \cup \{h\}$ and the monoid $\mathcal{POD}_n$ is generated by $\mathcal{PO}_n \cup \{h\}$ [36], where h is the order two permutation defined in the last section. Moreover, we can find generators of $\mathcal{O}_n$ (respectively, $\mathcal{PO}_n$) and relations on $\mathcal{OD}_n$ (respectively, $\mathcal{POD}_n$) satisfying the conditions of Theorem 1.1 (see [36]). Hence,

$$\mathsf{K}_{\mathsf{Ab}}(\mathcal{OD}_n) \subseteq \mathcal{O}_n \quad \text{and} \quad \mathsf{K}_{\mathsf{Ab}}(\mathcal{POD}_n) \subseteq \mathcal{PO}_n.$$

Since $\mathsf{K}_{\mathsf{Ab}}(\mathcal{O}_n) = \mathcal{O}_n$ and $\mathsf{K}_{\mathsf{Ab}}(\mathcal{PO}_n) = \mathcal{PO}_n$, we can state:

Theorem 4.2. $\mathsf{K}_{\mathsf{Ab}}(\mathcal{OD}_n) = \mathcal{O}_n$ *and* $\mathsf{K}_{\mathsf{Ab}}(\mathcal{POD}_n) = \mathcal{PO}_n$. *Moreover, the monoids* $\mathcal{OD}_n$ *and* $\mathcal{POD}_n$ *are solvable and* $\ell_{\mathsf{Ab}}(\mathcal{OD}_n) = 1 = \ell_{\mathsf{Ab}}(\mathcal{POD}_n)$. □

Now, let $\overline{\mathcal{PT}}_n$ be the submonoid of $\mathcal{PT}_n$ of all singular partial transformations together with the identity, i.e.

$$\overline{\mathcal{PT}}_n = \{s \in \mathcal{PT}_n \mid s = 1 \text{ or } |\operatorname{Im}(s)| \leq n-1\}.$$

The submonoid $\overline{\mathcal{T}}_n = \mathcal{T}_n \cap \overline{\mathcal{PT}}_n$ of $\overline{\mathcal{PT}}_n$ (and also of $\mathcal{T}_n$) is known to be generated by idempotents [41]. Since an element $s \in \overline{\mathcal{PT}}_n$ can be written as a product $s = et$, with e the restriction to $\operatorname{Dom}(s)$ of the identity map on X_n (whence an idempotent) and $t \in \overline{\mathcal{T}}_n$ any extension of s, we may conclude that $\overline{\mathcal{PT}}_n$ is also generated by idempotents.

Now, since the group of units of both monoids $\mathcal{T}_n$ and $\mathcal{PT}_n$ is the symetric group $\mathcal{S}_n$ whose derived subgroup is the alternating group $\mathcal{A}_n$, we have

$$\mathsf{K}_{\mathsf{Ab}}(\mathcal{T}_n) \cap \mathcal{S}_n = \mathcal{A}_n = \mathsf{K}_{\mathsf{Ab}}(\mathcal{PT}_n) \cap \mathcal{S}_n$$

by Corollary 1.1. Hence:

Theorem 4.3. $\mathsf{K}_{\mathsf{Ab}}(\mathcal{T}_n) = \overline{\mathcal{T}}_n \cup \mathcal{A}_n$ *and* $\mathsf{K}_{\mathsf{Ab}}(\mathcal{PT}_n) = \overline{\mathcal{PT}}_n \cup \mathcal{A}_n$. *Moreover, the monoids $\mathcal{T}_n$ and $\mathcal{PT}_n$ are not solvable.*

As $\overline{\mathcal{T}}_n$ and $\overline{\mathcal{PT}}_n$ are generated by idempotents, they are solvable (with null abelian kernel length) although both monoids contain non-solvable groups, for $n \geq 6$ (see [23]). This is not the case for the monoids $\mathcal{O}_n$, $\mathcal{PO}_n$, $\mathcal{OD}_n$ and $\mathcal{POD}_n$ considered above. In fact, it is easy to show that $\mathcal{O}_n$ and $\mathcal{PO}_n$ are aperiodic and $\mathcal{OD}_n$ and $\mathcal{POD}_n$ have only cyclic groups of order less than or equal to two.

Next, we turn our attention to the monoids $\mathcal{OP}_n$ and $\mathcal{POP}_n$ of orientation-preserving transformations.

Consider the monoids $\overline{\mathcal{OP}}_n = \mathcal{OP}_n \cap \overline{\mathcal{PT}}_n$ and $\overline{\mathcal{POP}}_n = \mathcal{POP}_n \cap \overline{\mathcal{PT}}_n$.

First, we prove that $\overline{\mathcal{OP}}_n$ and $\overline{\mathcal{POP}}_n$ are idempotent generated.

Let $e_1, e_2, \ldots, e_{n-1}, f_1, f_2, \ldots, f_{n-1}$ be the transformations of X_n defined by

$$e_i = \left(\begin{array}{ccc|c|ccc} 1 & \cdots & i-1 & i & i+1 & \cdots & n \\ 1 & \cdots & i-1 & i+1 & i+1 & \cdots & n \end{array}\right)$$

and

$$f_i = \left(\begin{array}{ccc|c|ccc} 1 & \cdots & n-i & n-i+1 & n-i+2 & \cdots & n \\ 1 & \cdots & n-i & n-i & n-i+2 & \cdots & n \end{array}\right),$$

for $1 \leq i \leq n-1$. Also, let g be the n-cycle as defined in the previous section. The set $\{e_1, e_2, \ldots, e_{n-1}, f_1, f_2, \ldots, f_{n-1}, g\}$ generates $\mathcal{OP}_n$ and Catarino [13] gave a presentation of $\mathcal{OP}_n$ in terms of these $2n-1$ generators and n^2+2n relations (see also [31]). Some of these relations, with particular interest for us, are the following:

$$g^n = 1; \tag{1}$$

$$e_i g = g e_{i+1}, \text{ for } 1 \leq i \leq n-2; \tag{2}$$

$$e_{n-1} g = g^2 f_{n-1} f_{n-2} \cdots f_1. \tag{3}$$

Notice also that $e_1, e_2, \ldots, e_{n-1}, f_1, f_2, \ldots, f_{n-1}$ are idempotents.

It was also observed in [13] that the transformations e_1 and g suffice to generate $\mathcal{OP}_n$. Since $e_1 g^{n-2} = g^{n-2}e_{n-1}$, by relations (2), we may deduce that $\{e_{n-1}, g\}$ also generates $\mathcal{OP}_n$, whence all elements of $\overline{\mathcal{OP}}_n$, except the identity, may be written in the form

$$g^{k_0}e_{n-1}g^{k_1}e_{n-1}\cdots g^{k_{t-1}}e_{n-1}g^{k_t},$$

with $0 \leq k_0, k_1, \ldots, k_t \leq n-1$ and $t \geq 1$, and so

$$\{g^k e_{n-1} g^\ell \mid 0 \leq k, \ell \leq n-1\}$$

is a set of generators of $\overline{\mathcal{OP}}_n$.

Next, notice that $g^k e g^\ell$, with $e \in \mathcal{OP}_n$ an idempotent and $0 \leq k, \ell \leq n-1$ such that $\ell + k \equiv 0(\operatorname{mod} n)$, is an idempotent. Moreover, if $e \in \mathcal{OP}_n$ is a product of idempotents then $g^k e g^\ell$, with $0 \leq k, \ell \leq n-1$ such that $\ell + k \equiv 0(\operatorname{mod} n)$, is a product of idempotents: in fact, if $e = a_1 a_2 \cdots a_t$ then $g^k e g^\ell = g^k a_1 g^\ell\, g^k a_2 g^\ell \,\cdots\, g^k a_t g^\ell$.

Let $f = f_{n-1} f_{n-2} \cdots f_1$. Then

$$e_{n-1} = e_{n-1}e_{n-1} = e_{n-1}gg^{n-1}e_{n-1} = g^2 f g^{n-1} e_{n-1}$$

and so:

(1) $g^{n-1}e_{n-1} = g^{n-1}g^2 f g^{n-1}e_{n-1} = gfg^{n-1} \cdot e_{n-1}$;
(2) $g^{n-2}e_{n-1} = g^{n-2}g^2 f g^{n-1}e_{n-1} = f \cdot g^{n-1}e_{n-1}$;
(3) $g^{n-i}e_{n-1} = g^{n-i}g^2 f g^{n-1}e_{n-1} = g^{n-i+2} f g^{i-2} \cdot g^{n-(i-1)}e_{n-1}$, for $3 \leq i \leq n-1$.

Thus $g^{n-i}e_{n-1}$ is a product of idempotents, for $1 \leq i \leq n-1$. On the other hand, we have

$$\begin{aligned} e_{n-1}g^{n-i} &= e_{n-1}\, e_{n-1} g\, g^{n-i-1} \\ &= e_{n-1}\, g^2 f\, g^{n-i-1} \\ &= e_{n-1} g^{n-(i-1)} \cdot g^{i+1} f g^{n-i-1}, \end{aligned}$$

for $1 \leq i \leq n-1$. Hence $e_{n-1}g^{n-i}$ is also a product of idempotents, for $1 \leq i \leq n-1$.

Now, since $g^k e_{n-1} g^\ell = g^k e_{n-1} \cdot e_{n-1} g^\ell$, for $0 \leq k, \ell \leq n-1$, we have:

Proposition 4.1. *The monoid $\overline{\mathcal{OP}}_n$ is generated by idempotents.* □

Next, with an argument similar to the one used to show that $\overline{\mathcal{PT}}_n$ is idempotent generated, we prove that $\overline{\mathcal{POP}}_n$ is also generated by idempotents.

Let $s \in \overline{\mathcal{POP}_n}$ and suppose that $\mathrm{Dom}(s) = \{i_1 < i_2 < \cdots < i_k\}$ $(0 \le k < n)$. Define e as the restriction to $\mathrm{Dom}(s)$ of the identity map on X_n. Then e is an idempotent belonging to $\overline{\mathcal{POP}_n}$. Next, define a full map t on X_n by

$$(i)t = \begin{cases} (i_1)s, \text{ if } 1 \le i < i_2 \\ (i_\ell)s, \text{ if } i_\ell \le i < i_{\ell+1},\ 2 \le \ell < k \\ (i_k)s, \text{ if } i_k \le i \le n. \end{cases}$$

Then $t \in \overline{\mathcal{OP}_n}$ and $s = et$. Hence, by Proposition 4.1, we have:

Proposition 4.2. *The monoid $\overline{\mathcal{POP}_n}$ is generated by idempotents.* □

Now, since the group of units of both monoids $\mathcal{OP}_n$ and $\mathcal{POP}_n$ is $\mathcal{C}_n$, the cyclic group of order n generated by g (see [13, 11, 36]), whose derived subgroup is trivial, we have

$$\mathsf{K}_{\mathsf{Ab}}(\mathcal{OP}_n) \cap \mathcal{C}_n = \{1\} = \mathsf{K}_{\mathsf{Ab}}(\mathcal{POP}_n) \cap \mathcal{C}_n$$

by Corollary 1.1. Hence:

Theorem 4.4. $\mathsf{K}_{\mathsf{Ab}}(\mathcal{OP}_n) = \overline{\mathcal{OP}_n}$ *and* $\mathsf{K}_{\mathsf{Ab}}(\mathcal{POP}_n) = \overline{\mathcal{POP}_n}$. *Moreover, the monoids $\mathcal{OP}_n$ and $\mathcal{POP}_n$ are solvable and $\ell_{\mathsf{Ab}}(\mathcal{OP}_n) = 1 = \ell_{\mathsf{Ab}}(\mathcal{POP}_n)$.* □

Finally, for $n \ge 3$, we consider the monoids $\mathcal{OR}_n$ and $\mathcal{POR}_n$ of orientation-preserving or orientation-reversing transformations. Analogously to $\mathcal{OD}_n$, the monoid $\mathcal{OR}_n$ is generated by $\mathcal{OP}_n \cup \{h\}$ and the monoid $\mathcal{POR}_n$ is generated by $\mathcal{POP}_n \cup \{h\}$ [11, 36], where h is the order two permutation defined in the previous section. Moreover, we can find generators of $\mathcal{OP}_n$ (respectively, $\mathcal{POP}_n$) and relations on $\mathcal{OR}_n$ (respectively, $\mathcal{POR}_n$) satisfying the conditions of Theorem 1.1 (see [11, 36]). Hence,

$$\mathsf{K}_{\mathsf{Ab}}(\mathcal{OR}_n) \subseteq \mathcal{OP}_n \quad \text{and} \quad \mathsf{K}_{\mathsf{Ab}}(\mathcal{POR}_n) \subseteq \mathcal{POP}_n.$$

On the other hand, in both cases the group of units is the order $2n$ dihedral group (generated by g and h [11, 36]), whose derived subgroup is the cyclic group $\langle g^2 \rangle$ generated by g^2. By applying Corollary 1.1 together with previous observations and Theorem 4.4, we obtain:

Theorem 4.5. $\mathsf{K}_{\mathsf{Ab}}(\mathcal{OR}_n) = \overline{\mathcal{OP}_n} \cup \langle g^2 \rangle$, $\mathsf{K}^{(2)}_{\mathsf{Ab}}(\mathcal{OR}_n) = \overline{\mathcal{OP}_n}$, $\mathsf{K}_{\mathsf{Ab}}(\mathcal{POR}_n) = \overline{\mathcal{POP}_n} \cup \langle g^2 \rangle$ *and* $\mathsf{K}^{(2)}_{\mathsf{Ab}}(\mathcal{POR}_n) = \overline{\mathcal{POP}_n}$. *Moreover, $\mathcal{OR}_n$ and $\mathcal{POR}_n$ are solvable monoids and $\ell_{\mathsf{Ab}}(\mathcal{OP}_n) = 2 = \ell_{\mathsf{Ab}}(\mathcal{POP}_n)$.* □

Acknowledgments

The first author gratefully acknowledges support of FCT through CMUP and the FCT and POCTI Project POCTI/32817/MAT/2000 which is funded in cooperation with the European Community Fund FEDER.

The second author gratefully acknowledges support of FCT and FEDER, within the POCTI project "Fundamental and Applied Algebra" of CAUL.

References

1. A.Ya. Aĭzenštat, The defining relations of the endomorphism semigroup of a finite linearly ordered set, *Sibirsk. Mat.* **3** (1962) 161-169 (Russian).
2. J. Almeida, Hyperdecidable pseudovarieties and the calculation of semidirect products, *Int. J. Algebra Comput.* **9** (1999) 241-261.
3. J. Almeida, Dynamics of implicit operations and tameness of pseudovarieties of groups, *Trans. Amer. Math. Soc.* **354** (2002) 387–411.
4. J. Almeida and M. Delgado, Sur certains systèmes d'équations avec contraintes dans un groupe libre, *Portugal. Math.* **56** (1999) 409–417.
5. J. Almeida and M. Delgado, Sur certains systèmes d'équations avec contraintes dans un groupe libre—addenda, *Portugal. Math.* **58** (2001) 379–387.
6. J. Almeida and M. Delgado, Tameness of the pseudovariety of Abelian groups, *Int. J. Algebra Comput.* To appear.
7. J. Almeida and B. Steinberg, On the decidability of iterated semidirect products and applications to complexity, *Proc. London Math. Soc.* **80** (2000) 50–74.
8. J. Almeida and B. Steinberg, Syntactic and global semigroup theory, a synthesis approach, in *Algorithmic Problems in Groups and Semigroups*, J.-C. Birget, S. Margolis, J. Meakin, M. Sapir, eds., Birkhäuser, 2000, 1–23.
9. J. Almeida and M.V. Volkov, The gap between partial and full, *Int. J. Algebra Comput.* **8** (1998) 399–430.
10. C.J. Ash, Inevitable graphs: a proof of the type II conjecture and some related decision procedures, *Int. J. Algebra Comput.* **1** (1991) 127–146.
11. R.E. Arthur and N. Ruškuc, Presentations for two extensions of the monoid of order-preserving mappings on a finite chain, *Southeast Asian Bull. Math.* **24** (2000) 1–7.
12. K. Auinger, A new proof of the Rhodes type II conjecture, *Int. J. Algebra Comput.* To appear.
13. P.M. Catarino, Monoids of orientation-preserving transformations of a finite chain and their presentations, *Semigroups and Applications,* eds. J.M. Howie and N. Ruškuc, World Scientific, (1998), 39–46.
14. P.M. Catarino and P.M. Higgins, The monoid of orientation-preserving mappings on a chain, *Semigroup Forum* **58** (1999) 190–206.
15. T. Coulbois, Free product, profinite topology and finitely generated subgroups, *Int. J. Algebra Comput.* **11** (2001) 171–184.

16. T. Coulbois, *Propriétés de Ribes-Zalesskiĭ, topologie profinie, produit libre et généralisations*, Ph. D. Thesis, Université Paris VII, 2000.
17. D.F. Cowan and N.R. Reilly, Partial cross-sections of symmetric inverse semigroups, *Int. J. Algebra Comput.* **5** (1995) 259–287.
18. M. Delgado, Abelian pointlikes of a monoid, *Semigroup Forum* **56** (1998) 339–361.
19. M. Delgado, On the hyperdecidability of pseudovarieties of groups, *Int. J. Algebra Comput.* **11** (2001) 753–771.
20. M. Delgado, Computing commutative images of rational languages and applications, *Theor. Inform. Appl.* **35** (2001) 419–435.
21. M. Delgado and J. Morais, A GAP [50] package on semigroups, *in preparation.*
22. M. Delgado and V.H. Fernandes, Abelian kernels of some monoids of injective partial transformations and an application, *Semigroup Forum* **61** (2000) 435–452.
23. M. Delgado and V.H. Fernandes, Solvable idempotent commuting monoids, Tech. Rep., CMUP: 2002-21. *Submitted.*
24. M. Delgado and V.H. Fernandes, Abelian kernels of monoids of order-preserving maps and of some of its extensions, *Semigroup Forum.* To appear.
25. M. Delgado, V.H. Fernandes, S. Margolis and B. Steinberg, On semigroups whose idempotent-generated subsemigroup is aperiodic, *Int. J. Algebra Comput.* To appear.
26. M. Delgado and P.-C. Héam, A polynomial time algorithm to compute the Abelian kernel of a finite monoid, *Semigroup Forum* **67** (2003) 97–110.
27. M.S. Dummit and R.M. Foote, *Abstract Algebra*, Englewood Clifs: Prentice-Hall, 1991.
28. V.H. Fernandes, Semigroups of order-preserving mappings on a finite chain: a new class of divisors, *Semigroup Forum* **54** (1997) 230–236.
29. V.H. Fernandes, Normally ordered inverse semigoups, *Semigroup Forum* **58** (1998) 418–433.
30. V.H. Fernandes, The monoid of all injective orientation preserving partial transformations on a finite chain, *Comm. Algebra* **28** (2000) 3401–3426.
31. V.H. Fernandes, The monoid of all injective order preserving partial transformations on a finite chain, *Semigroup Forum* **62** (2001) 178–204.
32. V.H. Fernandes, Presentations for some monoids of partial transformations on a finite chain: a survey, *Semigroups, Algorithms, Automata and Languages,* eds. G.M.S. Gomes, J.-E. Pin and P.V. Silva, World Scientific, (2002), 363–378.
33. V.H. Fernandes, A division theorem for the pseudovariety generated by semigroups of orientation preserving transformations on a finite chain, *Comm. Algebra* **29** (2001) 451–456.
34. V.H. Fernandes, Semigroups of order-preserving mappings on a finite chain: another class of divisors, *Izvestiya VUZ. Matematika* **3** (478) (2002) 51–59 (Russian).
35. V.H. Fernandes, G.M.S. Gomes and M.M. Jesus, Presentations for some monoids of injective partial transformations on a finite chain, *Southeast Asian Bull. Math.* To appear.

36. V.H. Fernandes, G.M.S. Gomes and M.M. Jesus, Presentations for some monoids of partial transformations on a finite chain, *Comm. Algebra.* To appear.
37. G.M.S. Gomes and J.M. Howie, On the ranks of certain semigroups of order-preserving transformations, *Semigroup Forum* **45** (1992) 272–282.
38. K. Henckell, S. Margolis, J.-E. Pin and J. Rhodes, Ash's type II theorem, profinite topology and Malcev products. Part I, *Int. J. Algebra Comput.* **1** (1991) 411–436.
39. B. Herwig and D. Lascar, Extending partial automorphisms and the profinite topology on free groups, *Trans. Amer. Math. Soc.* **352** (2000) 1985–2021.
40. P.M. Higgins, Divisors of semigroups of order-preserving mappings on a finite chain, *Int. J. Algebra Comput.* **5** (1995) 725–742.
41. J.M. Howie, The subsemigroup generated by the idempotents of a full transformation semigroup, *J. London Math. Soc.* **41** (1966) 707–716.
42. J.M. Howie, Product of idempotents in certain semigroups of transformations, *Proc. Edinburgh Math. Soc.***17** (1971) 223–236.
43. J.M. Howie, *Fundamentals of Semigroup Theory,* Oxford University Press, 1995.
44. G. Lallement, *Semigroups and Combinatorial Applications,* John Wiley & Sons, 1979.
45. D. McAlister, Semigroups generated by a group and an idempotent, *Comm. Algebra* **26** (1998) 515–547.
46. L.M. Popova, The defining relations of the semigroup of partial endomorphisms of a finite linearly ordered set, *Leningradskij gosudarstvennyj pedagogicheskij institut imeni A. I. Gerzena, Uchenye Zapiski* **238** (1962) 78–88 (Russian).
47. V.B. Repnitskiĭ and M.V. Volkov, The finite basis problem for the pseudovariety $\mathcal{O}$, *Proc. R. Soc. Edinb.*, Sect. A, Math. **128** (1998) 661–669.
48. L. Ribes and P.A. Zalesskiĭ, On the profinite topology on a free group, *Bull. London Math. Soc.* **25** (1993) 37–43.
49. B. Steinberg, Inevitable graphs and profinite topologies: Some solutions to algorithmic problems in monoid and automata theory, stemming from group theory, *Int. J. Algebra Comput.* **11** (2001) 25–71.
50. The GAP Group. *GAP – Groups, Algorithms, and Programming, Version 4.3*, 2002. (`http://www.gap-system.org`).
51. A. Vernitskii and M.V. Volkov, A proof and generalisation of Higgins' division theorem for semigroups of order-preserving mappings, *Izv.vuzov. Matematika*, No.1 (1995) 38–44 (Russian).

BRAIDS AND FACTORIZABLE INVERSE MONOIDS

DAVID EASDOWN
School of Mathematics and Statistics,
University of Sydney, NSW 2006, Australia
E-mail: de@math.usyd.edu.au

JAMES EAST
School of Mathematics and Statistics,
University of Sydney, NSW 2006, Australia,
E-mail: jamese@math.usyd.edu.au

D. G. FITZ-GERALD
School of Mathematics and Physics,
University of Tasmania,
GPO Box 252-37, Hobart, TAS 7001, Australia
E-mail: D.FitzGerald@utas.edu.au

What is the untangling effect on a braid if one is allowed to snip a string, or if two specified strings are allowed to pass through each other, or even allowed to merge and part as newly reconstituted strings? To calculate the effects, one works in an appropriate factorizable inverse monoid, some aspects of a general theory of which are discussed in this paper. The coset monoid of a group arises, and turns out to have a universal property within a certain class of factorizable inverse monoids. This theory is dual to the classical construction of fundamental inverse semigroups from semilattices. In our braid examples, we will focus mainly on the "merge and part" alternative, and introduce a monoid which is a natural preimage of the largest factorizable inverse submonoid of the dual symmetric inverse monoid on a finite set, and prove that it embeds in the coset monoid of the braid group.

1. Introduction

The motivation for this work comes from several directions. In Birman's theory of knot invariants [2], the singular braid monoid plays a role, in which strings are allowed to touch, creating "double points". One can attempt to simplify a knot or link by allowing one string to pass through another. The moment at which strings touch is a "double point" and a "singular" knot or link is created. These singular versions then feature in recursive formulae for invariants such as the Alexander and Jones polynomials. Braids close up to form knots and links (Alexander's Theorem), so it is useful to investigate means by which braids may be simplified or modified by some manipulation of the strings. This is the idea which lead to the *singular braid monoid*, first introduced by Baez [1], and developed by Birman [2]. The question of what happens when we snip one or more strings has been fully investigated by Easdown and Lavers [3], leading to a preimage of the symmetric inverse monoid of a finite set, known as the *inverse braid monoid*, exactly analogous to the relationship between the braid group and the symmetric group.

FitzGerald and Leech [5] use duality in category theory to introduce new classes of inverse monoids, a special case of which is $\mathcal{I}_X^*$, the *dual symmetric inverse monoid on a set* X, which comprises biequivalences on X and a binary operation involving composition and joins of equivalence relations. One may ask (and this remains unresolved) whether $\mathcal{I}_X^*$ has a natural preimage involving braids or a modification of braids. Below we study a candidate, the *merge and part braid monoid*, for what might be a useful preimage of $\mathcal{F}_X^*$, the largest factorizable inverse monoid of $\mathcal{I}_X^*$. The ingredients are braids and equivalence relations on strings, where equivalent strings are allowed to touch, merge and then part as reconstituted strings. Another possibility, not dealt with here in any detail, is to allow equivalent strings just to pass through each other. This leads to the *permeable braid monoid*, studied in detail by East [4], who also discusses decision problems and presentations for both types of braid monoids, and explores relationships with Coxeter groups in general.

As a simple illustration of these ideas, consider the pure braid β on 4 strings depicted in Figure 1. Certainly β cannot be continuously deformed into the identity braid, denoted by 1. This follows quickly from the fact that the pure braid group is an iterated semidirect product of free groups. However, to see this directly, without the theory of pure braids, one can argue as follows. Suppose β is equivalent to 1. If we ignore ("snip and shrivel") the second and third strings, then the configurations in Figure 2 are certainly also equivalent. But taking the projection onto a horizontal plane, the first of these produces a loop at the first point containing the fourth point, whilst the second produces a degenerate loop at the first point with the fourth point on the outside:

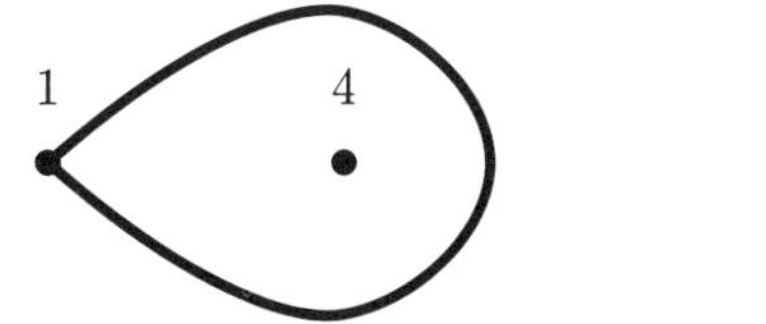

1 4

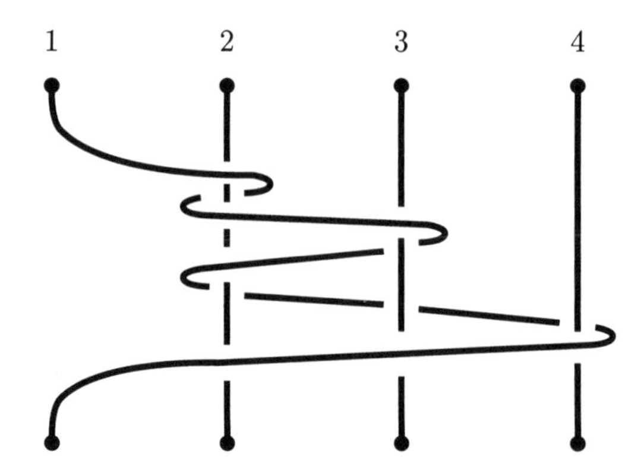

Figure 1. the pure braid β

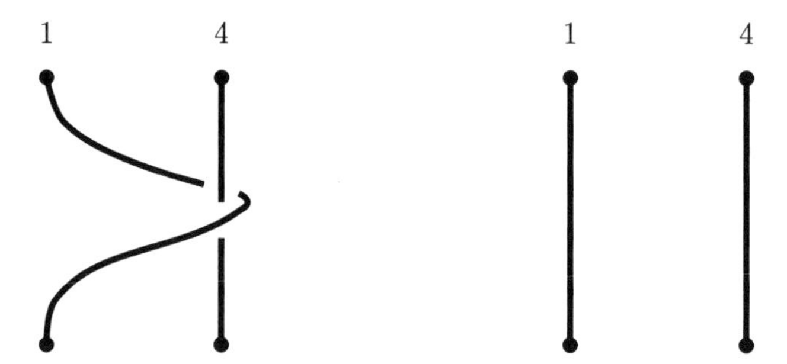

Figure 2. β and the identity braid simplify after removing the second and third strings

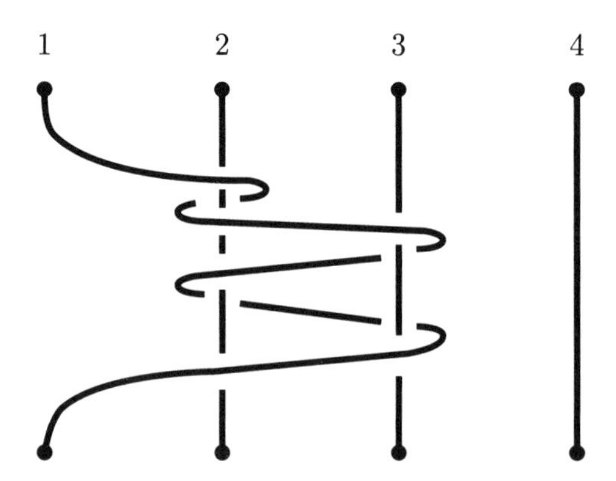

Figure 3. the pure braid β'

The fourth point remaining *inside* a loop at the first point is an invariant of a continous deformation of the first braid. This invariant fails in the horizontal projection of the final braid, which gives a contradiction.

What happens if we wish to simplify β by allowing the first and fourth strings only to pass through each other? (This is an instance working within the permeable braid monoid studied by East [4].) The first, second and third strings are not allowed to touch, as usual, during any continuous deformation. For example, β can become β' in Figure 3, which in turn can be represented by the configuration in Figure 4, where x and y are canonical generators of a free subgroup of the pure braid group on the first 3 strings. Hence there is no hope of transforming β' and hence also β into 1, since the commutator $xyx^{-1}y^{-1}$ is nontrivial in the free group.

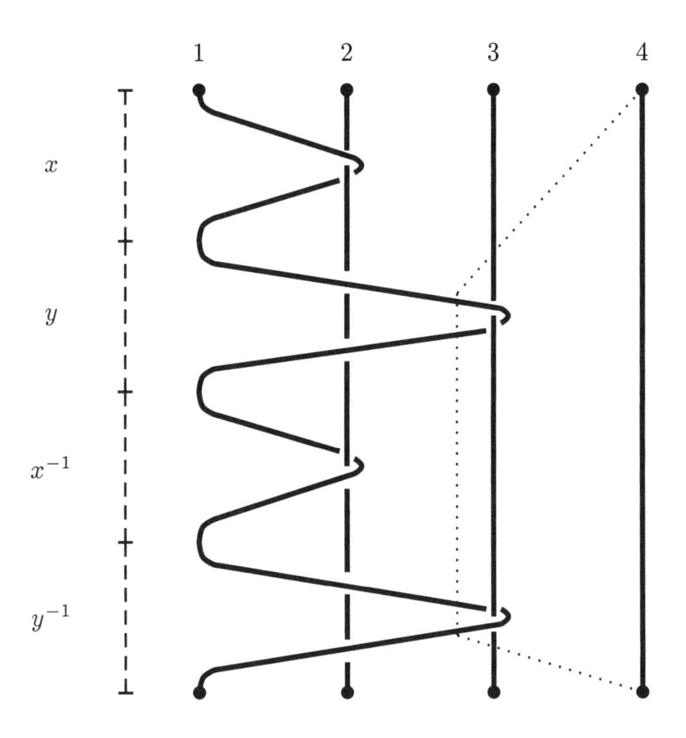

Figure 4. representing β' by a commutator

What can happen if we modify the rules further, and now allow the first and fourth strings to touch, and, at the moment of touching, forget where the respective parts of strings came from, and then part as newly constituted strings? In this case, one can see using Figures 4 and 5 that β

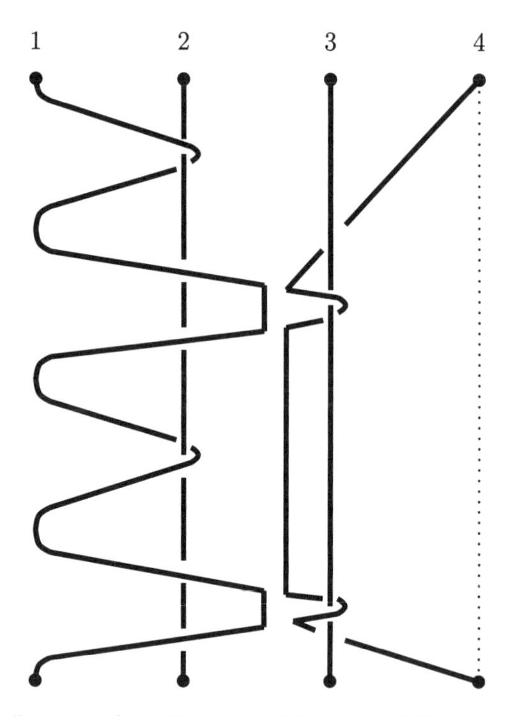

Figure 5. "merge and part" to unravel the braid of Figure 4

unravels completely! In Figure 4 a dotted line indicates where the fourth string can be stretched behind other strings and made to intersect with the first string, thus "merging". Parting as newly reconstituted strings, as indicated by Figure 5, creates a configuration which can easily be seen to unwrap, so represents the identity braid. These diagrams prove that in the merge and part monoid, defined in the next section, $xyx^{-1}y^{-1}$ is trivial. This verifies one of the relations in a presentation (see [4]).

2. The Merge and Part Braid Monoid

Let n be a positive integer which is fixed throughout. Denote by $B = B_n$ the braid group on n strings, and by E the set of all equivalences on $\{1, \ldots, n\}$, which is an upper semilattice under $\vee$. Denote the identity elements of B and E by 1 (which has to be read in context) and the zero of E by 0.

If $\beta \in B$ then $\overline{\beta}$ denotes the associated permutation, so that overline is a group homomorphism from B onto the symmetric group. If $\mathcal{E} \in E$ then put

$$\mathcal{E}^{\beta} \;=\; \{\,(i,j) \mid (i,j)\overline{\beta} \in \mathcal{E}\,\}\,,$$

where we define $(i,j)\overline{\beta} = (i\overline{\beta}, j\overline{\beta})$. If follows quickly that

$$\phi : B \to \operatorname{Aut} E\,, \beta \mapsto \beta\phi : \mathcal{E} \mapsto \mathcal{E}^{\beta}$$

is an antihomomorphism, so we get the semidirect product

$$E \rtimes B \;=\; E \rtimes_{\phi} B \;=\; \{\,(\mathcal{E},\beta) \mid \mathcal{E} \in E, \beta \in B\,\}$$

with multiplication $(\mathcal{E},\beta)(\mathcal{E}_0,\beta_0) \;=\; (\mathcal{E} \vee \mathcal{E}_0^{\beta}, \beta\beta_0)$. It is routine now to check that $E \rtimes B$ is a factorizable inverse semigroup with group of units $\{\,(1,\beta) \mid \beta \in B\,\} \cong B$ and set of idempotents $\{\,(\mathcal{E},1) \mid \mathcal{E} \in E\,\} \cong E$. (The definition and properties of *factorizability* are reviewed and developed in the next section.)

For each $i = 1, \ldots, n-1$ denote by σ_i the usual braid generator where the ith string crosses over the $(i+1)$th string. For each $\mathcal{E} \in E$ define the subgroup

$$B_{\mathcal{E}} \;=\; \langle\, \beta^{-1}\sigma_i\beta \mid i \in \{1,\ldots,n-1\},\; (i,i+1)\overline{\beta} \in \mathcal{E}\,\rangle$$

(which we interpret to be the trivial subgroup when $\mathcal{E}$ is the identity equivalence relation). The following facts are immediate from the definitions:

$$B_{\mathcal{E}} \;\subseteq\; B_{\mathcal{E}'} \quad \text{if} \quad \mathcal{E} \subseteq \mathcal{E}'\,,$$

$$\mathcal{E}^{\beta} \;=\; \mathcal{E} \quad \text{if} \quad \beta \in B_{\mathcal{E}}\,,$$

$$\gamma B_{\mathcal{E}}\gamma^{-1} \;=\; B_{\mathcal{E}^{\gamma}} \quad \text{for all} \quad \gamma \in B\,.$$

Define an equivalence $\sim$ on $E \rtimes B$ by

$$(\mathcal{E},\beta) \sim (\mathcal{E}_0,\beta_0) \quad \text{if and only if} \quad \mathcal{E} = \mathcal{E}_0 \quad \text{and} \quad \beta\beta_0^{-1} \in B_{\mathcal{E}}\,.$$

Lemma 2.1. *The equivalence $\sim$ is a congruence.*

Proof. Suppose $(\mathcal{E}_1,\beta_1) \sim (\mathcal{E}_1',\gamma_1)$ and $(\mathcal{E}_2,\beta_2) \sim (\mathcal{E}_2',\gamma_2)$. Then $\mathcal{E}_1 = \mathcal{E}_1'$, $\mathcal{E}_2 = \mathcal{E}_2'$, $\beta_1\gamma_1^{-1} \in B_{\mathcal{E}_1}$ and $\beta_2\gamma_2^{-1} \in B_{\mathcal{E}_2}$, so that

$$\begin{aligned}\beta_1\beta_2(\gamma_1\gamma_2)^{-1} \;&=\; (\beta_1\gamma_1^{-1})\gamma_1(\beta_2\gamma_2^{-1})\gamma_1^{-1} \\ &\in\; B_{\mathcal{E}_1}\gamma_1 B_{\mathcal{E}_2}\gamma_1^{-1} \;\subseteq\; B_{\mathcal{E}_1 \vee \mathcal{E}_2^{\gamma_1}}\,.\end{aligned}$$

Further, since $\gamma_1\beta_1^{-1} = (\beta_1\gamma_1^{-1})^{-1} \in B_{\mathcal{E}_1} \subseteq B_{\mathcal{E}_1 \vee \mathcal{E}_2^{\beta_1}}$, and $(\gamma_1\beta_1^{-1})\phi \in \operatorname{Aut} E$, we have

$$\begin{aligned}
\mathcal{E}_1 \vee \mathcal{E}_2^{\beta_1} &= (\mathcal{E}_1 \vee \mathcal{E}_2^{\beta_1})^{\gamma_1\beta_1^{-1}} \\
&= \mathcal{E}_1^{\gamma_1\beta_1^{-1}} \vee (\mathcal{E}_2^{\beta_1})^{\gamma_1\beta_1^{-1}} \\
&= \mathcal{E}_1 \vee \mathcal{E}_2^{(\gamma_1\beta_1^{-1}\beta_1)} \\
&= \mathcal{E}_1 \vee \mathcal{E}_2^{\gamma_1} .
\end{aligned}$$

This proves $(\mathcal{E}_1, \beta_1)(\mathcal{E}_2, \beta_2) \sim (\mathcal{E}_1', \gamma_1)(\mathcal{E}_2', \gamma_2)$. □

Define the *merge and part braid monoid on n strings* to be

$$\widetilde{B} = \widetilde{B}_n = (E \rtimes B)/\sim .$$

Denote the $\sim$-congruence class of $(\mathcal{E}, \beta)$ by $[\mathcal{E}, \beta]$. Clearly, the natural map $(\mathcal{E}, \beta) \mapsto [\mathcal{E}, \beta]$ is one-one on $\{ (1, \beta) \mid \beta \in B \}$ and on $\{ (\mathcal{E}, 1) \mid \mathcal{E} \in E \}$, so we get the following:

Proposition 2.1. *The inverse monoid $\widetilde{B}$ is factorizable with group of units $\{[1, \beta] \mid \beta \in B\} \cong B$ and semilattice of idempotents $\{[\mathcal{E}, 1] \mid \mathcal{E} \in E\} \cong E$.*

We now provide a geometric realization of $\widetilde{B}$ which justifies the manipulations of strings used in the examples in the first section. Let

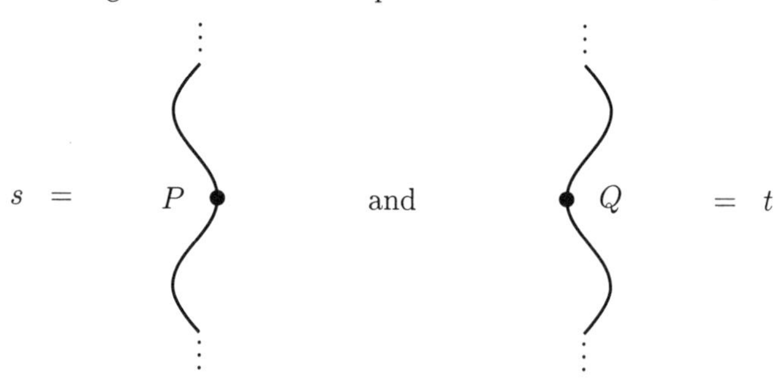

be two strings descending from fixed points on the upper plane to connect to fixed points on the lower plane. We say that a homotopy causes s and t to *merge and part* if, during the homotopy, s and t come together just once at say P and Q,

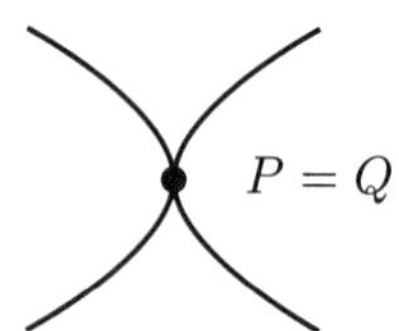

and then part, reconstituting as two strings made up of respective upper and lower strands. A catalogue of the possible configurations in the neighbourhood a moment before and after is given in Figure 6. Note that (2) and (3) can be interchanged using normal homotopy. Also an interchange between (6) and (7) can be achieved using two interchanges between types (4) and (5) in nearby neighbourhoods.

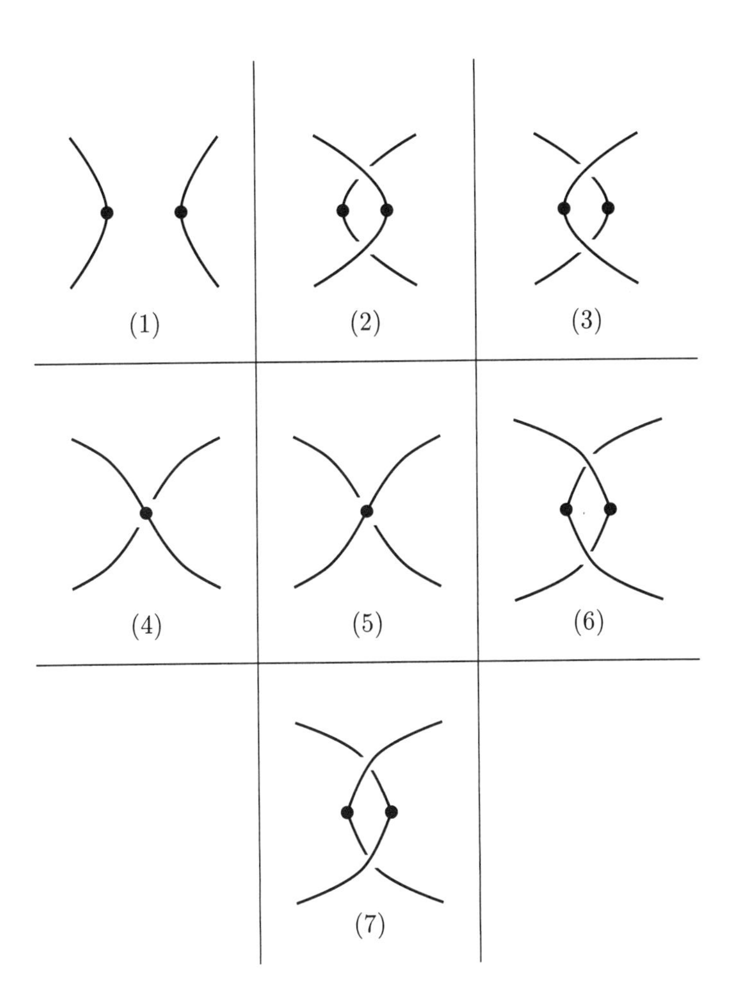

Figure 6. possibilities before and after "merge and part"

Consider a configuration of strings $s_1, \ldots, s_n$ emanating from points $1, \ldots, n$ respectively on the upper plane, and let $\mathcal{E} \in E$. We say s_i is *$\mathcal{E}$-equivalent* to s_j if $(i,j) \in \mathcal{E}$, and *$\mathcal{E}$-inequivalent* to s_j if $(i,j) \notin \mathcal{E}$.

Theorem 2.1. *Let $\mathcal{E}, \mathcal{E}_0 \in E$ and $\beta, \beta_0 \in B$. Then $(\mathcal{E}, \beta) \sim (\mathcal{E}_0, \beta_0)$ if and only if $\mathcal{E} = \mathcal{E}_0$ and there exists a homotopy from a representative of β to a representative of β_0 such that, in the course of the homotopy, $\mathcal{E}$-inequivalent strings never touch and $\mathcal{E}$-equivalent string are allowed to merge and part (one at a time, a finite number of times).*

Proof. ($\Rightarrow$) Suppose that $(\mathcal{E}, \beta) \sim (\mathcal{E}_0, \beta_0)$. Then $\mathcal{E} = \mathcal{E}_0$ and $\beta\beta_0^{-1} \in B_{\mathcal{E}}$. We prove there is a homotopy of the required type from a representative of β to a representative of β_0 by induction on the number of generators of $B_{\mathcal{E}}$ in the product forming $\beta\beta_0^{-1}$. If no generators are required then $\beta\beta_0^{-1} = 1$ so there is a homotopy in which no strings touch, which starts an induction. Thus, to prove the inductive step, it is sufficient to check that there is a homotopy of the required type between representatives of 1 and $\gamma^{-1}\sigma_i\gamma$ where $i \in \{1, \ldots, n-1\}$ and $(i, i+1)\overline{\gamma} \in \mathcal{E}$. Put

$$j = i\overline{\gamma}\,, \quad k = (i+1)\overline{\gamma}\,.$$

Let $\widehat{\gamma}$ be a representative of γ and $\widehat{1}$ the straight string representative of the identity braid 1. We get a homotopy from $\widehat{1}$ to a representative of $\gamma^{-1}\sigma_i\gamma$ as the composite of H_1 and H_2 displayed in Figure 7 (where $\widehat{\gamma}^{-1}$ is the reflection of $\widehat{\gamma}$ in a horizontal plane, and the representatives have been contracted in the second and third parts of the diagram). It is to be understood in Figure 7 that H_1 is a homotopy where no strings touch and H_2 is a homotopy in which only the jth and kth strings have merged and parted.

($\Leftarrow$) Suppose now $\mathcal{E} = \mathcal{E}_0$ and a homotopy H exists of the required type between a representative $\widehat{\beta}$ of β and a representative $\widehat{\beta_0}$ of β_0. If no strings touch then $\beta = \beta_0$, so certainly $\beta\beta_0^{-1} \in B_{\mathcal{E}}$, which starts an induction. Suppose H is the composite of homotopies H_1 and H_2 where during H_2 exactly one pair of strings merge and part. Let γ be the braid of the representative $\widehat{\gamma}$ resulting from applying H_1 to $\widehat{\beta}$. By an inductive hypothesis

$$\beta\gamma^{-1} \in B_{\mathcal{E}}\,.$$

The homotopy H_2 can be replaced (if necessary) by a composite $H_3H_4H_5$

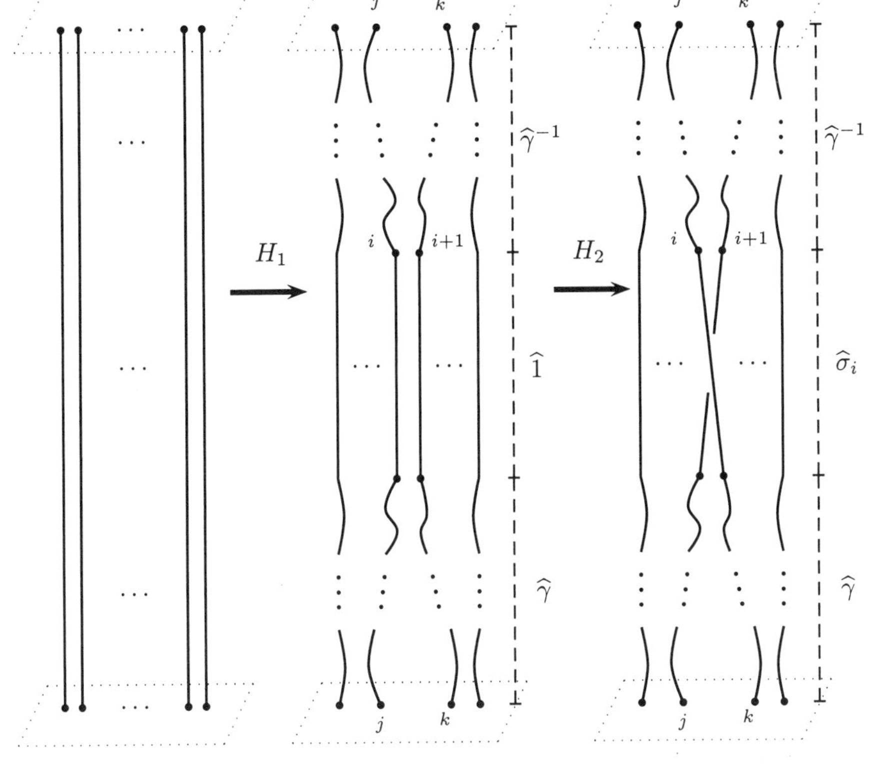

Figure 7. "merge and part" homotopy from $\widehat{1}$ to a representative of $\gamma^{-1}\sigma_i\gamma$

where $\gamma = \gamma_1\gamma_2$ for some $\gamma_1, \gamma_2 \in B$,

$$\widehat{\gamma} \xrightarrow{H_3} \left|\begin{array}{c}\widehat{\gamma}_1\\ \widehat{1}\\ \widehat{\gamma}_2\end{array}\right| \xrightarrow{H_4} \left|\begin{array}{c}\widehat{\gamma}_1\\ \widehat{\tau}\\ \widehat{\gamma}_2\end{array}\right| \xrightarrow{H_5} \widehat{\beta}_0 ,$$

and $\tau \in \{1, \sigma_i, \sigma_i^{-1}, \sigma_i^2, \sigma_i^{-2}\}$ (according to cases (1) to (7) catalogued in Figure 6), where H_3 and H_5 have no strings touching, and H_4 alters the $\widehat{1}$, contracted in the middle, by causing the ith and $(i+1)$th strings to merge and part. Because no strings touch during H_5 we have

$$\beta_0 = \gamma_1\tau\gamma_2 .$$

In order for H_4 to apply to the ith and $(i+1)$the strings, we need $(i, i+1)\overline{\gamma}_1^{-1} \in \mathcal{E}$. Thus $\gamma_1\tau\gamma_1^{-1} \in B_{\mathcal{E}}$, so

$$\beta\beta_0^{-1} = \beta\gamma_2^{-1}\tau^{-1}\gamma_1^{-1} = \beta\gamma^{-1}\gamma_1\tau^{-1}\gamma_1^{-1} \in B_{\mathcal{E}} ,$$

and the theorem is proved. □

In the final section of this paper we will return to discuss this monoid and prove it embeds in the coset monoid of the braid group.

3. Coset Monoids of Groups and Duality

In this section our aim is to use coset monoids of groups to dualize the following result of Munn [8]:

Theorem 3.1. *If S is an inverse semigroup with semilattice of idempotents E then there exists a homomorphism from S into T_E with kernel μ, the largest congruence contained in $\mathcal{H}$. Thus, if S is fundamental, then S embeds in T_E. Moreover every full inverse subsemigroup of T_E is fundamental.*

Coset monoids of groups were first studied by Schein [10], who discusses them in detail, and with generalizations to semigroups, in [11]. Other authors, such as McAlister [7], Leech [6], Nambooripad and Veeramony [9], have used them and generalizations in various contexts.

We will introduce the topic using the following simple example. Let G denote the symmetric group on a set of size 3, which has the presentation $\langle a, b \mid a^3 = b^2 = 1, a^b = a^{-1}\rangle$. If we identify G also with the group of symmetries of the triangle then we may list the elements of G as $\{1, a, a^2, b, ab, a^2b\}$, where $1, a, a^2$ form a normal subgroup of rotations and each of b, ab, a^2b is a reflection generating a nonnormal subgroup of order 2. Let $\mathcal{S}$ denote the lattice of subgroups of G. In particular $\mathcal{S}$ is a semilattice with respect to $\vee$ defined by $H \vee K = \langle H \cup K\rangle$ whose Hasse diagram may be depicted thus:

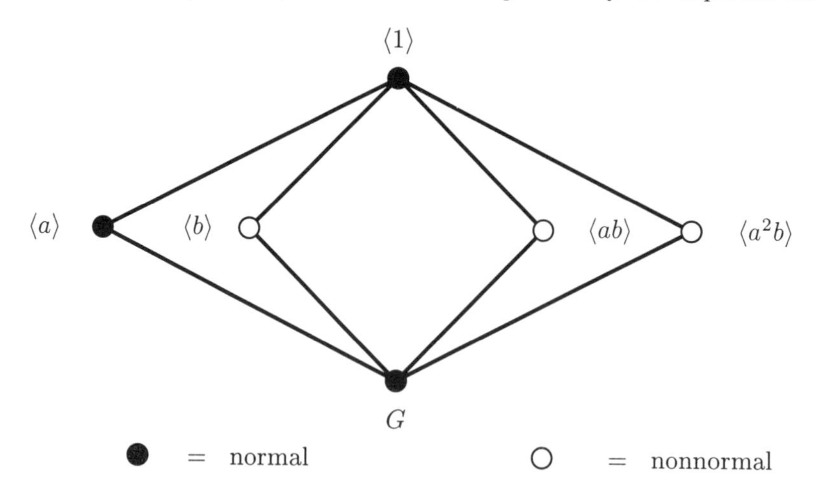

Then G acts on $\mathcal{S}$ by conjugation, fixing the normal subgroups and providing permutations of the nonnormal subgroups. (In this example the nonnormal subgroups provide a faithful permutation representation.) We may form

$$T = \mathcal{S} \rtimes G = \{(H,g) \mid H \leq G, g \in G\}$$

with multiplication $(H,g)(K,\ell) = (H \vee K^{g^{-1}}, g\ell)$. It is not hard to see (Figure 8) that the $\mathcal{D}$-classes of T correspond to conjugacy classes of subgroups of G, and that, in our example, the nonnormal subgroups lie in a single $\mathcal{D}$-class. Define a congruence $\sim$ on T by

$$(H,g) \sim (K,\ell) \quad \text{if and only if} \quad Hg = K\ell\,,$$

equality of cosets. As before, write $[H,g]$ for the congruence class of (H,g), so

$$T/\sim\, = \{\,[H,g] \mid H \leq G\,,\ g \in G\,\}$$

with multiplication $[H,g][K,\ell] = [H \vee K^{g^{-1}}, g\ell]$. But there is a bijection between $T/\sim$ and

$$\mathcal{C}(G) = \{\,Hg \mid H \leq G\,,\ g \in G\,\}\,,$$

the set of all cosets with respect to all subgroups of G. Thus $\mathcal{C}(G)$ inherits the multiplication

$$(Hg) * (K\ell) = (H \vee K^{g^{-1}})g\ell\,,$$

which one may show is the smallest coset containing the set product $HgK\ell$. We call $(\mathcal{C}(G), *)$ the *coset monoid* of G. (It is denoted by $K(G)$ by Schein and others, but we prefer our present notation because of a certain universal property with respect to a class $\mathcal{C}$ defined below.) In our example $\mathcal{C}(G) \cong T/\sim$ has 18 elements, displayed in Figure 9.

All of the preceding definitions of this section are now taken as read for any group G. Recall that an inverse monoid M is factorizable if $M = GE (= EG)$ where G is its group of units and E is its semilattice of idempotents, and it is standard to write G_e for the *stabilizer* of $e \in E$, that is,

$$G_e = \{\,g \in G \mid ge = e\,\} = \{\,g \in G \mid eg = e\,\}\,.$$

(The reader may easily check that, in the previous section, where B is the braid group and $\mathcal{E}$ is an equivalence on $\{1,\ldots,n\}$, the definition of $B_{\mathcal{E}}$ there gives precisely the definition of $B_{\mathcal{E}}$ here as a stabilizer.) Note that, for $g,h \in G$ and $e,f \in E$, using cancellation by a unit,

$$ge = f \quad\Longrightarrow\quad gege = ge\,,\ e = ege = ef = fe = f$$

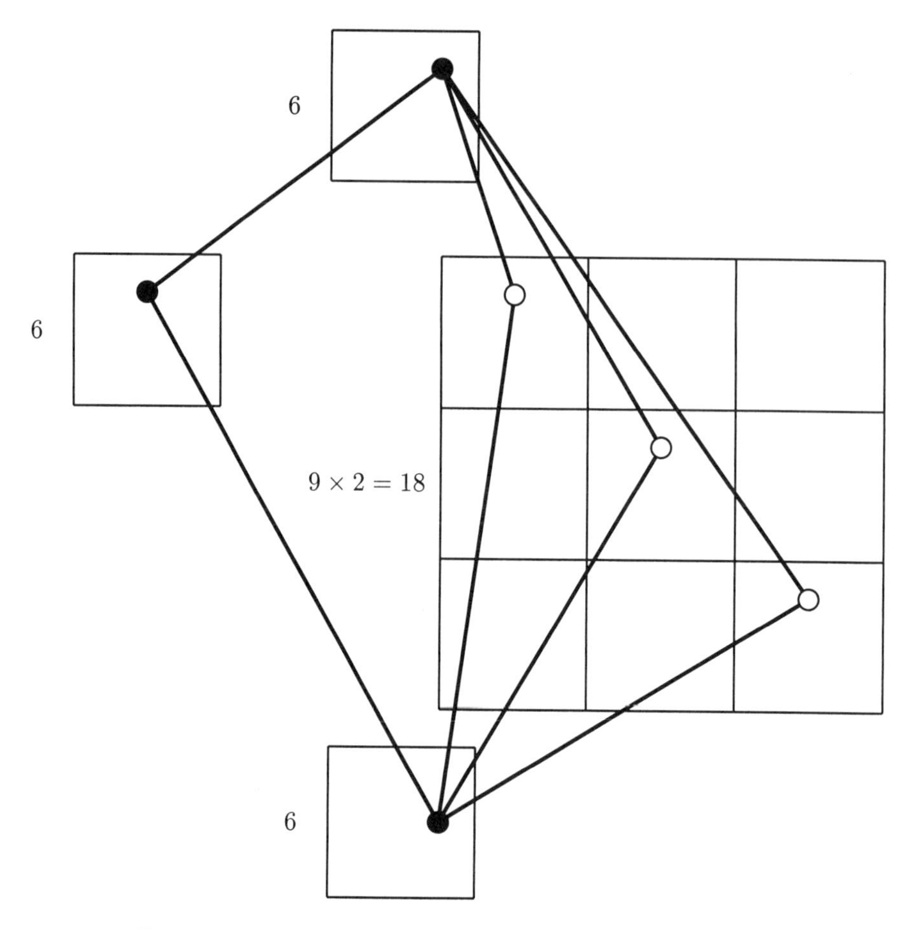

Figure 8. eggbox diagram for $T = \mathcal{S} \rtimes G$ containing 36 elements

and dually

$$eg = f \quad \Longrightarrow \quad e = f \ ,$$

from which it follows quickly that

$$eg \,\mathcal{R}\, fh \quad \Longleftrightarrow \quad e = f \quad \Longleftrightarrow \quad ge \,\mathcal{L}\, fh \ .$$

Now define $\mathcal{C}$ to be the class of factorizable inverse monoids $M = GE$ such that the mapping

$$e \mapsto G_e \ : \ E \to \mathcal{S}$$

respects joins, that is, $G_{ef} = \langle G_e \cup G_f \rangle$ for all $e, f \in E$.

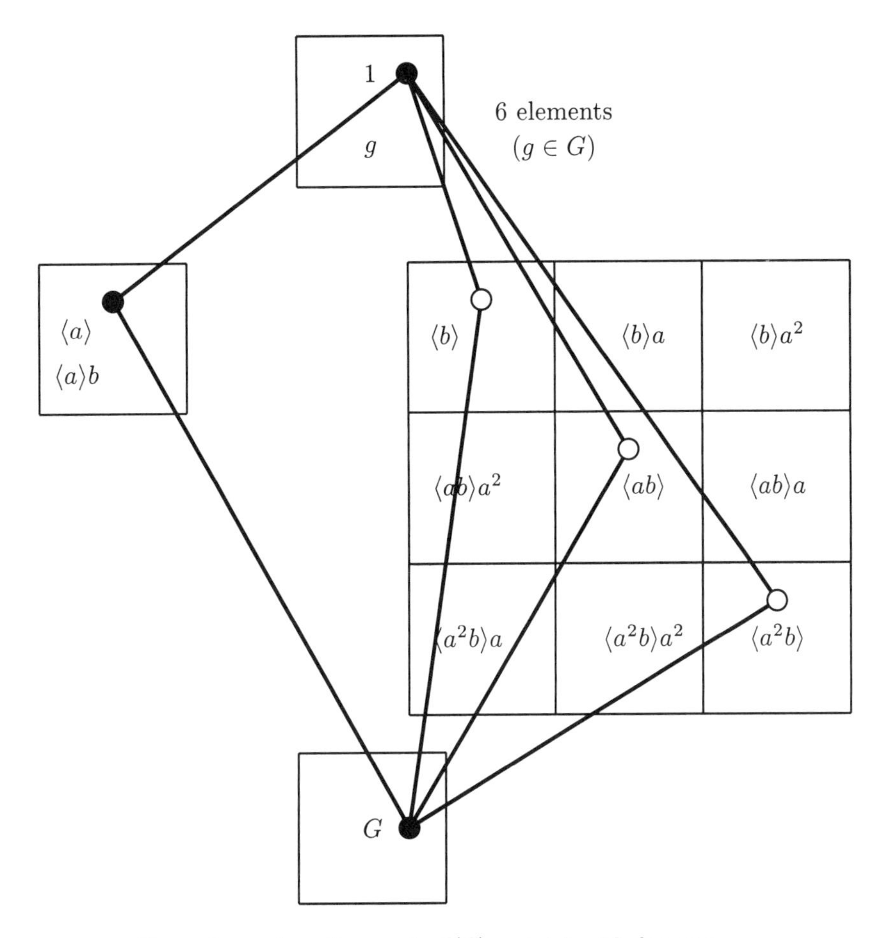

Figure 9. eggbox diagram for $\mathcal{C}(G)$ containing 18 elements

Clearly $\mathcal{C}(G)$ is factorizable with G as its group of units, and if $H, K \leq G$ then $G_H = H$, so that

$$G_{H*K} = G_{\langle H \cup K \rangle} = \langle H \cup K \rangle .$$

This verifies that

$$\mathcal{C}(G) \in \mathcal{C} .$$

Clearly also, since all stabilizers are trivial, $\mathcal{S} \rtimes G \in \mathcal{C}$. In fact, $\mathcal{C}(G)$ is a *cofundamental* image of $\mathcal{S} \rtimes G$ in the sense of Theorem 3.2 below, which dualizes Theorem 3.1.

Lemma 3.1. *Let M be any inverse semigroup and ρ a congruence on M. Then*

$$\rho \cap \mathcal{R} = 1 \quad \Longleftrightarrow \quad \rho \cap \mathcal{L} = 1 \, .$$

Proof. This follows quickly from the fact that inversion is an anti-isomorphism of M and if $x, y \in M$ then $(x, y) \in \rho$ if and only if $(x^{-1}, y^{-1}) \in \rho$. □

Lemma 3.2. *Let $M = GE \in \mathcal{C}$. Then*

$$\theta : M \to \mathcal{C}(G) \, , \; eg \mapsto G_e g \, [ge \mapsto gG_e] \, ,$$

is a homomorphism.

Proof. Let $g, h \in G$ and $e, f \in E$. Note θ is well-defined because if $eg = fh$ then $e = f$ and $egh^{-1} = e$, so $gh^{-1} \in G_e$, giving $G_e g = G_e h = G_f h$. Further, by definition of membership of $\mathcal{C}$,

$$\begin{aligned} (egfh)\theta &= (egfg^{-1}gh)\theta \;=\; G_{egfg^{-1}}gh \\ &= (G_e \vee G_{gfg^{-1}})gh \;=\; \langle G_e \cup G_{gfg^{-1}} \rangle gh \\ &= G_e g * G_f h \;=\; (eg)\theta * (fh)\theta \, . \end{aligned}$$

□

Lemma 3.3. *Let $M = GE \in \mathcal{C}$. Then there exists a largest congruence ν on M such that $\nu \cap \mathcal{R} = 1$ or (equivalently) $\nu \cap \mathcal{L} = 1$. Further there is a representation of M by $\mathcal{C}(G)$,*

$$\theta : M \to \mathcal{C}(G) \, , \; eg \mapsto G_e g \, [ge \mapsto gG_e] \, ,$$

whose kernel is ν.

Proof. By the previous lemma, θ is a representation. Clearly

$$\ker \theta \;=\; \{ \, (eg, fh) \mid G_e = G_f \quad \text{and} \quad gh^{-1} \in G_e \, \} \, .$$

Further,

$$(eg, fh) \in \ker \theta \cap \mathcal{R} \quad \Longrightarrow \quad e = f \text{ and } gh^{-1} \in G_e \quad \Longrightarrow \quad eg = eh \, ,$$

which proves $\ker \theta \cap \mathcal{R} = 1$.

Let ρ be any congruence on M such that $\rho \cap \mathcal{R} = 1$. To complete the proof it suffices to show $\rho \subseteq \ker \theta$. Suppose $(eg, fh) \in \rho$. Then $(g^{-1}e, h^{-1}f) \in \rho$ so

$$e \;=\; egg^{-1}e \, \rho \, fhh^{-1}f \, \rho \, f \, .$$

If $g \in G_e$ then

$$f \,\rho\, e \;=\; eg \,\rho\, fg$$

so $f = fg$, since $f \,\mathcal{R}\, fg$ and $\rho \cap \mathcal{R} = 1$, giving $g \in G_f$. Thus $G_e \subseteq G_f$ and similarly $G_f \subseteq G_e$ whence equality holds. But also

$$eg \,\rho\, fh \,\rho\, eh \quad \text{and} \quad eg \,\mathcal{R}\, eh$$

so $eg = eh$, whence $gh^{-1} \in G_e$. This proves $(eg, fh) \in \ker\theta$, so $\rho \subseteq \ker\theta$.□

Call $M \in \mathcal{C}$ *cofundamental* if $\nu = 1$, and call an inverse submonoid N of M *cofull* if $N \in \mathcal{C}$ and N has the same group of units as M.

Theorem 3.2. *If $M = GE \in \mathcal{C}$ then*

$$\theta : M \to \mathcal{C}(G)\,,\; eg \mapsto G_e g \; [ge \mapsto gG_e]$$

is a representation with kernel ν equal to the largest congruence on M such that

$$\nu \cap \mathcal{R} \;=\; 1_M\,.$$

Thus if M is cofundamental then M embeds in $\mathcal{C}(G)$ as a cofull inverse submonoid. Moreover every cofull inverse submonoid of $\mathcal{C}(G)$ is cofundamental.

Proof. In light of the previous lemmas, it remains to prove the final statement. Let M be a cofull inverse submonoid of $\mathcal{C}(G)$. If $(Hg, K\ell) \in \nu$ then

$$H \;=\; G_H \;=\; G_K \;=\; K \quad \text{and} \quad g\ell^{-1} \in G_H \;=\; H\,,$$

so $Hg = K\ell$. Thus $\nu = 1$ and M is cofundamental. □

The duality between Theorems 3.1 and 3.2 is perhaps unexpected. Observe that a congruence ρ on an inverse monoid M is idempotent-separating if and only if

$$\rho \subseteq \mathcal{H} \;=\; \mathcal{R} \cap \mathcal{L}\,,$$

that is,

$$\rho \subseteq \mathcal{R} \qquad \text{and} \qquad \rho \subseteq \mathcal{L}\,.$$

One's first guess at a dual property for ρ might be be to make it "antipodal" to $\mathcal{H}$:

$$\rho \cap \mathcal{H} \;=\; 1_M\,.$$

But this won't lead to a result like Theorem 3.2, because of the example M displayed in Figure 10, which is a cofull submonoid of $\mathcal{C}(G)$ (displayed in Figure 9) where G is the symmetric group on three letters. Observe that $\mathcal{H}$ is trivial on the ideal $M \backslash G$, and the Rees congruence ρ with respect to this ideal certainly is not trivial, yet $\rho \cap \mathcal{H} = 1_M$.

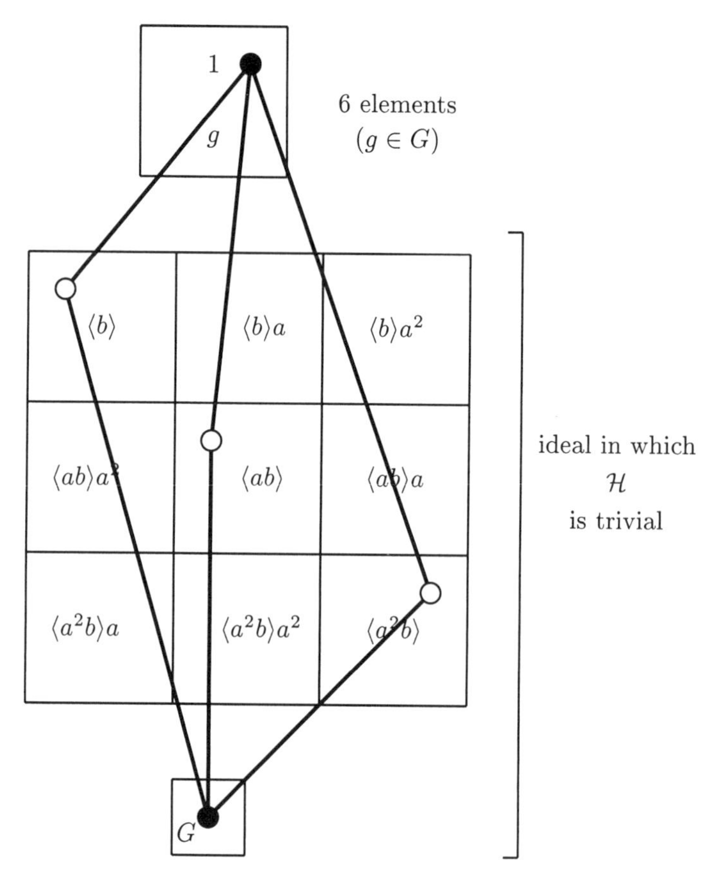

Figure 10. eggbox diagram for a cofull submonoid of $\mathcal{C}(G)$ where G is the symmetric group on 3 letters

One's second guess at a dual property for ρ might be be to make it "antipodal" to the "dual" of $\mathcal{H}$, which one might think of as $\mathcal{D}$:

$$\rho \cap \mathcal{D} \;=\; 1_M \,.$$

But then this won't lead to a result like Theorem 3.2 either, because for any group G with lattice of subgroups $\mathcal{S}$, the representation $\theta : \mathcal{S} \rtimes G \to \mathcal{C}(G)$ becomes projection onto the second coordinate $(e, g) \mapsto g$ (since all the stabilizers are trivial), and certainly in nontrivial examples, $\ker \theta \cap \mathcal{D} \neq 1_{\mathcal{S} \rtimes G}$.

The correct property turns out to first take the logical dual of the conjunction $\rho \subseteq \mathcal{R}$ and $\rho \subseteq \mathcal{L}$, which is a disjunction, and then make ρ "antipodal" to each alternative:

$$\rho \cap \mathcal{R} = 1_M \qquad \text{or} \qquad \rho \cap \mathcal{L} = 1_M .$$

But because of the equivalence of these two alternatives (Lemma 3.1) only one needs to be included in the statement of Theorem 3.2.

4. An Embedding in the Coset Monoid of the Braid Group

In this final section we prove that the merge and part braid monoid embeds in the coset monoid of the braid group. Let $X = \{1, \ldots, n\}$ where n is a fixed positive integer, and let E be the set of equivalence relations on X. Recall $B = B_n$ denotes the braid group on n strings and $\widetilde{B} = (E \rtimes B)/\sim$ is the merge and part monoid defined in the second section. The key step is the following:

Lemma 4.1. *Let $\mathcal{E}_1, \mathcal{E}_2 \in E$. Then $B_{\mathcal{E}_1} \vee B_{\mathcal{E}_2} = B_{\mathcal{E}_1 \vee \mathcal{E}_2}$.*

Proof. The forward set containment is obvious. Equality is also obvious if $\mathcal{E}_1$ and $\mathcal{E}_2$ are both the identity equivalence. Suppose then that at least one of these is not the identity equivalence. Let $\tau = \beta^{-1}\sigma_i\beta$ be a generator of $B_{\mathcal{E}_1 \vee \mathcal{E}_2}$ where $i \in X \backslash \{n\}$ and $\beta \in B$ such that

$$(i, i+1)\overline{\beta} \in \mathcal{E}_1 \vee \mathcal{E}_2 .$$

To complete the proof of the lemma, it suffices to show $\tau \in B_{\mathcal{E}_1} \vee B_{\mathcal{E}_2}$. Since $\mathcal{E}_1 \vee \mathcal{E}_2$ is the transitive closure of $\mathcal{E}_1 \cup \mathcal{E}_2$, without loss of generality we may suppose there exists a positive integer m and $x_1, \ldots, x_{2m} \in X$ such that

$$i\overline{\beta} = x_1 \; \mathcal{E}_1 \; x_2 \; \mathcal{E}_2 \; \ldots \; \mathcal{E}_1 \; x_{2m-2} \; \mathcal{E}_2 \; x_{2m-1} \; \mathcal{E}_1 \; x_{2m} = (i+1)\overline{\beta} .$$

For $a, b \in X$, $a < b$, put

$$\delta_{a,b} = \sigma_{a+1}\sigma_{a+2} \cdots \sigma_{b-1}$$

(interpreted as the identity braid if $b = a + 1$) and

$$\gamma_{a,b} = \gamma_{b,a} = \delta_{a,b}^{-1}\sigma_a\delta_{a,b} ,$$

so that $\overline{\gamma_{a,b}}$ is the transposition interchanging a and b and

$$(a, a+1)\overline{\delta_{a,b}} = (a, b) .$$

Thus, by definition, $\gamma_{c,d}$ is a generator of $B_{\mathcal{E}}$ for all $\mathcal{E} \in E$ and $(c, d) \in \mathcal{E}$ with $c \neq d$. Now put

$$\delta = \gamma_{x_1,x_2}\gamma_{x_2,x_3} \cdots \gamma_{x_{2m-2},x_{2m-1}}$$

(interpreted as the identity braid if $m = 1$) and

$$\gamma = \delta\gamma_{x_{2m-1},x_{2m}}\delta^{-1} .$$

Observe that

$$\gamma_{x_i,x_{i+1}} \in \begin{cases} B_{\mathcal{E}_1} & \text{if } i \text{ is odd} \\ B_{\mathcal{E}_2} & \text{if } i \text{ is even} \end{cases}$$

so that $\gamma \in \langle B_{\mathcal{E}_1} \cup B_{\mathcal{E}_2}\rangle$. Also

$$(i, i+1)\overline{\beta\gamma} = (x_1, x_{2m})\overline{\gamma} = (x_2, x_1)$$

so that $(\beta\gamma)^{-1}\sigma_i\beta\gamma \in B_{\mathcal{E}_1}$. Hence

$$\begin{aligned} \tau &= \gamma(\gamma^{-1}\beta^{-1}\sigma_i\beta\gamma)\gamma^{-1} \\ &\in \langle B_{\mathcal{E}_1} \cup B_{\mathcal{E}_2}\rangle B_{\mathcal{E}_1}\langle B_{\mathcal{E}_1} \cup B_{\mathcal{E}_2}\rangle \\ &\subseteq B_{\mathcal{E}_1} \vee B_{\mathcal{E}_2} . \end{aligned}$$

□

Lemma 4.2. *Let $\mathcal{E}_1, \mathcal{E}_2 \in E$ such that $\mathcal{E}_1 \neq \mathcal{E}_2$. Then $B_{\mathcal{E}_1} \neq B_{\mathcal{E}_2}$.*

Proof. Without loss of generality we have $x, y \in X$ such that $(x, y) \in \mathcal{E}_1$ but $(x, y) \notin \mathcal{E}_2$. Then, using the notation introduced in the proof of the previous lemma, $\gamma_{x,y} \in B_{\mathcal{E}_1}$ and $\overline{\gamma_{x,y}}$ is the transposition interchanging x and y. But, by a simple induction on the number of generators, if $\beta \in B_{\mathcal{E}_2}$ then $(x, x\overline{\beta}) \in \mathcal{E}_2$. Since $(x, x\overline{\gamma_{x,y}}) = (x, y) \notin \mathcal{E}_2$, we have that $\gamma_{x,y} \notin B_{\mathcal{E}_2}$, proving $B_{\mathcal{E}_1} \neq B_{\mathcal{E}_2}$. □

Theorem 4.1. *The merge and part monoid embeds in the coset monoid of the braid group.*

Proof. By Lemma 4.1, $\widetilde{B} \in \mathcal{C}$ so, by Theorem 3.2, $\theta : \widetilde{B} \to \mathcal{C}(B)$ is a representation. It follows quickly, by Lemma 4.2, that θ is faithful. □

The proofs in this section hold also for permutations (by overlining all braids), so we get the following result:

Corollary 4.1. *The largest factorizable inverse submonoid of the dual symmetric inverse monoid on a finite set embeds in the coset monoid of the symmetric group.*

Corollary 4.2. *The merge and part braid monoid and the largest factorizable inverse submonoid of the dual symmetric inverse monoid on a finite set are cofundamental.*

References

1. J. Baez, "Link invariants of finite type and perturbation theory," *Lett. Math. Phys.* **26** (1992), 43-51.
2. Joan S. Birman, "New points of view in knot theory," *Bull. Amer. Math. Soc.* **28** (1993), 253-287.
3. D. Easdown and T.G. Lavers, "The inverse braid monoid," *Advances in Mathematics*, to appear.
4. James East, "The factorisable braid monoid," in preparation.
5. D.G. FitzGerald and Jonathan Leech, "Dual symmetric inverse monoids and representation theory," *J. Austral. Math. Soc.* **64** (1998), 345-367.
6. Jonathan Leech, "Inverse monoids with a natural semilattice ordering," it Proc. London Math. Soc. **70** (1995), 146-182.
7. D.B. McAlister, "Embedding inverse semigroups in coset semigroups," *Semigroup Forum* **20** (1980), 255-267.
8. W.D. Munn, "Uniform semilattices and bisimple inverse semigroups," *Quart. J. Math. Oxford (2)* **17** (1966), 151-159.
9. K.S.S. Nambooripad and R. Veeramony, "Subdirect products of regular semigroups," *Semigroup Forum* **27** (1983), 265-307.
10. B.M. Schein, "Semigroups of strong subsets," *Volzhsky Matematichesky Sbornik* **4** (1966), 180-186.
11. B.M. Schein, "Cosets in groups and semigroups," *Semigroups with applications*, eds. Howie, Munn, Weinert, World Scientific, Singapore 1992, 205-221.

HYPERBOLIC GROUPS AND COMPLETELY SIMPLE SEMIGROUPS

JOHN FOUNTAIN

Department of Mathematics,
University of York,
Heslington, York YO10 5DD, U.K.
E-mail: jbf1@york.ac.uk

MARK KAMBITES

School of Mathematics & Statistics, Carleton University,
Herzberg Laboratories, 1125 Colonel By Drive,
Ottawa, Ontario K1S 5B6, Canada
E-mail: mkambite@math.carleton.ca

We begin with a brief introduction to the theory of word hyperbolic groups. We then consider four possible conditions which might reasonably be used as definitions or partial definitions of hyperbolicity in semigroups: having a hyperbolic Cayley graph; having hyperbolic Schützenberger graphs; having a context-free multiplication table; or having word hyperbolic maximal subgroups. Our main result is that these conditions coincide in the case of finitely generated completely simple semigroups.

This paper is based on a lecture given at the workshop on *Semigroups and Languages* held at the *Centro de Álgebra da Universidade de Lisboa* in November 2002. The aim of the paper, as of the talk, is to provide semigroup theorists with a gentle introduction to the concept of a word hyperbolic group (hereafter referred to as simply a 'hyperbolic group'), to discuss recent work of Gilman which makes it possible to introduce a notion of (word) hyperbolic semigroup and then to examine these ideas in the context of completely simple semigroups. The first three sections of the paper are expository while the fourth introduces new results. After briefly discussing Dehn's algorithm, we describe hyperbolic groups in terms of 'slim triangles', and then describe some properties of these groups which we need later in the paper. Hyperbolic groups were introduced by Gromov in [14]; readers who want to find out more can also consult [1], Chapter

III.Γ of [4], [11] and [22]. There are many characterisations of hyperbolic groups in terms of the geometry of Cayley graphs, but none of these extend naturally to semigroups because the algebraic properties of a semigroup are not closely related to the geometric properties of any notion of Cayley graph for a semigroup. However, Gilman [13] recently characterised hyperbolic groups in language theoretic terms; given a group G with finite generating set A and language L over A that projects onto G, he associated a language with G called the 'multiplication table relative to L' of the group, and showed that G is hyperbolic if and only if it has a context-free multiplication table relative to some rational language. This led Duncan and Gilman [9] to modify the definition of multiplication table for the semigroup setting and propose a definition of hyperbolic semigroup; using [13] they showed that for groups this is equivalent to the original definition. We give an account of this work in Section 3.

In Section 4 we examine completely simple semigroups which are hyperbolic in the sense of Duncan and Gilman. By definition, such a semigroup S is finitely generated, and so if S is isomorphic to the Rees matrix semigroup $M(G; I, \Lambda; P)$, then the sets I and Λ must be finite. Given this, one's first guess turns out to be correct, that is, S is hyperbolic if and only if G is hyperbolic. Moreover, in the case of a completely simple semigroup, the geometry of the Schützenberger graphs of the $\mathscr{R}$-classes is relevant. We show that the semigroup is hyperbolic if and only if each of these graphs is hyperbolic in the geometric sense defined in Section 1.

1. Hyperbolic Groups

Combinatorial group theory arose from the study of fundamental groups in the early years of the development of topology. It was given impetus and direction by Dehn's publication in 1912 of three basic decision problems: the word problem, the conjugacy problem and the isomorphism problem. We shall discuss only the word problem, and that only briefly.

We denote the free monoid on a non-empty set A by A^*, and the free semigroup by A^+. If A^{-1} is the set of symbols $\{a^{-1} : a \in A\}$, we define $(a^{-1})^{-1} = a$ for all $a \in A$, and for a non-empty word $w = x_1 \dots x_n$ where $x_i \in A \cup A^{-1}$, we define $w^{-1} = x_n^{-1} \dots x_1^{-1}$; for the empty word 1, we put $1^{-1} = 1$. A group G is *generated* by A (or A *generates* G) if there is a surjective homomorphism $\varphi : (A \cup A^{-1})^* \to G$ from the free monoid $(A \cup A^{-1})^*$ onto G with $a^{-1}\varphi = (a\varphi)^{-1}$ for all $a \in A$. If there is a finite set which generates G, then G is *finitely generated.* We also say that $\varphi : (A \cup A^{-1})^* \to G$ is a *choice of generators* for G, and speak of a *finite choice of generators* when A is finite. When X is a finite set which contains a formal inverse for each of its elements, we say that a surjective homomorphism $\varphi : X^* \to G$ is *symmetric* finite choice of generators for G. One wants to know when two words u, v in $(A \cup A^{-1})^*$ represent the same element of G, that is, when $u\varphi = v\varphi$. Equivalently, one is asking when $(uv^{-1})\varphi = 1$. A word r that represents the identity in G is called a *relator*, and there is a corresponding *relation* "$r = 1$". Let R be a set of relators and consider the congruence $\sim$ on $(A \cup A^{-1})^*$ generated by

$$\{(r, 1) : r \in R\} \cup \{(xx^{-1}, 1) : x \in A \cup A^{-1}\}.$$

If the congruences $\sim$ and $\ker \varphi$ coincide, then R is called a *set of relators for* G, and the pair $\langle A \mid R \rangle$ is a *presentation* for G. Rather than specifying a relator, one often specifies a relation $u = v$, meaning that uv^{-1} is a relator. The group G is *finitely presented* if it has a presentation $\langle A \mid R \rangle$ with both A and R finite. We mention the fact that if A and B are finite sets of generators for G and if G has a finite presentation $\langle A \mid R \rangle$, then it has a finite presentation $\langle B \mid S \rangle$ for some set S of words over $B \cup B^{-1}$.

A group G with a finite generating set A has *solvable word problem* if the set of words over A which represent the identity in G is recursive, that is, there is an algorithm which will decide which words in $(A \cup A^{-1})^*$ represent the identity and which do not.

Novikov [21], Boone [5] and Britton [6] independently proved that there are finitely presented groups with unsolvable word problem.

On the positive side, there are some presentations which lead to a fast algorithm for solving the word problem. Let A be a finite set of generators for a group G; recall that a word in $(A \cup A^{-1})^*$ is *freely reduced* if it has no factors of the form aa^{-1} or $a^{-1}a$ with $a \in A$, and that any word may be freely reduced by deleting such factors until none remain. Now suppose that we have a finite list of words $u_1, \ldots, u_n, v_1, \ldots, v_n$ such that, for each i, the words u_i and v_i represent the same element of G and $|v_i| < |u_i|$; suppose further that every non-empty freely reduced word that represents the identity of G contains at least one of the u_i as a factor. Then we have the following algorithm to solve the word problem for G. Given any word w, freely reduce it to obtain a word w'; if w' is not empty and does not have a factor u_i for any i, it does not represent the identity of G. If w' does have a factor u_i, replace it by v_i and freely reduce the resulting word to obtain w''. Note that $|w''| < |w'|$. Now repeat the procedure starting with w'' and continue until the procedure can be no longer applied.

Then w represents 1 in G if and only if the end result of the procedure is the empty word. Moreover, one can show that the algorithm is very fast. When a group admits this procedure for solving the word problem, we say that it has a *Dehn algorithm.* The algorithm leads to a *Dehn presentation* $\langle A \mid u_1v_1^{-1}, \ldots, u_nv_n^{-1}\rangle$ for G.

Dehn proved that Fuchsian groups admit Dehn presentations [8]. In the 1950s and 60s, various 'small cancellation' groups were shown to have Dehn presentations (see [20] for an account of this work). One motivation for the introduction of hyperbolic groups was to answer to the question of which groups admit a Dehn presentation. In his groundbreaking work [14], Gromov proved the following result among many other things.

Theorem 1.1. *A group is hyperbolic if and only if it admits a (finite) Dehn presentation.*

Having explained one of the reasons for studying hyperbolic groups, we now introduce some concepts which allow us to give one definition for this class of groups. First, let $\varphi : (A \cup A^{-1})^* \to G$ be a finite choice of generators for G. We consider the *(right) Cayley graph* $C_\varphi(G)$ of G *relative* to (A, φ). This is a (labelled) directed graph with vertex set G and an edge (labelled a) from g to $g(a\varphi)$ for each $g \in G$ and $a \in A$. Each edge is given a local metric in which it has unit length, and $C_\varphi(G)$ is turned into a metric space by defining the distance $d_\varphi(x, y)$ between two points x, y to be equal to the length of the shortest path joining them. When φ is understood, we sometimes refer to the 'Cayley graph relative to the generating set A'.

Notice that if g, h are vertices of $C_\varphi(G)$ (that is, elements of G), then $d_\varphi(g, h)$ is the length of the shortest word in $(A \cup A^{-1})^*$ representing $g^{-1}h$. Restricted to G, the metric d_φ is called the *word metric* with respect to (A, φ). We are interested only in finite generating sets A, but even in this case the metric on the Cayley graph and the word metric depend on the choice of A. However, if $\psi : (B \cup B^{-1})^* \to G$ is another finite choice of generators, then the metric spaces $(C_\varphi(G), d_\varphi)$, (G, d_φ), $(C_\psi(G), d_\psi)$ and (G, d_ψ) are closely related as we now explain.

Let (X, d) and (X', d') be metric spaces. A function $f : X \to X'$ is a *quasi-isometry* if there are constants $\lambda \geqslant 1$, $\epsilon \geqslant 0$ and $C \geqslant 0$ such that every point of X' lies in the closed C-neighbourhood of $f(X)$ and

$$\frac{1}{\lambda}d(x, y) - \epsilon \leqslant d'(f(x), f(y)) \leqslant \lambda d(x, y) + \epsilon$$

for all $x, y \in X$.

When there is a quasi-isometry from X to X', the spaces are said to be *quasi-isometric*. Quasi-isometric metric spaces have many geometric properties in common. We also mention that being quasi-isometric is an equivalence relation on the class of metric spaces.

The following are examples of quasi-isometries.

(1) The natural inclusion $\mathbb{Z} \to \mathbb{R}$ is a quasi-isometry. Take $\lambda = 1, \epsilon = 0$ and $C = \frac{1}{2}$. Similarly, the natural inclusion $\mathbb{Z}^n \to \mathbb{R}^n$ is a quasi-isometry.
(2) Let G be a group with finite choice of generators $\varphi : (A \cup A^{-1})^* \to G$ as above. Then the inclusion map $G \to C_\varphi(G)$ is a quasi-isometry from (G, d_φ) to $(C_\varphi(G), d_\varphi)$ with $\lambda = 1, \epsilon = 0$ and $C = \frac{1}{2}$.
(3) Let $\varphi : (A \cup A^{-1})^* \to G$ and $\psi : (B \cup B^{-1})^* \to G$ be finite choices of generators for a group G. For each member of $B\psi$ choose a word over $A \cup A^{-1}$ representing it, and similarly, for each element of $A\varphi$ choose a word over $B \cup B^{-1}$ which represents it. Let N be the maximum of the lengths of the chosen words. Then the identity map $G \to G$ is a quasi-isometry from (G, d_φ) to (G, d_ψ) and from (G, d_ψ) to (G, d_φ) with $\lambda = N$, $\epsilon = 0$ and $C = 0$.

We now introduce the ideas of geodesic and hyperbolic metric spaces. Let x, y be points in a metric space (X, d). A *geodesic segment* from x to y of length ℓ is the image of an isometric embedding i from the closed interval $[0, \ell] \subseteq \mathbb{R}$ with $i(0) = x$ and $i(\ell) = y$. Thus $d(i(s), i(t)) = |t - s|$ for all $s, t \in [0, \ell]$.

A metric space is a *geodesic (metric) space* if, for any two points of the space, there is a geodesic segment from one to the other. A (*geodesic*) *triangle* in a metric space consists of three points (the vertices) and, for each pair of these points, a choice of geodesic segment joining them. For a positive constant δ, a triangle in a metric space is δ-*slim* (following the terminology of [4]) if each edge of the triangle is contained in the closed δ-neighbourhood of the union of the other two edges.

A geodesic metric space X is *hyperbolic* if there is a global constant δ such that all triangles in X are δ-slim. The name comes from the fact that these spaces are generalisations of classical hyperbolic space $\mathbb{H}^n$ which has the slim triangles property with $\delta = 2$. Among other examples of hyperbolic metric spaces, we have the following.

(1) A bounded geodesic metric space is hyperbolic because if the distance between any two points in the space is at most M, then clearly

every triangle is M-slim.

(2) Any tree can be regarded as a metric space, and it is geodesic because any two points are connected by a shortest path. All triangles are 0-slim since any edge of a triangle is contained in the union of the other two, so that a tree is a hyperbolic metric space.

An important fact is that hyperbolicity is an invariant of quasi-isometry among geodesic spaces.

Theorem 1.2. *Let X and X' be quasi-isometric geodesic spaces. Then X is hyperbolic if and only if X' is hyperbolic.*

A group G with finite choice of generators $\varphi : (A \cup A^{-1})^* \to G$ is said to be *hyperbolic* if the metric space $(C_\varphi(G), d_\varphi)$ is hyperbolic. As $(C_\varphi(G), d_\varphi)$ is clearly a geodesic space for any finite choice of generators $\varphi : (A \cup A^{-1})^* \to G$, to say that G is hyperbolic means that, for some δ, all geodesic triangles in $(C_\varphi(G), d_\varphi)$ are δ-slim. In view of the theorem and our remarks above, this is independent of the choice of finite generating set for G. Some examples of hyperbolic groups are the following.

(1) Every finite group is hyperbolic because, relative to any finite set of generators, its Cayley graph is bounded.
(2) Every finitely generated free group is hyperbolic because, relative to a free set of generators, its Cayley graph is a tree.
(3) Let G and H be cyclic groups of order 3 with generators g and h respectively. Then the free product $G * H$ is hyperbolic. This can be seen by considering triangles in the Cayley graph relative to the generating set $\{g, h\}$.

The simplest example of a finitely generated group which is not hyperbolic is $\mathbb{Z} \times \mathbb{Z}$. We can see this by considering the Cayley graph relative to the generating set $\{(0,1),(1,0)\}$. For any positive integer N, consider the triangle with vertices $(0,0), (N,0)$ and (N,N); there is a unique geodesic segment joining $(0,0)$ to $(N,0)$, and a unique geodesic segment joining $(N,0)$ to (N,N), but there are several joining $(0,0)$ to (N,N). We choose the one, illustrated in Figure 1, passing through $(0,N)$. Then this point is outside the $(N-1)$-neighbourhood of the union of the edges from $(0,0)$ to $(N,0)$ and from $(N,0)$ to (N,N). Hence not all triangles are $(N-1)$-slim, and as this is true for every N, the Cayley graph (and so the group) is not hyperbolic.

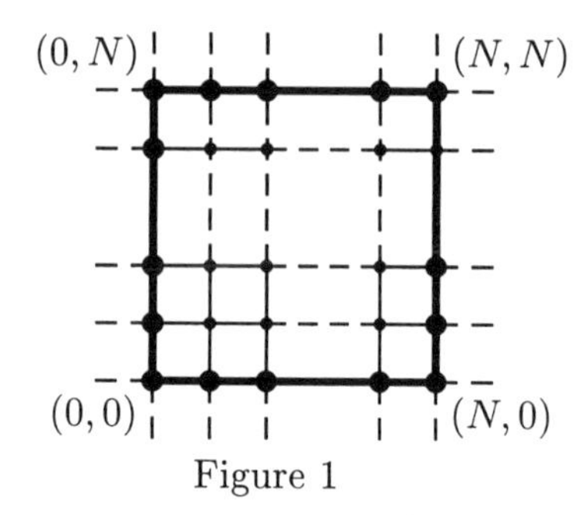

Figure 1

Next, we give a few important properties of hyperbolic groups. Proofs can be found in the books referred to above.

Theorem 1.3. *If a group G is hyperbolic, then:*

(1) *G is finitely presented;*
(2) *G has solvable word problem;*
(3) *G has solvable conjugacy problem;*
(4) *every subgroup of finite index in G is hyperbolic;*
(5) *every finite extension of G is hyperbolic;*
(6) *G does not contain a copy of $\mathbb{Z} \times \mathbb{Z}$ as a subgroup;*
(7) *G has only finitely many conjugacy classes of finite subgroups;*
(8) *G is biautomatic.*

We now describe a property of hyperbolic spaces which will be important for us in Section 4. Let (X, d) be a hyperbolic space and x, y, z be any three points in X, and consider a geodesic triangle with these points as vertices. The triangle inequality guarantees that there are unique non-negative numbers a, b, c such that $d(x, y) = a + b$, $d(x, z) = a + c$ and $d(y, z) = b + c$. Let p, q, r be the points on the edges from x to y, from x to z and from y to z respectively such that $d(x, p) = a$, $d(x, q) = a$ and $d(y, r) = b$ as shown in Figure 2.

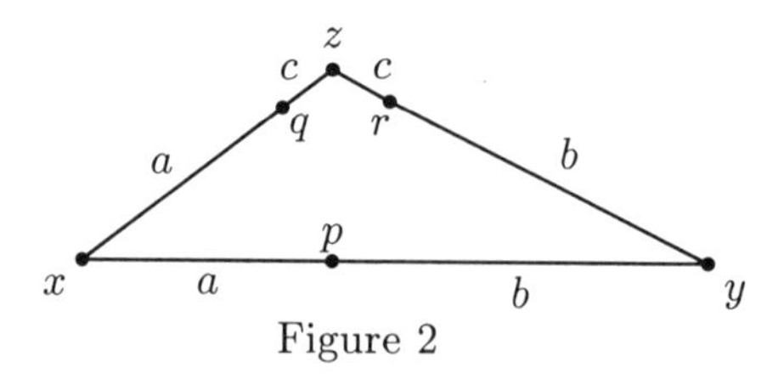

Figure 2

Let I_{xp} denote the geodesic segment from x to p which forms part of the edge of the triangle from x to y and so on. There are isometries $\alpha : I_{xp} \to I_{xq}$, $\beta : I_{yp} \to I_{yr}$ and $\gamma : I_{zq} \to I_{zr}$. It is known that the hyperbolicity of the space ensures that there is a global constant λ such that for any such triangle we have $d(u, \alpha(u)) < \lambda$, $d(v, \beta(v)) < \lambda$ and $d(w, \gamma(w)) < \lambda$ for all u, v, w in the domains of α, β, γ respectively. Following [4] we say that the triangles are λ-*thin*. In fact, the property that all geodesic triangles are λ-thin for some $\lambda \geqslant 0$ characterises hyperbolic spaces (see Proposition 1.17 of Chapter III.H in [4]). Since X is a geodesic space, there is a geodesic (of length less than λ) joining each pair of points which correspond under one of the three isometries α, β and γ.

When the space in question is the Cayley graph of a hyperbolic group and x, y, z are group elements, it is clear that either all the points p, q, r are group elements (vertices of the Cayley graph) or none of them are. Moreover, a group element on a side of the triangle corresponds to another group element under the appropriate isometry. Thus there are two possible configurations which we illustrate schematically in Figure 3 below where the dots represent group elements.

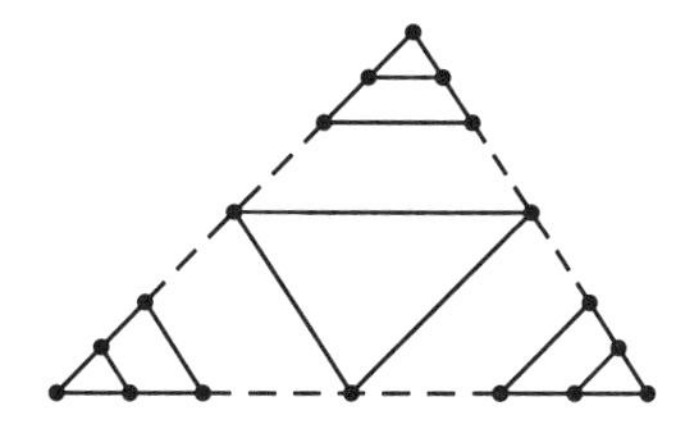
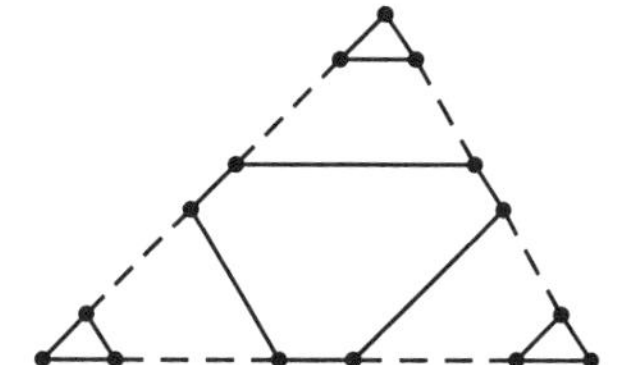

Figure 3

2. Multiplication Tables of Finitely Generated Groups

We now turn to Gilman's language theoretic characterisation of hyperbolic groups. We need some preliminaries about formal languages, and we begin by reminding the reader about context-free grammars, languages and rational transductions. For more on formal languages see [3],[15],[17], and for an introduction to the use of formal languages in group theory see [12].

Let A be a finite non-empty set. A *context-free grammar* over A is a 4-tuple $\Gamma = (V, A, P, S)$ where V is a finite set disjoint from A, P is a finite subset of $V \times (V \cup A)^*$ and S is a member of V. The elements of V are called *variables* or *non-terminals*, those of A are called *terminals*, those of P are called *productions* and S is called the *start symbol*.

If $(T, \alpha) \in P$, we write $T \to \alpha$. If $\beta, \gamma \in (V \cup A)^*$, then we write $\beta \Rightarrow \gamma$

if there exist δ, η in $(V \cup A)^*$ and a production $T \to \alpha$ such that $\beta = \delta T \eta$ and $\gamma = \delta \alpha \eta$. The reflexive transitive closure of $\Rightarrow$ is denoted by $\stackrel{*}{\Rightarrow}$. The *language generated by* Γ is defined to be

$$L(\Gamma) = \{w \in A^* : S \stackrel{*}{\Rightarrow} w\}.$$

A language $L \subseteq A^*$ is *context-free* if $L = L(\Gamma)$ for some context-free grammar Γ. We shall be interested in context-free languages in the free semigroup A^+, so it is worth mentioning the following standard result (where 1 is the empty word).

Lemma 2.1. *If $L \subseteq A^*$ is context-free, then so are $L \cup \{1\}$ and $L \setminus \{1\}$.*

Let X, Y be subsets of a monoid M. The product XY of X and Y is the set $\{xy : x \in X, y \in Y\}$, and the submonoid of M generated by X is denoted by X^*. The set $\mathrm{Rat}(M)$ of *rational subsets* of M is the smallest collection of subsets of M satisfying the following two conditions:

(1) every finite subset of M is rational;
(2) if $X, Y \subseteq M$ are rational, then so are $X \cup Y$, XY and X^*.

We obtain the definition of a *rational subset* of a semigroup S by replacing M by S and modifying (2) by replacing X^* by X^+, the subsemigroup generated by X. If S does not have an identity, S^1 denotes the monoid obtained from S by adjoining an identity. For $X \subseteq S$, we use X^* to denote the submonoid of S^1 generated by X. Clearly, $X^+ = XX^*$ and so any rational subset of S is also a rational subset of S^1.

A (nondeterministic) *finite automaton* over a monoid M is a finite directed graph (whose vertices are called *states*) with edges labelled by elements of M, a distinguished state called the *initial* state and a set of states called *accept* states.

A *path* in a finite automaton over M is a finite sequence of directed edges such that the terminal vertex of an edge in the sequence is the initial vertex of the next edge in the sequence. The *label* of a path is the product (in order) of the labels of its edges. A path is *successful* if it starts at the initial state of the automaton and ends at an accept state. The subset of M *accepted* by the automaton is the set of labels of successful paths. The significance of this idea is made clear by the following result. For a proof, see, for example, Theorem 2.6 of [12].

Proposition 2.1. *A subset of a monoid M is rational if and only if it is accepted by a finite automaton over M.*

A *rational transduction* $\rho : A^* \to B^*$ where A and B are finite is a rational subset of the direct product $A^* \times B^*$. Similarly, one also defines a *rational transduction* $\rho : A^+ \to B^+$ between finitely generated free semigroups. The rational transductions from A^+ to B^+ are precisely those rational transductions from A^* to B^* that are contained in $A^+ \times B^+$.

We observe that a homomorphism $\varphi : A^* \to B^*$ is an example of a rational transduction since the graph of φ is $\{(a, a\varphi) : a \in A\}^*$. Another example is provided by $\rho_L : A^* \to A^*$ where $L \subseteq A^*$ is rational and $\rho_L = \{(w, w) : w \in L\}$.

For any binary relation $\rho : X \to Y$ and subset L of X, we put

$$L\rho = \{y \in Y : (x, y) \in \rho \text{ for some } x \in L\}.$$

The basic properties of rational transductions are collected in the following result. Proofs can be found in Section III.4 of [3] or Section 5 of [12].

Proposition 2.2. *Let A, B, C be finite sets and suppose that $\rho : A^* \to B^*$, $\sigma : A^* \to B^*$, $\tau : B^* \to C^*$ are rational transductions. Then:*

(1) *$\rho \cup \sigma$, $\rho\sigma$ (the subset product of ρ and σ) and ρ^* are rational transductions;*
(2) *$\rho^{-1} : B^* \to A^*$ and $\rho \circ \tau : A^* \to C^*$ are rational transductions;*
(3) *if $L \subseteq A^*$ is rational, then so is $L\rho$;*
(4) *if $L \subseteq A^*$ is context-free, then so is $L\rho$.*

Of course, there is a corresponding result for rational transductions between free semigroups. As an immediate corollary of the proposition (and using the example ρ_L above), we have the following standard results about rational and context-free languages.

Corollary 2.1. *Let A, B be finite non-empty sets, L be a rational language over A and let $\varphi : A^* \to B^*$ be a homomorphism.*

(1) *If $K \subseteq A^*$ and $J \subseteq B^*$ are rational, then so are $K \cap L$, $K\varphi$ and $J\varphi^{-1}$.*
(2) *If $K \subseteq A^*$ and $J \subseteq B^*$ are context-free, then so are $K \cap L$, $K\varphi$ and $J\varphi^{-1}$.*

We conclude this section by explaining the notion of multiplication table for a finitely generated group and stating Gilman's theorem [13].

Let $\varphi : (A \cup A^{-1})^* \to G$ be a finite choice of generators for a group G, and let # be a symbol not in A. For $L \subseteq (A \cup A^{-1})^*$ with $L\varphi = G$, the language

$$M = \{u\#v\#w : u, v, w \in L \text{ and } (uvw)\varphi = 1\}$$

over $A \cup \{\#\}$ is called the *multiplication table of G relative to* (A, L, φ).

Theorem 2.1. *Let* $\varphi : (A \cup A^{-1})^* \to G$ *be a finite choice of generators for a group* G. *Then* G *is hyperbolic if and only if there is a rational language* L *over* A *with* $L\varphi = G$ *such that the multiplication table for* G *relative to* (A, L, φ) *is context-free.*

3. Word Hyperbolic Semigroups

Theorem 2.1 suggests a way of defining a notion of hyperbolicity for semigroups. This was made explicit by Duncan and Gilman in [9], who proposed the following definition which was subsequently further explored in [16]. A *finite choice of generators* for a semigroup S is a surjective homomorphism $\varphi : A^+ \to S$ where A is a finite alphabet. A *choice of representatives* for S is a triple (A, L, φ) where $\varphi : A^+ \to S$ is a finite choice of generators and $L \subseteq A^+$ is such that $L\varphi = S$. It is a *rational* choice of representatives if the language L is rational. There are, of course, corresponding notions for monoids. For a word $w = a_1 \ldots a_n$ where $a_i \in A$, the *reverse* of w is $w^r = a_n \ldots a_1$, and $1^r = 1$. Let $\#$ be a symbol not in A. Following [9], we say that S is *hyperbolic* if, for some finite choice of generators $\varphi : A^+ \to S$, there is a rational language L over A with $L\varphi = S$ such that the language $T = \{u\#v\#w^r : u, v, w \in L \text{ and } (uv)\varphi = w\varphi\}$ is context-free. The language T is called the *multiplication table* of S relative to (A, L, φ), and we say that S is hyperbolic *with respect to* the choice of generators $\varphi : A^+ \to S$.

If M is a monoid, then using a finite choice of monoid generators, we can define M to be 'hyperbolic as a monoid', and, of course, a group may be 'hyperbolic as a semigroup', 'hyperbolic as a monoid' or 'hyperbolic as a group'. However, Duncan and Gilman show that the various notions coincide (see Theorem 3.5 and Corollary 4.3 of 9), that is, we have the following result.

Theorem 3.1.

(1) *A monoid* M *is hyperbolic as a semigroup if and only if it is hyperbolic as a monoid;*

(2) *A group* G *is hyperbolic as a semigroup if and only if it is hyperbolic as a group.*

Crucial for this result and of interest in its own right is the following.

Theorem 3.2. *If the semigroup S is hyperbolic with respect one finite choice of generators, then it is hyperbolic with respect to every such choice.*

It is noted in [9] that finite semigroups are hyperbolic, and it is also shown that the bicyclic monoid $M = \langle a, b \mid ab = 1 \rangle$ is hyperbolic. The latter shows that, when applied to inverse semigroups, the definition of hyperbolic we are using gives a different class from that arising from a geometric definition of hyperbolic inverse semigroup suggested in [25]. To describe this approach, we need the notion of the Schützenberger graph of an $\mathscr{R}$-class R of a semigroup S. These graphs were first defined by Stephen [26] in the context of inverse semigroups, but can be defined quite generally. Let $\varphi : A^+ \to S$ be a finite choice of generators for a semigroup S, and let R be an $\mathscr{R}$-class of S. The *Schützenberger graph* $\Gamma_\varphi(R)$ of R (relative to φ) is a (labelled) directed graph with vertex set R and an edge (labelled a) from s to $s(a\varphi)$ for each $s \in R$ and each $a \in A$ such that $s(a\varphi)\mathscr{R}s$. Note that $\Gamma_\varphi(R)$ is strongly connected; it is made into a metric space (denoted by $(\Gamma_\varphi(R), d_\varphi)$) in just the same way that the Cayley graph of a group is. If $\theta : B^+ \to S$ is another finite choice of generators, then (just as in the case of Cayley graphs of groups) $\Gamma_\varphi(R)$ and $\Gamma_\theta(R)$ are quasi-isometric. Again, as in the group case, we may consider the discrete metric space (R, d_φ) obtained by restricting the metric to R.

In the case where S is an inverse semigroup, one considers a choice of generators of the form $\varphi : (A \cup A^{-1})^+ \to S$ where, as in the group case, A is an alphabet, $A^{-1} = \{a^{-1} \mid a \in A\}$ is an alphabet disjoint from A and in one-one correspondence with A, and such that for every $a \in A$ we have $a^{-1}\varphi = (a\varphi)^{-1}$. In this case, we observe that whenever $\Gamma_\varphi(R)$ contains an edge from s to t labelled a, it contains also an edge from t to s labelled a^{-1}.

Note that if G is a group, then there is only one $\mathscr{R}$-class, namely G itself, and the Schützenberger graph is the right Cayley graph.

In [25], an inverse semigroup S is defined to be hyperbolic in a way which is equivalent to: S is finitely generated and, for each $\mathscr{R}$-class R, the Schützenberger graph of R is hyperbolic and R contains only finitely many $\mathscr{H}$-classes. Clearly, the bicyclic monoid is not hyperbolic in this sense.

4. Completely Simple Semigroups

Recall that a semigroup S is *completely simple* if it has no two-sided ideals other than itself, and it possesses minimal one-sided ideals. It is well known (see, for example, Chapter 3 of [18]) that S is completely simple if and only if it is isomorphic to a Rees matrix semigroup $M = M(G; I, J; P)$ over a group G where I, J are non-empty sets and $P = (p_{ji})$ is a $J \times I$ matrix over G and M is the set $I \times G \times J$ with multiplication given by

$$(i, g, j)(\ell, h, k) = (i, gp_{\ell j}h, k).$$

We now analyse when such a semigroup is hyperbolic. To be hyperbolic, S must be finitely generated. It is straightforward to show (and is also immediate from a more general result of Ayik and Ruškuc [2]) that this is equivalent to G being finitely generated, and I and J being finite. In other words, the maximal subgroups of S (which are all isomorphic) are finitely generated, and S has finitely many $\mathscr{R}$- and $\mathscr{L}$-classes. It turns out that adapting the Steinberg [25] definition to completely simple semigroups does give precisely those semigroups that are hyperbolic in the sense of [9].

Following [9], for any semigroup S, we define the Cayley graph $C_\sigma(S)$ relative to a choice of generators $\sigma : A^+ \to S$ to have vertices $S \cup \{*\}$ where $* = 1$ if S is a monoid, and $*$ is a symbol not in S otherwise. For each $a \in A$, there is a directed edge with label a from $*$ to $a\sigma$, and, for each $s \in S$, a directed edge with label a from s to $s(a\sigma)$. The distance between two vertices is the length of the shortest undirected path joining the vertices. Then $C_\sigma(S)$ is made into a metric space $(C_\sigma(S), d_\sigma)$ in just the same way that the Cayley graph of a group is. Again, as in the group case, a change of generators leads to a quasi-isometric space.

Our main result is the following.

Theorem 4.1. *Let S be a finitely generated completely simple semigroup isomorphic to the Rees matrix semigroup $M(G; I, J; P)$. Then the following are equivalent:*

(1) *S is hyperbolic;*
(2) *G is hyperbolic;*
(3) *for any choice of generators for S, the Schützenberger graph of each $\mathscr{R}$-class of S is hyperbolic;*
(4) *for any choice of generators for S, the Cayley graph of S is hyperbolic.*

First we establish some notation which will be used throughout the section. Let S be a finitely generated completely simple semigroup with $\mathscr{R}$-classes $R_1, \ldots, R_m$ and $\mathscr{L}$-classes $L_1, \ldots, L_n$, and put $R_i \cap L_j = H_{ij}$. Each H_{ij} is a maximal subgroup of S and we denote its identity by e_{ij}. Let

$R = R_1$, $H = H_{11}$ and $e = e_{11}$. Note that each R_i is a subsemigroup with a single $\mathscr{R}$-class.

We now consider the equivalence of (2), (3) and (4).

Let $X \subseteq H$ be a *symmetric* (that is, closed under inverses) finite set of generators for H; we assume that $e_{11} \in X$. Then, as is well known and easy to see, $Y = X \cup \{e_{12}, \ldots, e_{1n}\}$ is a generating set for R, and $Z = Y \cup \{e_{21}, \ldots, e_{m1}\}$ is a generating set for S. In each case the choice of generators is given by the inclusion map, so there is no ambiguity in writing $(C_X(H), d_X)$ for the Cayley graph of H relative to (X, ι) regarded as a metric space, and we can use similar notation for the other graphs involved.

Lemma 4.1. *The spaces* $(C_X(H), d_X)$, $(\Gamma_Y(R), d_Y)$, $(\Gamma_Z(R), d_Z)$ *and* $(C_Z(S), d_Z)$ *are quasi-isometric.*

Proof. It suffices to consider the discrete spaces (H, d_X), (R, d_Y), (R, d_Z) and $(S \cup \{*\}, d_Z)$. First, note that $e_{1j}x \in H$ for each $j \in \{2, \ldots, n\}$ and each $x \in X$ so that $e_{1j}x$ can be expressed as a word over X. Choose one such word for each $e_{1j}x$ and let M be the greatest of the lengths of the chosen words. Next, observe that if $r \in R$, then $r = he_{1j}$ for some $h \in H$ and some $j \in \{1, \ldots, n\}$. Hence $d_Y(H, r) \leqslant 1$ and so the inclusion map $H \to R$ is a quasi-isometry from (H, d_X) to (R, d_Y) with $\lambda = M$, $\epsilon = 0$ and $C = 1$.

Now consider (R, d_Z) and let $r, s \in R$ be such that there is an edge from r to s labelled by e_{i1}. Suppose that $r \in H_{1j}$ and put $h_{ji} = e_{1j}e_{i1}$. Note that h_{ji} is in H (as is s) and that $uh_{ji} = ue_{i1}$ for all $u \in H_{1j}$. Let w_{ji} be a word over X which represents h_{ji}. Then, since $rh_{ji} = re_{i1} = s$, there is an edge path from r to s in (R, d_Y) labelled by w_{ji}. Let N be the length of the longest of the words w_{ji}. Then the identity map $R \to R$ is a quasi-isometry from (R, d_Y) to (R, d_Z) with $\lambda = N$, $\epsilon = 0$ and $C = 0$.

Finally, note that there is no ambiguity in the notation because the d_Z-metric on R is the restriction to R of the d_Z-metric on S. Thus all we have to do is note that every element of $S \cup \{*\}$ has distance at most 1 from an element of R so that (R, d_Z) and $(S \cup \{*\}, d_Z)$ are indeed quasi-isometric.

If S is isomorphic to the Rees matrix semigroup $M(G; I, J; P)$, then G is isomorphic to each maximal subgroup of S. Hence, in view of the fact that Cayley graphs of a group (or semigroup) relative to different finite generating sets are quasi-isometric, as are Schützenberger graphs of an $\mathscr{R}$-class of a semigroup relative to different finite generating sets for the semigroup, it follows from Lemma 4.1 and Theorem 1.2 that (2), (3) and

(4) of Theorem 4.1 are equivalent.

Similarly, the fact that (1) implies (2) in Theorem 1.2 is a consequence of the next proposition.

Proposition 4.1. *If S is hyperbolic, then so is H.*

Proof. Since S is finitely generated we can retain the notation for the $\mathscr{R}$-classes, *etc.*, and since S is hyperbolic, there is a choice of representatives (A, L, σ) for S where L is rational and the multiplication table T of S relative to (A, L, σ) is context-free.

First, we define a choice of generators for H. Put

$$B = \{b_{ja} : a \in A,\ j \in J\}$$

and define a homomorphism $\rho : B^+ \to H$ by $b_{ja}\rho = e_{1j}(a\sigma)e$. Clearly B is finite. Also ρ maps onto H, for if $h \in H$, then $h = (a_1\sigma)\dots(a_n\sigma)$ for some $a_1, \dots, a_n \in A$. Suppose that $a_k\sigma \in H_{i(k),j(k)}$. Clearly, $i(1) = 1$ and $j(n) = 1$ so that, writing $a(i)$ for a_i when it occurs as a subscript, we have

$$\begin{aligned} h &= (a_1\sigma)\dots(a_n\sigma) \\ &= e(a_1\sigma)e.e_{1j(1)}(a_2\sigma)e.e_{1j(2)}(a_3\sigma)e.\dots e_{1j(n-1)}(a_n\sigma)e \\ &= (b_{1a(1)}b_{j(1)a(2)}\dots b_{j(n-1)a(n)})\rho. \end{aligned}$$

Next, we find a rational choice of representatives for H over the alphabet B. Let $K = H\sigma^{-1}$ and observe that $K = K'A^*K'' \cup (K' \cap K'')$ where $K' = \{a \in A : a\sigma \in R\}$ and $K'' = \{a \in A : a\sigma \in L_1\}$. Now K' and K'' are finite so that K is rational and hence so is $L \cap K$.

Now define a function $\varphi : K \to B^+$ by

$$(a_1 \dots a_n)\varphi = b_{1a(1)}b_{j(1)a(2)}\dots b_{j(n-1)a(n)}$$

where $a_i \in A$ and $a_k\sigma \in H_{i(k)j(k)}$ as above. We have just seen that $w\varphi\rho = w\sigma$ for all $w \in K$. Since $L\sigma = S$, we have $(L \cap K)\varphi\rho = (L \cap K)\sigma = H$ and thus $(B, (L\cap K)\varphi, \rho)$ is a choice of representatives for H. To show that this choice is rational, we show that φ is a rational transduction. We define a finite automaton M over the semigroup $A^+ \times B^+$. The state set of M is $\{q_0\} \cup S/\mathscr{L} \cup A$ where we assume the union is disjoint. The initial state is q_0 and the unique accept state is L_1. The edge set is

$$\begin{aligned} &\{(q_0, (a, b_{1a}), L_{a\sigma}) : a\sigma \in R\} \\ &\cup \{(L_j, (1, b_{ja}), a) : a \in A,\ j \in \{1, \dots, n\}\} \\ &\cup \{(a, (a, 1), L_{a\sigma}) : a \in A\}. \end{aligned}$$

It is straightforward to verify that M accepts the graph of φ, so that φ is rational. By (3) of Proposition 2.2, $(L \cap K)\varphi$ is rational and so we do have a rational choice of representatives.

The multiplication table of H relative to $(B, (L \cap K)\varphi, \rho)$ is

$$T' = \{u\#v\#w^r : u, v, w \in (L \cap K)\varphi \text{ and } (u\rho)(v\rho) = w\rho\}.$$

Thus, since $\varphi\rho = \sigma$, the word $u\#v\#w^r$ is in T' if and only if $u = u'\varphi, v = v'\varphi, w = w'\varphi$ for some $u', v', w' \in L \cap K$ such that $(u'\sigma)(v'\sigma) = w'\sigma$, that is,

$$u'\#v'\#w'^r \in T \cap (K\{\#\}K\{\#\}K^r).$$

Recall that the reversal φ^r of φ is defined to be

$$\{(x^r, y^r) \in K^r \times K^r : (x, y) \in \varphi\},$$

and that since φ is rational, so is φ^r (see, for example, page 66 of [3]). Next we define a function $\theta : K\{\#\}K\{\#\}K^r \to B^+\{\#\}B^+\{\#\}B^+$ by

$$(x\#y\#z^r)\theta = (x\varphi)\#(y\varphi)\#(z^r\varphi^r).$$

so that $T' = (T \cap (K\{\#\}K\{\#\}K^r))\theta$. Clearly, the graph of θ is $\varphi\{(\#,\#)\}\varphi\{(\#,\#)\}\varphi^r$ and so we have that θ is a rational subset of $(A \cup \{\#\})^+ \times (B \cup \{\#\})^+$. Thus θ is a rational transduction. Now T is context-free and $K\{\#\}K\{\#\}K^r$ is clearly rational so that, in view of Corollary 2.1(2), $T \cap (K\{\#\}K\{\#\}K^r)$ is context-free, and hence, by Proposition 2.2(4), so is T'. Thus, regarded as a semigroup, H is hyperbolic, and so, by Theorem 3.1, it is a hyperbolic group.

We now embark on proving that condition (2) of Theorem 4.1 implies condition (1). We start by recalling the following result (see, for example, Corollary 2.20 of Chapter III.Γ in [4]).

Proposition 4.2. *Let G be a hyperbolic group and $\sigma : (A \cup A^{-1})^* \to G$ be a finite choice of generators. Then the set P of words in $(A \cup A^{-1})^*$ which label geodesics in the Cayley graph $C_\sigma(G)$ is a rational language.*

We call the set P the *language of geodesics* of G relative to σ. The following lemma and its proof are inspired by Section 4 of [13].

Lemma 4.2. *Let $\sigma : A^* \to G$ be a symmetric finite choice generators for a hyperbolic group G, and let P be the language of geodesics. Then the language*

$$Q = \{u\#v\#w\#x : u, v, w, x \in P \text{ and } (uvwx)\sigma = 1\}$$

over $A \cup \{\#\}$ is context-free.

Proof. First, we note that $C = P\{\#\}P\{\#\}P\{\#\}P$ is a rational subset of $(A\cup\{\#\})^*$. We produce a context-free grammar $\mathscr{G}$ such that $Q = L(\mathscr{G}) \cap C$, and hence Q is context-free as desired.

Since G is hyperbolic, there is a non-negative constant λ such that all geodesic triangles in the Cayley graph $C_\sigma(G)$ are λ-thin. Define a context-free grammar $\mathscr{G}$ as follows. The terminal alphabet is $A \cup \{\#\}$ and the set of variables is

$$V = \{X_z : z \in (A \cup \{\#\})^* \text{ and } |z| \leqslant 2\lambda\}.$$

Let $W = A \cup \{\#\} \cup V$ and extend $\sigma : A^* \to G$ to a homomorphism $\sigma : W^* \to G$ by defining $X_z\sigma = z\sigma$ and $\#\sigma = 1$.

In this proof we denote the empty word by ϵ; the start symbol of the grammar is X_ϵ and the set of productions consists of all pairs (X_z, α) where $\alpha \in W^*$, $|\alpha| \leqslant 7$ and $X_z\sigma = \alpha\sigma$.

Clearly, if $X_z \overset{*}{\Rightarrow} \beta$, then $X_z\sigma = \beta\sigma$ and so the grammar generates a context-free language consisting of certain words which are mapped to 1 by σ. To complete the proof, it is clearly enough to show that $Q \subseteq L(\mathscr{G})$.

Let $u\#v\#w\#x \in Q$ so that u, v, w, x label geodesics in $C_\sigma(G)$ and $(uvwx)\sigma = 1$. Note that $X_\epsilon \to \epsilon\#\epsilon\#\epsilon\#\epsilon$ is a production, so that certainly $\epsilon\#\epsilon\#\epsilon\#\epsilon \in L(\mathscr{G})$. If some but not all of u, v, w, x are empty, then the words are labels of the sides of a geodesic triangle or a degenerate geodesic triangle in $C_\sigma(G)$, and the argument in Section 4 of [13] shows that $u\#v\#w\#x$ is in $L(\mathscr{G})$. Thus we may assume that $u = a_1 \dots a_p, v = b_1 \dots b_q, w = c_1 \dots c_r$ and $x = d_1 \dots d_s$ with a_i, b_j, c_k, d_h being letters in A. Then in $C_\sigma(G)$ there is a geodesic quadrilateral with vertices $1, u\sigma, (uv)\sigma$ and $(x\sigma)^{-1}$ and geodesic sides labelled u, v, w and x. There is at least one geodesic segment joining $(x\sigma)^{-1}$ and $u\sigma$. Consider one such; then this segment is the common side of two geodesic triangles, as illustrated in Figure 4.

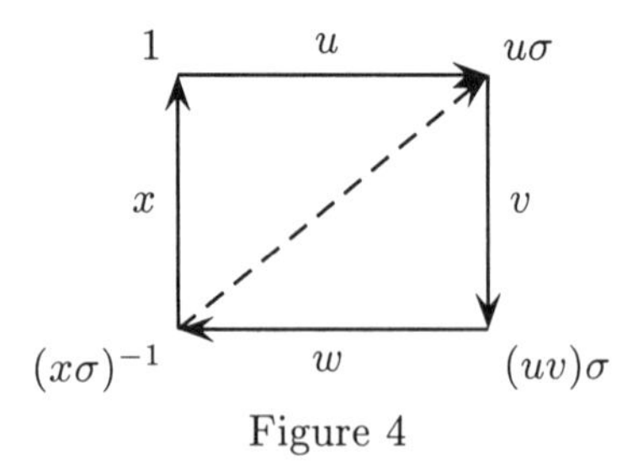

Figure 4

Since each triangle in the diagram is λ-thin, we can pair the points on their perimeters as described at the end of Section 1 in such a way that each pair of matched points is joined by a path of length at most λ. Joining the paths which meet at the common side of the two triangles gives a matching between points on the sides of the quadrilateral, and matched points are joined by a path of length at most 2λ. Moreover, a point which is a group element is matched with one or more group elements.

Let L_u, L_v, L_w, L_x be the sides of the quadrilateral with labels u, v, w, x respectively in Figure 4, and let L be the side joining $(x\sigma)^{-1}$ to $u\sigma$. Let Δ be the triangle with sides L_u, L_x and L, and Δ' be that with sides L_v, L_w and L. Consider a group element g on L_u. In Δ, the point g is matched with at most one point on L_x and at most one point on L. If it is matched with a point on L, then the latter point is matched with at most one point on L_v and with at most one point on L_w. Thus g cannot be matched with more than one point on a specific side of the quadrilateral.

The matching described can occur in one of several ways. A group element on a particular side may be matched with:

(*i*) a group element on an adjacent side;
(*ii*) a group element on an opposite side;
(*iii*) a group element on an adjacent side and a group element on an opposite side;
(*iv*) group elements on each of the adjacent sides;
(*v*) group elements on each of the other three sides.

Which of these occur depends on the nature of the configuration obtained when we put together paths joining matching points in Δ and Δ' which meet on the side L. The configuration is determined by where the central triangle or hexagon in Δ and that in Δ' meet the common side L. This can occur in one of several (actually 16) ways. The following diagrams illustrate three of the possibilities; the assiduous reader can easily draw the remaining ones.

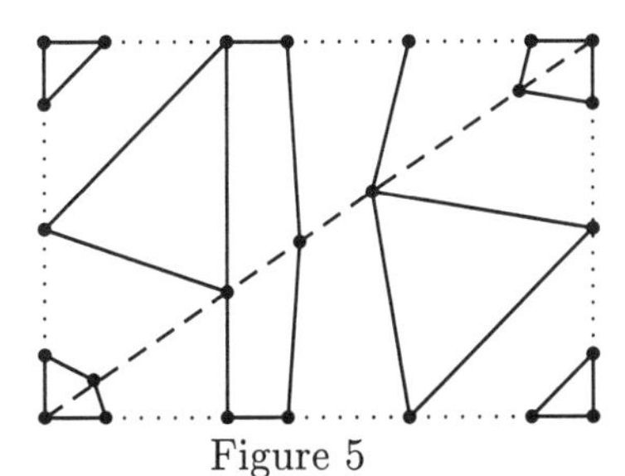

Figure 5

In Figure 5 both geodesic triangles have central triangles; the central triangle of Δ meets the common side to the left of the central triangle in Δ'. Two of the other possible configurations are illustrated below in Figure 6. In these diagrams some of the edges and some of the group elements are labelled for later reference.

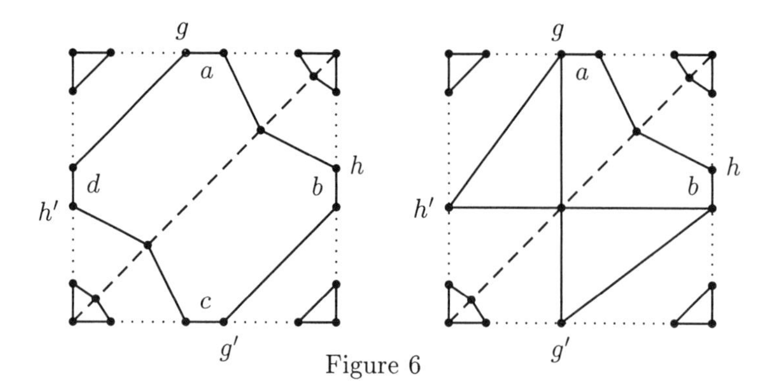

Figure 6

In the left hand quadrilateral in Figure 6, the two triangles Δ and Δ' have central hexagons and the hexagons have a side in common. Deleting the common edge L of Δ and Δ' (and so, in particular, removing the common side of the central hexagons), we see that we are left with a quadrilateral with a central octagon.

In the right hand quadrilateral of Figure 6, Δ and Δ' both have central triangles which meet at a common point. This is the only configuration in which possibility (v) above holds, that is, the only one in which there is a group element matched with group elements on each of the other three sides.

We obtain a derivation of $u\#x\#w\#x$ by using the quadrilateral and the labels on the paths joining matching pairs of group elements on the sides of the quadrilateral. The precise derivation we get depends on which of the sixteen possible configurations is involved. We illustrate the process for obtaining the derivation when the configuration in question is that of Figure 5, and subsequently comment on what happens when we use the two configurations in Figure 6. We adopt the notational convention that if w is one of u_i, y_i, z_i or t_i, then X_w will be denoted by U_i, Y_i, Z_i or T_i respectively.

When we dispense with the common side of the two triangles in Figure 5 and label the paths we obtain the following diagram.

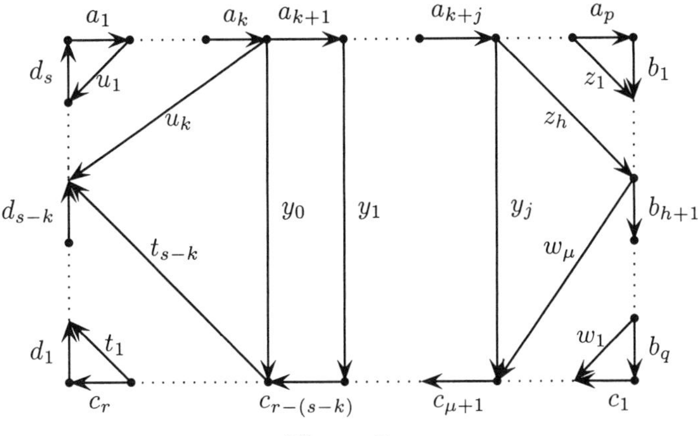

Figure 7

In this diagram, $(a_1 \dots a_k)\sigma$ is the group element on L_u which is a vertex of the central triangle in Δ. We also have $h = p - (k + j)$ and $\mu = q - h$.

We use the diagram to get a derivation in the following way. We start with the production $X_\epsilon \to a_1 U_1 d_s$ and then apply productions determined by internal quadrilaterals to obtain the derivation:

$$X_\epsilon \Rightarrow a_1 U_1 d_s \Rightarrow a_1 a_2 U_2 d_{s-1} d_s \Rightarrow \cdots \Rightarrow a_1 \dots a_k U_k d_{s-(k-1)} \dots d_s.$$

Now, corresponding to the internal triangle labelled by y_0, t_{s-k} and u_k, apply the production $U_k \to Y_0 T_{s-k}$. From Y_0, by applying productions determined by internal quadrilaterals, we obtain

$$Y_0 \Rightarrow a_{k+1} Y_1 c_{r-(s-k)} \Rightarrow \cdots \Rightarrow a_{k+1} \dots a_{k+j} Y_j c_{\mu+1} \dots c_{r-(s-k)},$$

and from T_{s-k}, by applying productions determined by internal quadrilaterals and the triangle labelled by t_1, c_r and d_1 we get

$$\begin{aligned} T_{s-k} &\Rightarrow c_{r+1-(s-k)} T_{s-(k+1)} d_{s-k} \Rightarrow \cdots \Rightarrow c_{r+1-(s-k)} \dots c_{r-1} T_1 d_2 \dots d_{s-k} \\ &\Rightarrow c_{r+1-(s-k)} \dots c_r \# d_1 \dots d_{s-k}. \end{aligned}$$

Now we apply the production $Y_j \to Z_h W_\mu$ and in a similar way obtain derivations

$$W_\mu \Rightarrow b_{h+1} W_{\mu-1} c_\mu \Rightarrow \cdots \Rightarrow b_{h+1} \dots b_q \# c_1 \dots c_\mu$$

and

$$Z_h \Rightarrow a_{k+j+1} Z_{h-1} b_h \Rightarrow \cdots \Rightarrow a_{k+j+1} \dots a_p \# b_1 \dots b_h.$$

Putting all this together we see that

$$X_\epsilon \overset{*}{\Rightarrow} a_1 \ldots a_p \# b_1 \ldots b_q \# c_1 \ldots c_r \# d_1 \ldots d_s$$

as desired.

The other configurations give rise to corresponding derivations in a similar way. As we have noted, when we dispense with the common side of the the two triangles in the left hand diagram in Figure 6 there is a central octagon. If the labels of the paths joining g to $h'(d\sigma)$, $g(a\sigma)$ to h, $h(b\sigma)$ to g' and $g'c$ to h' are u, y, w and t respectively, we need a production of the form $U \to aZbWcTd$. This explains why we need to allow words of length up to 7 on the right hand side of productions.

In the right hand diagram in Figure 6, there are group elements matched with others on the opposite side of the quadrilateral for both pairs of opposite sides. We do not use the path joining h and h' in constructing the derivation from this configuration. If the labels of the paths joining g to h', g to g', g' to h', $g(a\sigma)$ to h and $h(b\sigma)$ to g' are u, y, t, z and w respectively, we use productions $U \to YT$ and $Y \to aZbW$ as well as ones arising from internal quadrilaterals and triangles as in the above derivations.

Thus we obtain $Q \subseteq L(\mathscr{G})$ as required.

We will actually use the following easy consequence of the lemma.

Corollary 4.1. *Let $\sigma : A^* \to G$ be a symmetric finite choice of generators for a hyperbolic group G, and let P be the language of geodesics. Let g be an element of G. Then the language*

$$T_g = \{u\#w\#x : u, w, x \in P \text{ and } (u\sigma)g(wx)\sigma = 1\}$$

is context-free.

Proof. First, note that by Corollary 2.1(1), $g\sigma^{-1} \cap P$ is rational so that

$$P_g = P\{\#\}(g\sigma^{-1} \cap P)\{\#\}P\{\#\}P$$

is also rational. Hence by Corollary 2.1(2), the language $Q_g = Q \cap P_g$ is context-free.

We define a finite automaton M over the monoid $(A \cup \{\#\})^* \times (A \cup \{\#\})^*$. The state set of M is $\{q_0, q_1, q_2\}$ with initial state q_0 and unique accept state q_2. The edge set is

$$\begin{aligned}&\{(q_0, (a, a), q_0) : a \in A\} \cup \{(q_0, (\#, \epsilon), q_1)\} \\ &\quad \cup \{(q_1, (a, \epsilon), q_1) : a \in A\} \cup \{(q_1, (\#, \#), q_2)\} \\ &\quad \cup \{((q_2, (a, a), q_2) : a \in A\} \cup \{(q_2, (\#, \#), q_2)\},\end{aligned}$$

where again we use ϵ to denote the empty word. We illustrate M in Figure 8 where we use the convention that if an arrow has several labels, then it is labelled by the set of those labels. We denote the diagonal $\{(a,a) : a \in A\}$ by D, and write B for $D \cup \{\#, \#\}$.

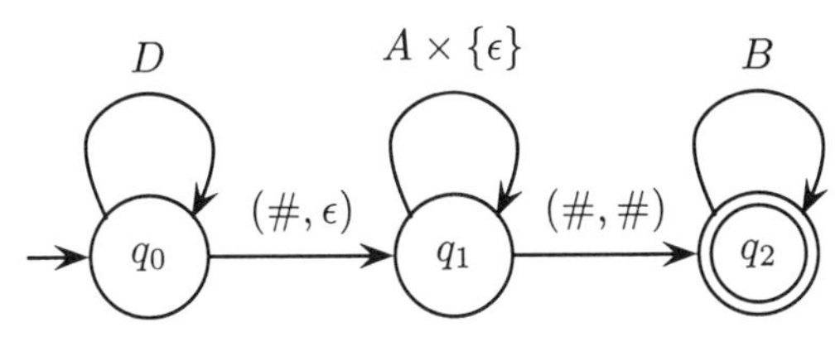

Figure 8

Let θ be the rational transduction accepted by M. Then, by (4) of Proposition 2.2, $Q_g\theta$ is context-free. But $Q_g\theta = T_g$.

We are now ready to complete the proof of Theorem 4.1 which is a consequence of the next result. We revert to the notation introduced at the beginning of the section for the $\mathscr{R}$-classes, *etc.* of a finitely generated completely simple semigroup S. In particular, we have $\mathscr{H}$-classes H_{ij} for $i \in \{1, \ldots, m\}$ and $j \in \{1, \ldots, n\}$, we write H for H_{11}, the idempotent in H_{ij} is e_{ij}, $X \subseteq H$ is a symmetric finite set of generators for H and $e = e_{11} \in H$.

For clarity, we let $A, B = \{b_i : i = 1, \ldots, m\}, C = \{c_j : j = 1, \ldots, n\}$ be disjoint sets and $\varphi : A \to X$ be a bijection. Then, extending φ to a homomorphism $\sigma : A^* \to H$ gives a symmetric choice of generators for H, and extending φ to $\rho : (A \cup B \cup C)^+ \to S$ by defining $b_i\rho = e_{i1}$ and $c_j\rho = e_{1j}$ gives a choice of (semigroup) generators for S.

Proposition 4.3. *Let S be a finitely generated completely simple semigroup with a maximal subgroup H. If H is hyperbolic, then S is hyperbolic.*

Proof. Using the above notation, let P be the language of geodesics for H relative to σ. Let $h_{ji} = e_{1j}e_{i1}$ and note that $h_{ji} \in H$. Then, by Corollary 4.1, the language

$$T_{ji} = \{u\#w\#x : u, w, x \in P \text{ and } (u\sigma)h_{ji}(wx)\sigma = 1\}$$

is context-free. Since X (and so also A) is symmetric, we have $x \in P$ if and only if $x^{-1} \in P$. Thus

$$T_{ji} = \{u\#w\#x^{-1} : u, w, x \in P \text{ and } (u\sigma)h_{ji}(w\sigma) = x\sigma\}.$$

Define a finite automaton M over the semigroup $(A \cup \{\#\})^+ \times (A \cup \{\#\})^+$ as follows: the state set is $\{q_0, q_1, q_2\}$ where q_0 is the initial state and q_2 is the unique accept state, and the edge set is

$$\begin{aligned}&\{(q_0,(a,a),q_0) : a \in A\} \cup \{(q_0,(\#,\#),q_1)\}\\&\quad \cup \{(q_1,(a,a),q_1) : a \in A\} \cup \{(q_1,(\#,\#),q_2)\}\\&\quad \cup \{((q_2,(a^{-1},a),q_2) : a \in A\}.\end{aligned}$$

The automaton is illustrated in Figure 9 where D denotes the diagonal $\{(a,a) : a \in A\}$ and C denotes the set $\{(a^{-1},a) : a \in A\}$.

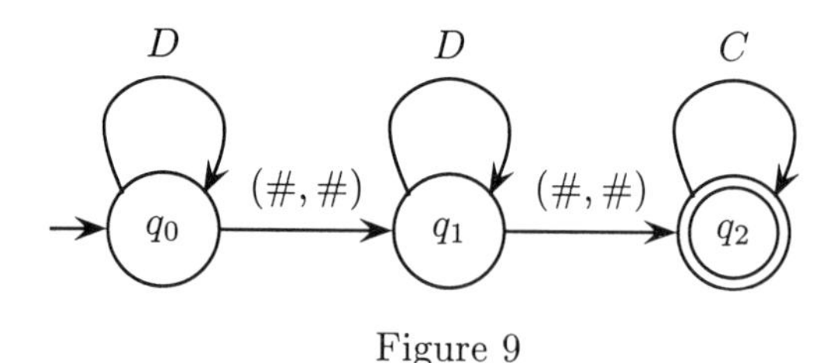

Figure 9

Let θ be the rational transduction accepted by M. Then, by (4) of Proposition 2.2, $T_{ji}\theta$ is context-free, that is,

$$T'_{ji} = \{u\#w\#x^r : u,w,x \in P \text{ and } (u\sigma)h_{ji}(w\sigma) = x\sigma\}$$

is context-free.

We have already pointed out that $\rho : (A \cup B \cup C)^+ \to S$ is surjective. In fact, every element of S can be written as $e_{i1}he_{1j}$ for some i,j and some $h \in H$ (see, for example, Chapter 3 of [18]). Hence $S = L\rho$ where $L = BPC$. Clearly, L is a rational subset of $(A \cup B \cup C)^+$, so to show that S is hyperbolic, it is enough to show that the multiplication table T_L of S relative to $(A \cup B \cup C, L, \rho)$ is context-free.

Let $X = A \cup B \cup C \cup \{\#\}$. For $i,k \in \{1,\ldots,m\}$ and $j,\ell \in \{1,\ldots,n\}$, we define an automaton $M_{ijk\ell}$ over the monoid $X^* \times X^*$ as follows. There are five states $q_0,\ldots,q_4$ with q_0 being the initial state and q_4 the unique accept state. The edge set is

$$\begin{aligned}&\{(q_0,(1,b_i),q_1)\} \cup \{(q_1,(a,a),q_1) : a \in A\}\\&\cup \{(q_1,(\#,c_j\#b_k),q_2)\} \cup \{(q_2,(a,a),q_2) : a \in A\}\\&\cup \{(q_2,(\#,c_\ell\#c_\ell),q_3)\} \cup \{(q_3,(a,a),q_3) : a \in A\}\\&\cup \{(q_3,(1,b_i),q_4)\}.\end{aligned}$$

The automaton is illustrated in Figure 10 where, as before, D denotes the diagonal $\{(a,a) : a \in A\}$.

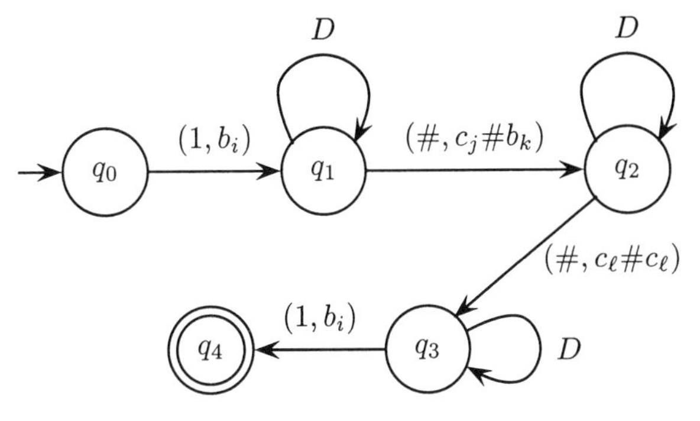

Figure 10

Let $\psi_{ijk\ell}$ be the rational transduction accepted by $M_{ijk\ell}$, and put $T_{ijk\ell} = T'_{jk}\psi_{ijk\ell}$. Then

$$T_{ijk\ell} = \{b_i u c_j \# b_k w c_\ell \# (b_i x c_\ell)^r : u, w, x \in P \text{ and } (u\sigma)h_{jk}(w\sigma) = x\sigma\},$$

and by Proposition 2.2(4), $T_{ijk\ell}$ is context-free.

Let $\alpha, \beta, \gamma \in (A \cup B \cup C)^+$. Then $\alpha\#\beta\#\gamma^r \in T_L$ if and only if for some $i, k \in \{1, \ldots, m\}$, $j, \ell \in \{1, \ldots, n\}$ and $u, w, x \in P$ we have $\alpha = b_i u c_j$, $\beta = b_k w c_\ell$, $\gamma = b_i x c_\ell$ and

$$(b_i u c_j)\rho(b_k w c_\ell)\rho = (b_i x c_\ell)\rho,$$

that is,

$$e_{i1}(u\sigma)h_{jk}(w\sigma)e_{1\ell} = (e_{i1}(u\sigma)e_{1j})(e_{k1}(w\sigma)e_{1\ell}) = e_{i1}(x\sigma)e_{1\ell}.$$

The latter holds if and only if $(u\sigma)h_{jk}(w\sigma) = x\sigma$. Hence $T_L = \bigcup T_{ijk\ell}$ where the indices i, k range over $1, \ldots, m$ and j, ℓ range over $1, \ldots, n$. Thus T_L is a finite union of context-free languages, and so is itself context-free as required.

We conclude with a straightforward corollary of Theorem 4.1. First, we recall the notion of an 'automatic semigroup'. Let A be a finite alphabet and \$ be a symbol not in A. We define a function $\delta : (A^+ \times A^+) \to (A^\$ \times A^\$)^+$ where $A^\$ = A \cup \{\$\}$ by

$$(v, w)\delta = \begin{cases} (a_1, b_1)\ldots(a_m, b_m) & \text{if } m = n \\ (a_1, b_1)\ldots(a_m, b_m)(\$, b_{m+1})\ldots(\$, b_n) & \text{if } m < n \\ (a_1, b_1)\ldots(a_n, b_n)(a_{n+1}, \$)\ldots(a_m, \$) & \text{if } m > n \end{cases}$$

where $v = a_1 \ldots a_m$ and $w = b_1 \ldots b_n$.

Let $\varphi : A^+ \to S$ be a finite choice of generators for the semigroup S. A triple (A, L, φ) is an *automatic structure* for S if L is a rational subset of

A^+ with $L\varphi = S$ such that the sets

$$L_= = \{(v, w)\delta : v, w \in L \text{ and } v\varphi = w\varphi\}$$

and

$$L_a = \{(v, w)\delta : v, w \in L \text{ and } (va)\varphi = w\varphi\}$$

are rational subsets of $(A^\$ \times A^\$)^+$ for every $a \in A$. A semigroup is *automatic* if it has an automatic structure, and it is *prefix-automatic* if it has an automatic structure (A, L, φ) with L prefix-closed. We remark that this definition is equivalent to that used in [23], [24] (see, for example, Proposition 7.1 of [19]).

It follows from Theorem 2.5.9 of [10] that an automatic group is prefix-automatic. However, whether or not every automatic semigroup is prefix-automatic is an open question.

Examples of hyperbolic monoids which are not automatic are given in [16], but it is known that every hyperbolic group is automatic (see, for example Corollary 2.20 of Chapter III.Γ in [4]). For completely simple semigroups, the situation is similar to that for groups. First, we quote part of Theorem 7.9 of [19].

Proposition 4.4. *Let M be a finitely generated Rees matrix semigroup (without zero) over a group G. Then M is prefix-automatic if and only if G is automatic.*

Corollary 4.2. *Every hyperbolic completely simple semigroup is prefix-automatic.*

Proof. Let S be a hyperbolic completely simple semigroup. Then S is finitely generated and isomorphic to a Rees matrix semigroup $M(G; I, J; P)$. By Theorem 4.1, G is hyperbolic, and hence, by the results quoted above, G is prefix-automatic. Now, by Proposition 4.4, S is prefix-automatic.

Acknowledgements

The first author would like to thank the organisers (M. Branco, V. H. Fernandes and Gracinda M. S. Gomes) for the invitation to speak at the *Semigroups and Languages* workshop.

The work was carried out while the second author was a research student at the University of York supported by an EPSRC studentship. He would like to thank Kirsty for all her support and encouragement.

References

1. J. Alonso, T. Brady, D. Cooper, V. Ferlini, M. Lustig, M. Mihalik, M. Shapiro and H. Short, Notes on word hyperbolic groups, in *Group Theory from a Geometrical Viewpoint* (E. Ghys, A. Haefliger and A. Verjovsky, eds.), World Scientific, (1991), 3–63.
2. H. Ayik and N. Ruškuc, Generators and relations of Rees matrix semigroups, *Proc. Edinburgh Math. Soc.* **42** (1999), 481–495.
3. J. Berstel, *Transductions and context-free languages*, Teubner, 1979.
4. M. Bridson and A. Haefliger, *Metric spaces of non-positive curvature*, Springer-Verlag, 1999.
5. W. W. Boone, The word problem, *Ann. of Math.* **2** (1959), 207–265.
6. J. L. Britton, The word problem for groups, *Proc. London Math. Soc.* **8** (1958), 493–506.
7. M. Dehn, Über unendliche diskontinuierliche Gruppen, *Math. Ann.* **71** (1912), 116–144.
8. M. Dehn, Transformationen der Kurven auf zweitseitigen Flächen, *Math. Ann.* **72** (1912), 413–421.
9. A. J. Duncan and R. H. Gilman, Word hyperbolic semigroups, *Math. Proc. Camb. Phil. Soc.*, to appear.
10. D. B. A. Epstein et al., *Word processing in groups*, Jones and Bartlett, 1992.
11. E. Ghys and P. de la Harpe (eds.), *Sur les groupes hyperboliques d'apres Mikhael Gromov*, Birkhäuser, 1990.
12. R. H. Gilman, Formal languages and infinite groups, in *Geometrical and Computational Perspectives on Infinite Groups.* DIMACS Ser. Discrete Math. Theoret. Comput. Sci. **25**, Amer. Math. Soc., (1996), 27–51.
13. R. H. Gilman, On the definition of word hyperbolic groups, *Math. Zeitschr.* **242** (2002), 529–541.
14. M. Gromov, Hyperbolic groups, in *Essays in Group Theory* (S. M. Gersten, ed.), Springer-Verlag, MSRI Publ. **8** (1987), 75–263.
15. M. Harrison, *Introduction to Formal Language Theory*, Addison-Wesley, 1978.
16. M. Hoffmann, D. Kuske, F. Otto and R. M. Thomas, Some relatives of automatic and hyperbolic groups, *Proc. of the Thematic Term on Semigroups, Automata, Algorithms and Languages* (G. M. S. Gomes, J.-E. Pin and P. V. Silva, eds.), World Scientific, (2002), 379–406.
17. J. E. Hopcroft and J. D. Ullman, *Introduction to Automata Theory, Languages and Computation*, Addison-Wesley, 1979.
18. J. M. Howie, *Fundamentals of Semigroup Theory*, Oxford University Press, 1995).
19. M. E. Kambites, *Combinatorial aspects of partial algebras*, Ph.D. thesis, University of York, 2003.
20. R. C. Lyndon and P. E. Schupp, *Combinatorial group theory*, Springer-Verlag, 1977.
21. P. S. Novikov, On the algorithmic unsolvability of the word problem in group theory, (Russian), *Trudy Mat. Inst. Steklov* **44** (1955), 143 pp.

22. K. Ohshika, *Discrete groups*, Amer. Math. Soc., 2002.
23. P. V. Silva and B. Steinberg, Extensions and submonoids of automatic monoids, *Theor. Comp. Sci.* **289** (2002), 727–754.
24. P. V. Silva and B. Steinberg, A geometric characterization of automatic monoids *Quart. J. Math.*, to appear.
25. B. Steinberg, A topological approach to inverse and regular semigroups, *Pacific J. Math.* **208** (2003), 367–396.
26. J. B. Stephen, Presentations of inverse monoids, *J. Pure Appl. Algebra* **63** (1990), 81–112.

AN INTRODUCTION TO E^*-UNITARY INVERSE SEMIGROUPS — FROM AN OLD FASHIONED PERSPECTIVE

DONALD B. MCALISTER*

Department of Mathematical Sciences,
Northern Illinois University,
DeKalb, Illinois 60115, U.S.A.
E-mail: don@math.niu.edu

Thirty years ago, in 1972, H.E. Scheiblich published the first usable description of the free inverse semigroup on a set. This led to a flowering of research on inverse semigroups, including the structure of E-unitary inverse semigroups. In a similar fashion, the publication of the book "Inverse semigroups: the theory of partial symmetry" by Mark V. Lawson, and, in particular, the chapters on applications of inverse semigroups to tilings have led to another flowering of research; this time into the theory of E^*-unitary inverse semigroups. This talk will give an introduction to the later theory in the light of the earlier perspective that grew from Scheiblich's result.

1. Introduction and pre-history

2002 is a very good year in the history of semigroups. It is precisely 50 years after the publication of Wagner's seminal paper [45] on inverse semigroups, "Generalized Groups", Doklady Akademiĭ Nauk SSSR 84 (1952), 1119-1122. It is also 30 years after the publication of another seminal paper in the theory of inverse semigroups, Scheiblich's Semigroup Forum paper on the structure of free inverse semigroups, Semigroup Forum 4, (1972), 351-359; [39], [40]. The period between the publication of these two papers, in particular the latter part of the 1960's, had seen a considerable development of information about the structure and properties of inverse semigroups. There were the papers of Munn [27],[29] on fundamental inverse semigroups; there were the beautiful theorems of Reilly [34], and Kochin [11] and Munn [28], on the structure of regular (inverse) semigroups whose idempotents

*Author's second address: Centro de Álgebra da Universidade de Lisboa, Av. Prof. Gama Pinto, 2, 1649–003 Lisboa, Portugal.

form an ω-chain; the structure of more general, but still special, classes of inverse semigroups was investigated, especially by Warne (see for example [46], [47]) and Schreier type extension theorems were developed, for example by Coudron [4] and D'Alarcão [5] for idempotent separating extensions, and by Saitô [36] for group coextensions of inverse semigroups. In principle, it was possible to say that the structure of inverse semigroups was known. But there was a problem with these structure theorems.

The building blocks, full subsemigroups of isomorphisms between principal ideals of a semilattice, and semilattices of groups, were simple and natural. The interrelations between the building blocks were not. The conditions required to describe how extensions were put together were intricate, to say the least. They involved group actions and factor sets which had to obey complicated and stringent interrelated conditions. It was even worse when a congruence wasn't involved. Frankly, the structure theory of inverse semigroups had reached the point of diminishing returns. One could reasonably say that each new theorem resulted in a gain of information but a loss of insight. A good — or would it be more accurate to say bad — example of this situation can be found in one of my own papers "0-bisimple inverse semigroups" [19] which appeared in the Proceedings of the London Mathematical Society in 1974 although the work had been done before I left Ireland in 1970.

It was clear that some other approach was needed. I, personally, had been trying to find new constructive ways of describing inverse semigroups, which would avoid factor sets and their complicated interactions, by using cones in partially ordered structures to build inverse semigroups. The idea had been to try to generalize the Clifford [3] and Reilly [35] theory of bisimple inverse semigroups. An example of this was given in the paper [18] which generalized Eberhardt and Sheldon's insightful characterization of one parameter inverse semigroups [6]. It is interesting that generalizations of both of these approaches have found their way into the modern theory of inverse semigroups, in particular of E^*-unitary inverse semigroups, due to Lawson; see for example [14], [15].

But it was Scheiblich's construction that provided the new direction. The description which he gave of the free inverse semigroup was so implicitly simple, but so powerful, that it led to a whole new way of looking at inverse semigroups. Earlier structure theorems had taken two, relatively, simple objects — a kernel (semilattice of groups) and an image (a fundamental inverse semigroup) — and tried to tie them together by means of a Schreier type extension theory. But now that we knew the free inverse semigroup was an essentially — at least conceptually — simple object it became possible to consider a dual approach. On the theory that if an object is simple so are its homomorphic images we could try to construct some conceptually simple inverse semigroups which would have all inverse semigroups as nice (idempotent separating, so that the kernels were nice) homomorphic images. So Scheiblich's result led directly to the theory of E-unitary inverse semigroups. The construction of these semigroups reflected Scheiblich's construction of the free inverse semigroups but the class of semigroups was already known and a Schreier type extension theory had been given for it earlier by T. Saitô [36] who called these semigroups proper. That's where the P in E-unitary inverse semigroups comes from.

2. History: E-unitary inverse semigroups.

Let's recall Scheiblich's construction [39, 40] of the free inverse monoid $FIM(X)$ on a set X. Its elements consist of pairs (A, g) where A is a finite set of reduced words in the free group $FG(X)$ on X which is closed under prefixes, and g is an element of the free group whose reduced form belongs to A. Multiplication is defined by

$$(A, g)(B, h) = (A \cup gB, gh).$$

Note that, since A is closed under prefixes, $1 \in A$.

There are a couple of points to be made here. The first relates to the multiplication. This involves translates by elements of the free group of prefix closed subsets. What are these translates? Well to understand this we need to think geometically. The free group $FG(X)$ is a geometric object or, rather, its Cayley graph is. In this sense, a finite prefix closed subset B is just a convex connected neighborhood of the identity and so, because the group acts by automorphisms on the Cayley graph, gB is just a convex connected neighborhood of any of its members, in particular of g. The fact that g belongs to both A and to gB means that $A \cup gB$ is connected and so multiplication is well-defined.

The second remark is a philosophical or cultural one. There has always been something unsatisfying about this setup for someone like me, who is a child of the twentieth century and who grew up believing that Einstein's relativity theory was the most important scientific development of the 1900's. We can summarize the latter by saying that the Lorentz group acts on space-time and all points which lie on same orbit are equivalent. Well, in Scheiblich's theorem, we have a group acting but we insist on localizing everything at 1. Relativity would say that we should regard equally all points of the free group. That is we should look at triples (g, A, h) where A is a neighborhood of g and h The triple (g, A, h) with $g, h \in A$ is then equivalent, under the obvious group action, to $(1, g^{-1}A, g^{-1}h)$ and so multipliciation on equivalence classes of triples can be made to mirror the multiplication of pairs. We shall return to this remark later but notice that an equivalence class of the triples admits an easy interpretation as a directed graph over X with a pair of identified vertices. That is, it corresponds to a Munn tree and what we have is an interpretation of Munn's graphical version [30], [31] of the free inverse semigroup on X.

In any case, the philosophy behind the original study of E-unitary inverse semigroups was, as I have said, the following: construct a family of inverse semigroups from simple, familiar, naturally related objects in such a way that every inverse semigroup is a nice homomorphic image of a member of the family. Thus there are two theorems. Before stating them, we recall the definition of an E-unitary inverse semigroup:

Definition 2.1. Let S be an inverse semigroup. Then S is said to be *E-unitary* if $ab = b$ implies $a^2 = a$.

This isn't the usual definition — $e \leq a$ where e is idempotent, implies $a^2 = a$ — but is easily seen to be equivalent to it. [Recall that, in an inverse semigroup, the natural partial order is defined by $a \leq b$ if and only if $a = fb$ for some idempotent f.]

Theorem 2.1. *Let $\mathcal{X}$ be a down directed partially ordered set with $\mathcal{Y}$ an order ideal and subsemilattice of $\mathcal{X}$. Also let G be a group which acts on $\mathcal{X}$ by order automorphisms in such a way that $\mathcal{X} = G\mathcal{Y}$. Then*

$$P = P(G, \mathcal{X}, \mathcal{Y}) = \{(a, g) \in \mathcal{Y} \times G : g^{-1}a \in \mathcal{Y}\}$$

is an E-unitary inverse semigroup under the multiplication

$$(a, g)(b, h) = (a \wedge gb, gh).$$

Conversely, if S is an E-unitary inverse semigroup then $S \approx P(G, \mathcal{X}, \mathcal{Y})$ for a unique (up to isomorphism and equivalence of group actions) G,$\mathcal{X}$, and $\mathcal{Y}$.

Note that we have here both an existence and a uniqueness theorem for E-unitary inverse semigroups. All three components G, $\mathcal{X}$, and $\mathcal{Y}$ are intrinsic to the semigroup S. The group G is (isomorphic to) the maximum group homomorphic image of S, the semilattice $\mathcal{Y}$ is (isomorphic to) the semilattice of idempotents of S. The role of $\mathcal{X}$ is not so clear but it is none the less intrinsic to S.

Theorem 2.2. *Let S be an inverse semigroup. Then S is an idempotent separating homomorphic image of an E-unitary inverse semigroup.*

I was very proud of these theorems, and particularly the first one, when I obtained them. I wrote to Al Clifford telling him about it and asking him if he would like to see a copy of the manuscript before I submitted it for publication. As I remember, it was some time before he responded but when he did, the answer was short. "That can't possibly be true", he wrote. Well that was a bit of a blow to the ego after a lot of hard work over quite a period of time!

But I can understand his reaction. The structure of these inverse semigroups appeared to be much more simple than one would have expected on the basis of other structure theorems for special classes of inverse semigroups; for example the structure theorem for 0-bisimple inverse semigroups [19] which I mentioned earlier. I was fortunate I hadn't sent Al some of the early drafts of the proof. Then he would never have believed the result. But I am pretty sure that it is true.

There have been several different proofs since that first one. My personal favorite is the proof due to Douglas Munn [32]. The main part of it consists in determining the partially ordered set $\mathcal{X}$ since the group G and the semilattice $\mathcal{Y}$ have natural interpretations. In the original proof of the theorem, $\mathcal{X}$ was constructed by a painstaking analysis of how the maximum group homomorphic image acts on $\mathcal{R}$-classes.

Munn's approach was much more external than my internal approach. Let us review it. From the requirement that $\mathcal{X} = G\mathcal{Y}$ we know that each element of $\mathcal{X}$ has the form ga for some $g \in G$ and $a \in \mathcal{Y}$. Of course the expression of the elements ga may not be unique and the problem is to determine when $ga = hb$. More precisely, we want to know when $ga \leq hb$ in order to be able to describe the partial order on $\mathcal{X}$. Well, let's see, suppose that $S = P(G, \mathcal{X}, \mathcal{Y})$. We have

$$\begin{aligned} ga \leq hb \quad &\text{if and only if} \quad h^{-1}ga \leq b \\ &\text{if and only if} \quad s = \left(a, g^{-1}h\right) \in P \text{ and } s^{-1}s \leq b. \end{aligned}$$

Thus, for an E-unitary inverse semigroup S, we define a quasi-order $\preceq$ on $G \times E$ by

$$(g, a) \preceq (h, b) \iff \exists\ s \in S \text{ with } ss^{-1} = a,\ s^{-1}s \leq b \text{ and } s\sigma^{\natural} = g^{-1}h,$$

where $\sigma^\natural$ denotes the canonical homomorphism of S onto its maximum group homomorphic image $G = S/\sigma$. It is clear that G acts on this quasi-ordered set by automorphisms through multiplication of first coordinates, so G also acts by order automorphisms on the quotient partially ordered set

$$\mathcal{X} = \{[g, a] : g \in G, a \in E\}.$$

$\mathcal{X}$ is down directed and has $\mathcal{Y} = \{[1, a] : a \in E\}$ as an order ideal isomorphic to E and $S \approx P(G, \mathcal{X}, \mathcal{Y})$

In my opinion, this proof is a gem. It is so clear and insightful. That's not a surprise; it is just a trade mark of all of Douglas Munn's work. But there is more, this is a totally constructible proof with powerful applications. To see this, I want to turn to the second theorem. It states that inverse semigroups have E-unitary covers. But there is no uniqueness to them and the problem arises of how they may be constructed. Norman Reilly and I [26] gave a answer to this question by using the notion of a prehomomorphism from one inverse semigroup to another.

Definition 2.2. Let S and T be inverse semigroups. Then a function $\theta : S \to T$ is a *prehomomorphism* if $st\theta \leq s\theta t\theta$ for all $s, t \in S$.

It is easy to see that prehomomorphisms preserve inverses and therefore map idempotents to idempotents. We say that θ is *idempotent pure* if *only* idempotents are mapped to idempotents.

Suppose that G is a group. Then Schein [37] has shown that the set $\mathcal{K}(G)$ of all cosets of G, modulo subgroups of G, is an inverse semigroup under the multiplication $*$ where $X * Y$ is the smallest coset that contains the subset XY. The semigroup $\mathcal{K}(G)$ has many interesting properties. For example, the semilattice of idempotents is anti-isomorphic to the lattice of subgroups of G; the units are the one element cosets $\{g\}$, $g \in G$, and the zero is G. Like the symmetric inverse semigroup, the semigroups $\mathcal{K}(G)$ are universal for inverse semigroups in the sense that any inverse semigroup can be embedded in one of the semigroups $\mathcal{K}(G)$; [23].

Theorem 2.3. *Let S be an inverse semigroup and let $\theta : S \to \mathcal{K}(G)$ be an idempotent pure prehomomorphism from S into the inverse semigroup $\mathcal{K}(G)$ of all cosets of a group G. Then*

$$T = \{(s, g) \in S \times G : g \in s\theta\}$$

is an E-unitary cover of S with maximum group homomorphic image isomorphic to $H = \{g \in G : s\theta \leq g$ for some $s \in S\}$; the fact that θ is a *prehomomorphism implies that H is a subgroup of G . (Let us say that θ is* surjective *if $H = G$.)*

Conversely, each E-unitary cover of S with maximum group homomorphic image G is isomorphic to one of this form for some idempotent pure surjective prehomomorphism of S into $\mathcal{K}(G)$.

Because T is explicitly given in terms of θ, Munn's construction can be immediately applied to it and we get the following proposition which must be classified as folklore. At least, I've known about it since the middle 1970's:

Proposition 2.1. *Let $\theta : S \to \mathcal{K}(G)$ be an idempotent pure surjective prehomomorphism. Define a quasi-order $\preceq$ on $G \times E$ by*

$$(g, e) \preceq (h, f) \iff \exists s \in S \text{ with } ss^{-1} = e,\ s^{-1}s \leq f \text{ and } h \in g(s\theta).$$

Then $T \approx P(G, \mathcal{X}, \mathcal{Y})$ where $\mathcal{X}$ is the corresponding partially ordered set and $\mathcal{Y} = \{[1, e] : e \in E\} \approx E$.

The final result of this section really belongs to the present of the subject and not its history. It is included in this section because it deals directly with E-unitary inverse semigroups and fits nicely with the historical thread which I have been describing here. As I remarked above, there was something unnatural about the localization restriction in Scheiblich's description of the free semigroups. Of course, the same kind of complaint can be voiced about the old-fashioned description of E-unitary inverse semigroups because it is an exact analog to Scheiblich's. We saw that this localization restriction could be removed in the case of the free inverse semigroup. When this was done we obtained a very natural geometrical interpretation of Munn's description of the free inverse semigroup by labelled graphs.

Suppose we follow the same pattern in the more general situation. That is, given G, $\mathcal{X}$, $\mathcal{Y}$, we consider the triples $(g, x, h) \in G \times \mathcal{X} \times G$ with the idea that this triple corresponds to the pair (e, k) where $e = g^{-1}x$ and $k = g^{-1}h$; thus $(g, x, h) = (g, ge, gk)$. The restrictions e, $k^{-1}e \in \mathcal{Y}$ require that $g^{-1}x$ and $h^{-1}x = h^{-1}ge = k^{-1}e \in \mathcal{Y}$. We see then that two of the triples (g_1, x_1, h_1) and (g_2, x_2, h_2) correspond to the same element $(e, k) \in P(G, \mathcal{X}, \mathcal{Y})$ if and only if

$$x_1 = g_1 e,\ h_1 = g_1 k,\ x_2 = g_2 e,\ h_2 = g_2 k$$

That is, they both correspond to the same element (e, k) if and only if

$$g_2 = k' g_1,\ x_2 = k' x_1,\ h_2 = k' h_1$$

where $k' = g_2 g_1^{-1} = h_2 h_1^{-1}$. Thus we get a set in one-to-one correspondence with the elements of $P(G, \mathcal{X}, \mathcal{Y})$ if we start with the set of all triples $(g, x, h) \in G \times \mathcal{X} \times G$ such that $g^{-1}x,\ h^{-1}x \in \mathcal{Y}$ and define an equivalence relation by

$$(g_1, x_1, h_1) \sim (g_2, x_2, h_2) \Leftrightarrow \exists k' \in G \text{ with } g_2 = k' g_1,\ x_2 = k' x_1,\ h_2 = k' h_1.$$

Multiplication of two classes is constructed in such a way as to mimic the multiplication in $P(G, \mathcal{X}, \mathcal{Y})$. This can easily be done because

$$(g_1, x_1, h_1) \sim (1, g_1^{-1}x_1, g_1^{-1}h_1) \text{ and } (g_2, x_2, h_2) \sim (1, g_2^{-1}x_2, g_2^{-1}h_2)$$

so the product should be

$$(1, g_1^{-1}x_1 \wedge g_1^{-1}h_1 g_2^{-1}x_2, g_1^{-1}h_1 g_2^{-1}h_2) \sim (kg_1, kx_1 \wedge x_2, h_2) \text{ where } k = g_2 h_1^{-1}.$$

This form of the product has a natural interpretation in terms of the evident partial product on the set of triples (g, x, h) if we consider x as an arrow from the vertex g to the vertex h:

$$(a, x, b)(b, z, c) = (a, x \wedge z, c).$$

For $(g_1, x_1, h_1) \sim (kg_1, kx_1, kh_1) = (kg_1, kx_1, g_2)$ if we take $k = g_2 h_1^{-1}$ and the partial product

$$(kg_1, kx_1, g_2)(g_2, x_2, h_2) = (kg_1, kx_1 \wedge x_2, x_2).$$

Note that (kg_1, kx_1, kh_1) is the only member of the class of (g_1, x_1, h_1) which ends where (g_2, x_2, h_2) begins. Thus the product is well-defined.

Given the triple $(G, \mathcal{X}, \mathcal{Y})$ define $E(G, \mathcal{X}, \mathcal{Y})$ to be the set of all equivalence classes of triples $(g, x, h) \in G \times \mathcal{X} \times G$ such that $g^{-1}x,\ h^{-1}x \in \mathcal{Y}$ under the multiplication defined above. Then we have the following relativistic structure theorem for E-unitary inverse semigroups:

Theorem 2.4. *(Steinberg [42]) Let $\mathcal{X}$ be a partially ordered set, $\mathcal{Y}$ an ideal and subsemilattice of $\mathcal{X}$, and G a group which acts on $\mathcal{X}$ in such a way that $\mathcal{X} = G\mathcal{Y}$. Then $E(G, \mathcal{X}, \mathcal{Y})$ is an E-unitary inverse semigroup. Conversely, each E-unitary inverse semigroup is isomorphic to $E(G, \mathcal{X}, \mathcal{Y})$ for a unique G, $\mathcal{X}$, $\mathcal{Y}$ (up to isomorphism and equivalence of group actions).*

As a final remark in this section, let me point out that the process involved in Steinberg's relativity theorem is really a very familiar one. It is one that we expect the students in our beginning linear algebra classses to be completely comfortable with — although often they are not. They come into our classes knowing very well what a vector is — a directed line segment — and how to add directed line segments using the parallelogram law. We mathematicians insist in localizing everything at one particular point (the origin) and we come up with vector spaces. Steinberg's theorem for E-unitary inverse semigroups just reverses the process! Actually, it shows that E-unitary inverse semigroups are not really more complicated than vector spaces. The parallelogram law is just the special case where $\mathcal{X} = \mathcal{Y}$ is trivial and G is the group of translations of the plane!

3. The Present: E^*-unitary inverse semigroups

Most inverse semigroups have zeros, E-unitary ones usually do not; an E-unitary inverse semigroup with a zero is a semilattices. For this reason Szendrei [44] defined an *E^*-unitary inverse semigroup* to be an inverse semigroup S with zero in which $0 \neq e = e^2 \leq a$ implies $a^2 = a$. Clearly S is E-unitary if and only if S^0 is E^*-unitary and, more generally, any Rees factor semigroup of an E-unitary inverse semigroup is E^*-unitary. Later Bullman-Fleming, Fountain, and Gould [2] defined an inverse semigroup $S = S^0$ with zero to be *strongly E^*-unitary* if and only if there is a function $\theta : S \to G^0$ with the following properties

(1) $x\theta = 0$ if and only if $x = 0$;
(2) $x\theta = 1$ if and only if $x^2 = x$;
(3) if $xy \neq 0$ then $(xy)\theta = x\theta y\theta$.

for some group G.

In terms of the language introduced earlier, such a function θ is just a 0-restricted idempotent pure prehomomorphism of S into G^0. Further, θ becomes an idempotent pure prehomomorphism of S into $\mathcal{K}(G)$ if we interpret $s\theta$ as the one element coset $\{s\theta\}$ for $s \neq 0$ and 0θ as G, the zero of $\mathcal{K}(G)$. Thus θ gives rise to an E-unitary cover of S. Since G is the zero of $\mathcal{K}(G)$, this prehomomorphism is surjective.

Bullman-Fleming, Fountain, and Gould show that any strongly E^*-unitary inverse semigroup is E^*-unitary and that any Rees factor semigroup of an E-unitary inverse semigroup is strongly E^*-unitary. Indeed, if $S = P(G, \mathcal{X}, \mathcal{Y})$ is E-unitary then the function $\theta : S/I \to G^0$ defined by

$$(e,g)\theta = \begin{cases} g \text{ for } (e,g) \notin I \\ 0 \text{ for } (e,g) \in I \end{cases}$$

is a zero restricted idempotent pure prehomomorphism of S/I into G^0.

In fact, as noted independently by Margolis and Steinberg (see [43]) and McAlister [24], the converse is also true. For suppose that $\theta : S \to G^0$ is a 0-restricted idempotent pure prehomomorphism of $S = S^0$ into a group with zero G^0. Then

$$\begin{aligned} T &= \{(s,g) \in S \times G : g \in s\theta\} \\ &= \{(s,g) \in S \times G : g = s\theta \text{ if } s \neq 0\} \cup \{(0,g) : g \in G\} \end{aligned}$$

is an E-unitary cover for S. It is easy to see that S is isomorphic to the Rees factor semigroup T/I with $I = \{(0,g) : g \in G\}$. This means that the structure theorems for E-unitary inverse semigroups also apply to strongly E^*-unitary inverse semigroups.

Theorem 3.1. *Let $\mathcal{X}$ be a partially ordered set with zero, $\mathcal{Y}$ an ideal and subsemilattice of $\mathcal{X}$, and G a group which acts on $\mathcal{X}$ in such a way that $\mathcal{X} = G\mathcal{Y}$. Let*

$$P^*(G,\mathcal{X},\mathcal{Y}) = \{(e,g) \in \mathcal{Y} \times G : g^{-1}e \in \mathcal{Y} \text{ and } e \neq 0\} \cup \{0\}.$$

Then $P^*(G,\mathcal{X},\mathcal{Y})$ is a strongly E^*-unitary inverse semigroup under the multiplication

$$(e,g)(f,h) = (e \wedge gf, gh) \text{ if } e \wedge gf \neq 0$$

and other products equal to 0. Each strongly E^*-unitary inverse semigroup is isomorphic to one of this form for some G, $\mathcal{X}$, $\mathcal{Y}$ as above.

Theorem 3.2. *Let $\mathcal{X}$ be a partially ordered set with zero, $\mathcal{Y}$ an ideal and subsemilattice of $\mathcal{X}$ and G a group which acts on $\mathcal{X}$ in such a way that $\mathcal{X} = G\mathcal{Y}$. Let $E^*(G,\mathcal{X},\mathcal{Y})$ be the set of equivalence classes of triples (g,x,h) with $g^{-1}x$, $h^{-1}x \in \mathcal{Y}\backslash\{0\}$, together with 0 under the multiplication*

$$[g_1,x_1,h_1][g_2,x_2,h_2] = [kg_1, kg_1 \wedge g_2, h_2] \text{ if } kg_1 \wedge g_2 \neq 0$$

where $k = g_2h_1^{-1}$, and with all other products equal to 0. Then $E^*(G,\mathcal{X},\mathcal{Y})$ is a strongly E^*-unitary inverse semigroup. Each strongly E^*-unitary inverse semigroup is isomorphic to one of this form.

Since the strongly E^*-unitary inverse semigroup $P^*(G,\mathcal{X},\mathcal{Y})$ is a Rees factor semigroup of the E-unitary inverse semigroup $P(G,\mathcal{X},\mathcal{Y})$, the structure of ideals, maximal subgroups, and congruences can be "read off" from the corresponding results for E-unitary semigroups. And in a similar way we can get the analogous structures for $E(G,\mathcal{X},\mathcal{Y})$ and $E^*(G,\mathcal{X},\mathcal{Y})$ by translating the results for P-semigroups.

The partially ordered set $\mathcal{X}$ which appears in each of these theorems can be given explicitly, thanks to Munn's proof of the, so-called, P-theorem in terms of the idempotent pure prehomomorphism $\theta : S \to G^0$. However, neither the group G nor the prehomomorphism θ need be unique. Thus, in distinction to the situation with E-unitary inverse semigroups we do not have a unique representation for E^*-unitary inverse semigroups.

The global structure theorem in the form I have described it here has, as I indicated earlier, a natural geometric interpretation. We have a set of vertices V $(= G)$ and a set of admissible arrows (u, x, v) between the vertices; in the case above (g, x, h) is admissible if both $g^{-1}x$ and $h^{-1}x$ belong to $\mathcal{Y}$. The labels on the arrows admit a partial semilattice multiplication, and we have a group G which acts on both the vertices and edges; the action on vertices is fix-point free. Because of the partial multiplication on the labels, there is a partial multiplication on arrows which are, in the obvious sense, consecutive. We have an equivalence relation on arrows:

$$(u_1, x_1, v_1) \sim (u_2, x_2, v_2) \ \Leftrightarrow \exists\ g \in G \text{ with } u_2 = gu_1,\ x_2 = gx_1,\ v_2 = gv_1$$

and a product is defined on the equivalence classes by translating the first arrow, if possible, so that its end is the beginning of the second one and then taking the equivalence class of product of the consecutive classes, if it exists. Otherwise the product is 0.

This is precisely the situation which pertains to the tiling semigroups which were introduced by Kellendonk [8],[9], and Kellendonk and Lawson [10]; see also Lawson's book [13] . Lawson [16] has recently axiomatized the structure that is needed to carry out this program. He calls it a *mosaic*, and the resulting semigroup he calls a *mosaic semigroup*. I shall not describe in detail the axioms — you should read them in Lawson's paper or in Steinberg's [41]. Indeed, Lawson points out that his mosaics are a reformulation of what Steinberg calls an 'abstract geometric representation'. As I said above, the ingredients are:

(1) a directed graph whose edges are labeled by the non-zero elements of a partially ordered set $\mathcal{X}$ with zero;
(2) a group G which acts partially by order automorphisms on $\mathcal{X}$ and in a fix-point free manner on the set V of vertices of the graph; this gives an obvious action of the group on the graph;
(3) a relation, (which for analogy with tiling semigroups, Lawson denotes by $\in$), that determines the admissible arrows. The arrow (triple) (a, x, b) is admissible if both $a \in x$ and $b \in x$;
(4) there are connections between the relation $\in$ and the order on $\mathcal{X}$; for example, for each $a \in A$, the set $\{x \in \mathcal{X} : a \in x\} \cup \{0\}$ is subsemilattice and order ideal of $\mathcal{X}$.

The set of orbits of admissible arrows, together with zero, is turned into a semigroup under the multiplication:

$$[a, x, b][c, y, d] = [ga, gx \wedge hy, hd]$$

if there exist $g, h \in G$ with $gb = hc$ and $ga, hd \in gx \wedge hy \neq 0$, and with all other products equal to zero.

So we have shown that every strongly E^*-unitary inverse semigroup is a mosaic semigroup. It is shown in Lawson [16] that the converse is also true. Note that proving this requires the existence of a 0-restricted idempotent pure prehomomorphism from the mosaic semigroup into a group with zero. In order to produce this, Lawson makes use of a construction of Fountain.

A natural question arises. Is every E^*-unitary inverse semigroup a strongly E^*-unitary inverse semigroup? Frankly, this was my first guess when I started thinking about such semigroups after hearing, I believe at the workshop here six years ago, Gomes and Howie's structure theorem,[7] for E^*-unitary inverse semigroups which are categorical at zero, in terms of Brandt semigroups acting on partially ordered sets. Well, I was right for those semigroups, but not in general. Bullman-Fleming, Fountain and Gould [2] have given a lovely example to show that an inverse semigroup may be E^*-unitary without being strongly E^*-unitary.

Example 3.1. (Bullman-Flemming, Fountain and Gould) Let $\mathcal{Y}$ be the free semilattice on three generators α, β, γ and form the semilattice of groups S over $\mathcal{Y}$ where each group, except $G_{\alpha\beta\gamma}$ which is $\{0\}$, is infinite cyclic. The linking homomorphism $\theta_{\alpha,\alpha\beta}$ is the squaring function while all the other homomorphisms are identities except those into $G_{\alpha\beta\gamma}$. Thus S has the diagram below:

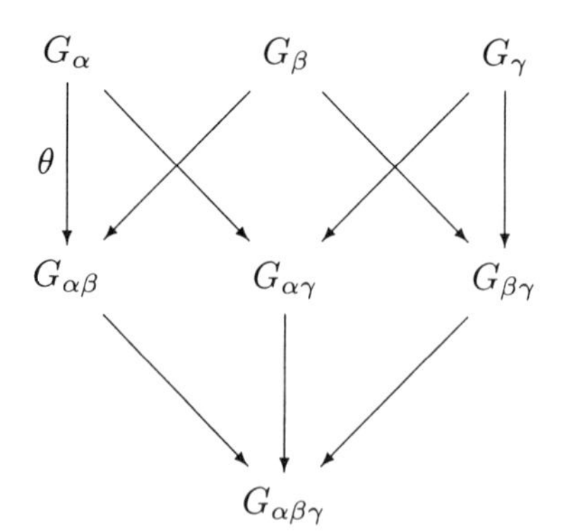

where θ is the squaring function, and the other maps are identical, or zero. Then S is E^*-unitary but not strongly E^*-unitary.

Proof. Let us denote the generators of the groups G_* by a_* and suppose that ϕ is a 0-restricted prehomomorphism of S into some group G. Then $a_\beta a_{\beta\gamma} = a_{\beta\gamma}^2 = a_\gamma a_{\beta\gamma}$ gives $a_\beta\phi = a_{\beta\gamma}\phi = a_\gamma\phi$ and similarly $a_\alpha a_{\alpha\gamma} = a_{\alpha\gamma}^2 = a_\gamma a_{\alpha\gamma}$ gives $a_\alpha\phi = a_{\alpha\gamma}\phi = a_\gamma\phi$ so that $a_\alpha\phi = a_\beta\phi = a_\gamma\phi \neq 0$. But also $a_\alpha a_{\alpha\beta} = a_{\alpha\beta}^3$ while $a_\beta a_{\alpha\beta} = a_{\alpha\beta}^2$ which gives $a_\alpha\phi = (a_\beta\phi)^2$. Thus $a_\alpha\phi = (a_\alpha\phi)^2$ is idempotent while a_α is not. Hence ϕ cannot be idempotent pure and consequently S is not strongly E^*-unitary. □

However, it is true that many natural examples of E^*-unitary inverse semigroups are strongly E^*-unitary; that is, they are Rees factor semigroups of E-unitary inverse semigroups.

(1) Tiling semigroups;

(2) E^*-unitary inverse semigroups which are categorical at zero; Gomes and Howie [7], and Bulman-Flemming, Fountain, and Gould [2];

(3) Almost factorizable E^*-unitary inverse semigroups [24];

(4) Toplitz inverse semigroups; these were introduced by Nica [33] and studied by him and by Lawson [15], in his summer school lecture notes;

(5) Inverse semigroups of graphs; these were introduced by Ash and Hall [1]. Lawson devotes a section of his 2001 Summer School lectures to these semigroups; it is interesting that these semigroups, like one-dimensional tiling semigroups of Kellendonk and Lawson which are a special case, can be also shown simply and directly to be strongly E^*-unitary by using the relationship between Scheiblich and Munn's representations of the free inverse semigrup; see McAlister [25];

(6) Inverse semigroups which are separated over a naturally quasisemilatticed 0-cancellative subsemigroup,which are partially embeddable in groups with zero. Naturally quasisemilatticed semigroups were introduced by McAlister [18] and are related to the semigroups in 5; examples of such semigroups are Rees factor semigroups of cancellative semigroups whose principal left and right ideals form semilattices under intersection, in particular of free semigroups;

(7) 0-bisimple monoids are E^*-unitary if and only if their right unit subsemigroup is cancellative; they are strongly E^*-unitary if and only if their right unit subsemigroup can be embedded in a group; [14]. If the semigroup does not have a zero then its right unit

subsemigroup obeys Ore's conditions and so, if it is cancellative, it is embeddable in a group. But if there is a zero this does not apply and we cannot conclude that in this case E^*-unitary is equivalent to strongly E^*-unitary.

So how does one tell if nice classes of E^*-unitary inverse semigroups are strongly E^*-unitary without having to treat each situation separately? In fact, one can give an answer to this question in terms of a universal construction.

Let S be an inverse semigroup with zero and denote by $S^* = \{\bar{s} : s \neq 0\}$ a set in one-to-one correspondence with the set of non-zero elements of S. Now let G be the group $FG(S^*)/N$ where N is the normal subgroup generated by the elements

$$\bar{s}^{-1}\overline{st}\bar{t}^{-1} \text{ where } s,\ t,\ st \text{ are non-zero elements of } S$$

and let η be the map $S \to G^0$ defined by

$$s\eta = \begin{cases} N\bar{s} \text{ when } s \neq 0 \\ 0 \text{ when } s = 0 \end{cases}.$$

Then η is easily seen to be a 0-restricted prehomomorphism of S into G^0 and we have the following proposition:

Proposition 3.1. *Let θ be a 0-restricted prehomomorphism of S into a group with zero H^0. Then there is a unique 0-restricted homomorphism ϕ of G^0 into H^0 such that the diagram*

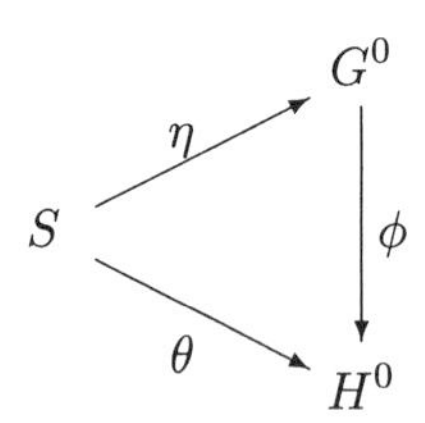

commutes. Conversely, if ϕ is 0-restricted homomorphism of G^0 into H^0 then $\theta = \eta\phi$ is a 0-restricted prehomomorphism of S into H^0.

Corollary 3.1. *Let S be an inverse semigroup with zero. Then S is strongly E^*-unitary if and only if $\bar{s} \in N$ implies $s^2 = s$.*

The criterion in the corollary, though it looks neat, merely shifts the original problem into a problem involving the group N; which doesn't really solve it. The examples above indicate that the answer to this question is closely related to the problem of knowing when a semigroup with zero admits a partial embedding into a group with zero. The following beautiful theorem of Steinberg [43] provides some parameters. More details will be found in Steinberg's paper elsewhere in these proceedings.

Theorem 3.3. *Let $\mathcal{C}$ be a pseudovariety of groups. Then the following conditions are equivalent:*

(1) the uniform word problem for $\mathcal{C}$ is decidable;
(2) it is decidable whether a finite inverse graph embeds in the Cayley graph of a group in $\mathcal{C}$;
(3) it is decidable whether a finite inverse graph is a Schützenberger graph of an E-unitary inverse semigroup with maximum group homomorphic image in $\mathcal{C}$;
(4) it is decidable whether a finite inverse semigroup is a Rees quotient of an E-unitary inverse semigroup with a maximum group homomorphic image in $\mathcal{C}$;
(5) it is decidable whether a finite inverse semigroup with zero admits a 0-restricted idempotent pure prehomomorphism into G^0 for some group G in $\mathcal{C}$.

Since the uniform word problem is known to be undecidable for the pseudovariety of all groups, the theorem has the following consequence.

Theorem 3.4. *It is undecidable whether a finite E^*-unitary inverse semigroup is strongly E^*-unitary.*

With this I will stop. I have only scratched the surface of the topic in this introduction. I would urge you to read the papers of Lawson and Steinberg. There is much of interest to be found in them. And, of course, there is the obvious question of what can one say about the structure of E^*-unitary inverse semigroups which are not strongly E^*-unitary!

Acknowledgments

I wish to thank Mark V. Lawson and Benjamin Steinberg, on whose work much of this article is based, for making available to me several preprints of their work. I also wish to thank them for giving me permission to use the results contained in those preprints.

This article was prepared while I was a visitor in the Centro de Álgebra da Universidade de Lisboa (CAUL) during the academic year 2002-2003.

I thank the Center and, in particular, its Director Gracinda M.S. Gomes for hospitality and the Fundação Luso-Americana para Desenvolvimento (FLAD) for generous financial support through Bolsa 338/2002.

References

1. Ash, C.J. and Hall, T.E. *Inverse semigroups on graphs*, Semigroup Forum **11** (1975), 140-145.
2. Bullman-Fleming, S., Fountain, J., and Gould, V. *Inverse semigroups with zero: covers and their structure*, J. Australian Math. Soc. (A) **67** (1999), 15-30.
3. Clifford, A.H., *A class of d-simple semigroups*, American J. Math. **75** (1953), 547-556.
4. Coudron, A. *Sur les extensions de demigroupes reciproques*, Bull. Soc.Royal Liege **37** (1965), 409-419.
5. D'Alarcão, H. *Idempotent separating extensions of inverse semigroups*, J. Australian Math. Soc. **9** (1969), 211-217.
6. Eberhart, C. and Shelden, J. *One parameter inverse semigroups*, Trans. American Math. Soc. **168** (1972), 53-66.
7. Gomes, G.M.S. and Howie, J.M. *A P-theorem for inverse semigroups with zero*, Portugalae Math. **55** (1996), 257-278.
8. Kellendonk, J. *The local structure of tilings and their integer group of invariants*, Comm. Math. Phys. **187** (1997), 115-157.
9. Kellendonk, J. *Topological equivalence of tilings*, J. Math. Physics **38** (1997), 1823-1842.
10. Kellendonk, J. and Lawson, M.V. *Tiling semigroups*, J. Algebra **224** (2000), 140-150.
11. Kochin, B.P. *The structure of inverse ideally simple ω-semigroups*, Vestnik Leningradskogo Universiteta **23** (1968), 41-50.
12. Lausch, H. *Cohomology of inverse semigroups*, J. Algebra **36** (1975), 273-303.
13. Lawson, M.V., "Inverse semigroups: the theory of partial symmetries", World Scientific, Singapore, 1998.
14. Lawson, M.V. *The structure of 0-E-unitary inverse semigroups, I: the monoid case*, Proc. Edinburgh Math. Soc. **42** (1999), 497-520.
15. Lawson, M.V. *E^*-unitary inverse semigroups*, Lecture Notes for Thematic Term on Semigroups, etc., Coimbra, 2001.
16. Lawson, M.V. *Mosaics*, Preprint, September 2002.
17. Lawson, M.V. *Some properties of one-dimensional tiling semigroups*, Preprint, September 2002.
18. McAlister, D.B. *Inverse semigroups separated over a subsemigroup*, Trans. American Math. Soc. **182** (1973), 85-117.
19. McAlister, D.B. *0-bisimple inverse semigroups*, Proc. London Math. Soc (3) **28** (1974), 193-221.
20. McAlister,D.B. *Groups, semilattices, and inverse semigroups*, Trans. American Math. Soc. **192** (1974), 193-221.

21. McAlister, D.B. *Groups, semilattices, and inverse semigroups, II*, Trans. American Math. Soc. **196** (1974), 251-270.
22. McAlister, D.B. *A random ramble through inverse semigroups*, in "Semigroups" (eds. T.E. Hall et al.) Academic Press (1980), 1-20.
23. McAlister, D.B. *Embedding inverse semigroups in coset semigroups.* Semigroup Forum **20** (1980), 255-267.
24. McAlister, D.B. *A note on 0-E-unitary inverse semigroups*, unpublished preprint.
25. McAlister, D.B. *Finite nilpotent semigroups, Ash's theorem, and one-dimensional tilings*, ms. of talk given at CAUL, October 2001, NIU, July 2002.
26. McAlister, D.B. and Reilly, N.R. *E-unitary covers for inverse semigroups*, Pacific J. Math. **68** (1977), 161-174.
27. Munn, W.D. *Uniform semilattices and bisimple inverse semigroups*, Quarterly Journal of Mathematics Oxford (2) **17** (1966),151-159.
28. Munn, W.D. Regular ω-semigroups, Glasgow Math. Journal **9** (1968), 46-66.
29. Munn, W.D. *Fundamental inverse semigroups*, Quarterly Journal of Mathematics Oxford (2) **21** (1970), 157-170.
30. Munn, W.D. *Free inverse semigroups*, Semigroup Forum **5** (1973), 262-269.
31. Munn, W.D. *Free inverse semigroups*, Proc. London Math. Soc. (3) **29** (1974), 385-404.
32. Munn, W.D. *A note on E-unitary inverse semigroups*, Bull. London Math. Soc. **8** (1976), 71-76.
33. Nica, A. *On a groupoid construction for actions of certain inverse semigroups*, International Journal of Math. **5** (1994), 349-372.
34. Reilly, N.R. *Bisimple ω-semigroups*, Proc. Glasgow Math. Assoc. **7** (1966), 160-167.
35. Reilly, N.R. *Bisimple inverse semigroups*, Trans. American Math. Soc. **132** (1968),101-114.
36. Saitô, T. *Proper ordered inverse semigroups*, Pacific J. Math. **15** (1965), 273-303.
37. Schein, B.M. *Semigroups of strong subsets*, Volzhskii Matematicheskiĭ Sbornik **4** (1966), 180-186.
38. Schein, B.M., *Review of Lawson's book "Inverse Semigroups: The Theory of Partial Symmetries"*, Semigroup Forum **65** (2002), 149-158.
39. Scheiblich, H.E. *Free inverse semigroups*, Semigroup Forum **4** (1972), 351-359.
40. Scheiblich, H.E. *Free inverse semigroups*, Proc. American Math. Soc. **38** (1973), 1-7.
41. Steinberg,B. *Building inverse semigroups from group actions*, Preprint, September 1999.
42. Steinberg, B. *McAlister's P-theorem via Schützenberger graphs*, Communications in Algebra **31** (2003), 4387-4392.
43. Steinberg, B. *The uniform word problem for groups and finite Rees quotients of E-unitary inverse semigroups.* Journal of Algebra **266** (2003), 1-13.

44. Szendrei, M.B. *A generalization of McAlister's P-theorem for E-unitary regular semigroups*, Acta. Sci. Math. (Szeged) **57** (1987), 229-249.
45. Wagner, W.W., *Generalized groups*, Doklady Akad. Nauk. SSSR **84** (1952), 1119-1112.
46. Warne, R.J. *I-bisimple semigroups*, Trans. American Math.Soc. **130** (1968), 367-386.
47. Warne, R.J. *E-bisimple semigroups*, J. Natur. Sci. and Math. **11** (1971), 51-81.

TURÁN'S GRAPH THEOREM AND MAXIMUM INDEPENDENT SETS IN BRANDT SEMIGROUPS

J. D. MITCHELL

Centro de Álgebra da Universidade de Lisboa,
Av. Prof. Gama Pinto, 2,
1649-003 Lisboa, Portugal
E-mail: mitchell@cii.fc.ul.pt

Let $\mathfrak{B}_n$ be an aperiodic Brandt semigroup $\mathcal{M}^0[G; n, n; P]$ where G is the trivial group, $n \in \mathbb{N}$ and $P = (a_{ij})_{n\times n}$ with $a_{ij} = 1_G$ if $i = j$ and $a_{ij} = 0$ otherwise. The maximum size of an independent set in $\mathfrak{B}_n$ is known to be $\lfloor n^2/4 \rfloor + n$, where $\lfloor n^2/4 \rfloor$ denotes the largest integer not greater than $n^2/4$. We reprove this result using Turán's famous graph theorem. Moreover, we give a characterization of all independent sets in $\mathfrak{B}_n$ with size $\lfloor n^2/4 \rfloor + n$.

1. Introduction

A subset X of a semigroup S is called *independent* if every element x is not contained in the subsemigroup generated by the other elements of X. More precisely, $x \notin \langle X \setminus \{x\} \rangle$ for every $x \in X$. This definition of independence is equivalent, in some sense, to the usual definition of independence in linear algebra. However, it is easy to find examples of semigroups where the maximum cardinality of any independent set is strictly larger than the minimum cardinality of any generating set. The cyclic group of order 6 is such an example. Furthermore, not every independent set in a semigroup S is necessarily a generating set for S. This distinction gives rise to three different types of set. That is, generating, independent generating and independent sets. These three kinds of set in turn provide us with three rank properties of S:

- **The lower rank:** the minimum cardinality of any generating set; denoted by $\text{rank}_L(S)$;

- **The intermediate rank:** the maximum cardinality of any independent generating set; denoted by $\text{rank}_I(S)$;
- **The upper rank:** the maximum cardinality of any independent set; denoted by $\text{rank}_U(S)$.

We note that the intermediate rank of a semigroup need not exist. For example, the group $\mathbb{Z}_{p^\infty}$ (p prime) of all p^nth roots of unity ($n \in \mathbb{N}$) is nonfinitely generated and the maximum cardinality of an independent set is 1. The upper rank of a semigroup need not exist either. For example, the additive monoid of natural numbers contains independent sets of any finite size, but no infinite independent sets. On the other hand, a countably infinite direct sum of the additive group of integers contains (countably) infinite independent sets. Nonetheless, whenever the three ranks are defined, for a semigroup S, the following inequalities hold:

$$\text{rank}_L(S) \leq \text{rank}_I(S) \leq \text{rank}_U(S).$$

In particular, these inequalities hold for all finite semigroups.

For brevity we call an independent set with maximum cardinality, if it exists, a *maximum independent set*. Likewise, we call a generating set with minimum cardinality a *minimum generating set* and an independent generating set with maximum cardinality a *maximum independent generating set*.

The most extensively studied of the three ranks is the lower rank (commonly called the *rank*), which has been considered for many examples of groups and semigroups, see, for example, [2] or [11]. The lower, intermediate and upper ranks were first considered together, in the context of semigroup theory, in [8]. Recently, there has been some interest in the study of the upper and intermediate ranks of some examples of groups. In [13] the intermediate rank of the symmetric group $\mathcal{S}_n$ over an n element set ($n \in \mathbb{N}$) was found to be $n-1$. All independent generating sets of this cardinality in $\mathcal{S}_n$ were determined in [5]. Continuing on from [13], the intermediate rank of the projective special linear groups $PSL_2(q)$, where q is a prime power, is given in [14].

Here we shall consider the upper, intermediate and lower rank of a particular type of Rees zero matrix semigroup. Let $S = \mathcal{M}^0[G; n, n; P]$ with index set $\{1, 2, \ldots, n\}$ ($n \in \mathbb{N}$), group G and diagonal matrix $P = (a_{ij})_{n \times n}$ with $a_{ij} = 1_G$, the identity of G, if $i = j$ and $a_{ij} = 0$, otherwise. Such semigroups S are known as *Brandt semigroups*. When G is the trivial group the semigroup S is called an *aperiodic Brandt semigroup.* We shall denote the aperiodic Brandt semigroup with index sets of size $n \in \mathbb{N}$ by $\mathfrak{B}_n$. It is well-known that a semigroup S is both completely 0-simple and an inverse semigroup if and only if it is a Brandt semigroup, see Theorem 5.1.8 in [9] for details. The semigroup $\mathfrak{B}_n$ can also be described as the set of elements $(\{1, 2, \ldots, n\} \times \{1, 2, \ldots, n\}) \cup \{0\}$ with multiplication given by:

$$(i, j)(k, l) = \begin{cases} (i, l) & j = k \\ 0 & j \neq k \end{cases}$$

and $(i, j)0 = 0(i, j) = 0^2 = 0$. For more information about semigroups see [9].

In the main result of this paper we find the cardinality of a maximum independent set in any aperiodic Brandt semigroup. The upper rank of these semigroups was first determined by Howie and Ribeiro in [8], where the authors remark:

'...[our work] is reminiscent of the theorem due to Turán in [12] concerning the maximum number of edges of a triangle-free graph with n vertices...[here] it does not seem possible to quote his work directly.'

We give a new proof of Theorem 3.3 in [8] by directly quoting Turán's famous theorem. Before doing this we first introduce some terminology relating to graphs. We refer to undirected graphs as *graphs* and directed graphs as *digraphs.* Throughout we assume that our digraphs do not have multiple edges (i.e. two distinct edges with the same direction from one vertex u to another vertex v) but may have at most one loop at each vertex. A digraph Γ contains a 2-*cycle* if there exist vertices u and v such that (u, v) and (v, u) are edges in Γ. We call the graph we obtain from a digraph Γ by replacing every directed edge by an undirected edge the *underlying graph of* Γ. Note that by our definition a digraph Γ containing a 2-cycle has no multiple edges, but the underlying graph of Γ does. A graph Γ is called *triangle-free* if no subgraph of Γ is isomorphic to the complete graph with three vertices (a triangle!). For a rational r we denote by $\lfloor r \rfloor$ the largest integer not greater than r and by $\lceil r \rceil$ the smallest integer not less than r.

Proposition 1.1. *Let Γ be a triangle-free graph with n vertices, no multiple edges and at most one loop at each vertex. Then the maximum number of edges in Γ is $\lfloor n^2/4 \rfloor + n$.*

For a proof see Theorem 2.3 in [7] or Chapter 29 in [1].

Proposition 1.1 is a special case of a more general theorem, due to Turán, which gives the maximum number of edges in a graph with no subgraph isomorphic to the complete graph on n vertices. This result initiated extremal graph theory. See [3] or [4] for more details.

For a graph, or digraph, Γ we shall denote by $V(\Gamma)$ the vertex-set of Γ and by $E(\Gamma)$ the edge-set of Γ. The degree of a vertex v in a graph Γ is the number of edges incident to v; we denote this number by $d(v)$. Note that a loop incident to a vertex is counted as two incident edges. For a vertex v in some graph (or digraph) Γ we denote by $\Gamma \setminus \{v\}$ the graph obtained from Γ by removing v and all its incident edges. A graph Γ is called *bipartite* if the vertices of Γ can be partitioned into two sets A and B where $E(\Gamma) \subseteq (A \times B) \cup \{ (v,v) : v \in V(\Gamma) \}$. We shall refer to a bipartite graph with $|A| = m$, $|B| = n$ and $E(\Gamma) = (A \times B) \cup \{ (v,v) : v \in V(\Gamma) \}$ as a *complete m,n-bipartite* graph. For more further information about graphs see [7].

2. The main result

In this section, we present our main result. We start by giving the lower rank of $\mathfrak{B}_n$ for $n > 1$. The following result is an immediate consequence of Theorem 3.3 in [6] (see Proposition 3.1 below also).

Proposition 2.1. *The minimum cardinality of a generating set for $\mathfrak{B}_n$ is n $(n > 1)$.*

Note that when $n = 1$ every element in $\mathfrak{B}_n$ is idempotent and hence $\operatorname{rank}_L(\mathfrak{B}_1) = \operatorname{rank}_I(\mathfrak{B}_1) = |\mathfrak{B}_1| = 2$. In [8] the intermediate rank of $\mathfrak{B}_n$ was determined:

Proposition 2.2. *The maximum cardinality of an independent generating set in $\mathfrak{B}_n$ is $2n - 2$ $(n > 1)$.*

Now, we make a connection between subsets of $\mathfrak{B}_n$ and digraphs. For a set $A \subseteq \mathfrak{B}_n$ define a digraph Γ_A with vertex-set $V(\Gamma_A) = \{1, 2, \ldots, n\}$ and edge-set $E(\Gamma_A) = A \setminus \{0\}$. Note that for any $A \subseteq \mathfrak{B}_n$, the digraph Γ_A contains no multiple edges but may contain loops.

Let Γ be a digraph. Then we call Γ *independent* if for every pair of adjacent vertices u and v any path from u to v contains their common incident edge. Note that u and v need not be distinct, and so we suppose that a vertex is adjacent to itself if and only if there is a loop at that vertex. The following routine lemma shows that independence of subsets of $\mathfrak{B}_n$ is equivalent to independence in the related digraph.

Lemma 2.3. *A subset $A \subseteq \mathfrak{B}_n$ is independent if and only if the related digraph Γ_A is independent.*

We give an upper bound for the cardinality of any independent set in $\mathfrak{B}_n$. For $n > 1$, consider the complete $\lceil n/2 \rceil, \lfloor n/2 \rfloor$-bipartite digraph with every edge, that is not a loop, directed from one vertex-set to the other. If $n = 6$ then this digraph is:

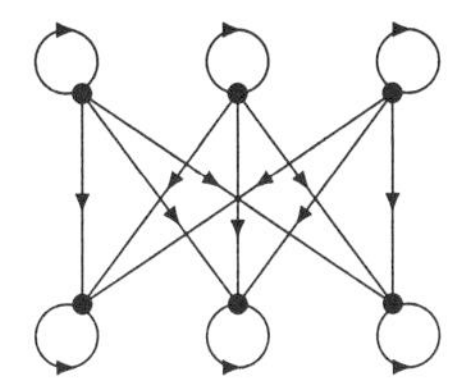

It is evident that digraphs of this type are independent, and so

$$\operatorname{rank}_U(\mathfrak{B}_n) \geq \lceil n/2 \rceil \lfloor n/2 \rfloor + n = \lfloor n^2/4 \rfloor + n. \tag{1}$$

Note that, for $n > 1$, this bound implies that the zero element 0 (in $\mathfrak{B}_n$) is never contained in a maximum independent set. However, for $\mathfrak{B}_1$ the entire semigroup forms an independent set and so $\operatorname{rank}_U(\mathfrak{B}_1) = 2$.

To show that inequality (1) is sharp it is sufficient, by Turán's theorem, to prove that the underlying graph of any independent digraph, with maximum number of edges, has no 2-cycles and is triangle-free. It is clear that any digraph containing a sequence of edges of the form $(v_1, v_2), (v_2, v_3), (v_1, v_3)$ is dependent. If an independent digraph contains a 3-cycle $(v_1, v_2), (v_2, v_3), (v_3, v_1)$ then none of the vertices v_1, v_2 or v_3 has an incident loop. Therefore to prove that an independent graph, with maximum number of edges, has a loop at every vertex is equivalent to proving that its underlying graph is triangle-free. As a first step we prove that every independent digraph with at least one 2-cycle contains a vertex of 'small' degree.

Lemma 2.4. *Let Γ_X be an independent digraph with n ($n \geq 4$) vertices that contains a 2-cycle. Then there is at least one vertex in Γ_X with degree at most $\lfloor n/2 \rfloor + 1$.*

Proof. Assume that $n = 2k+1$ and $d(v) \geq k+2 = \lfloor n/2 \rfloor + 2$ for all vertices v. Suppose that for every pair of vertices v and v' such that $(v, v'), (v', v) \in E(\Gamma_X)$ there exists a vertex $v''(\neq v', v)$ such that $(v', v''), (v'', v') \in E(\Gamma_X)$. Then there exists a directed cycle

$$(v_0, v_1), (v_1, v_2), \ldots, (v_{k-1}, v_k), (v_k, v_0)$$

of distinct vertices in Γ_X such that the reverse cycle

$$(v_0, v_k), (v_k, v_{k-1}), \ldots, (v_2, v_1), (v_1, v_0)$$

is also in Γ_X. But the path $(v_0, v_k), (v_k, v_{k-1}), \ldots, (v_2, v_1)$ does not contain the edge (v_0, v_1), a contradiction. It follows that there exists a pair of distinct vertices v and v', connected to only one adjacent vertex by 2 edges, such that there is a directed path of distinct vertices

$$(v, u_1), (u_1, u_2), \ldots, (u_{r-1}, u_r), (u_r, v'),$$

from v to v' and the reverse path is also in Γ_X. Note that u_1 may be equal to v'. There is no loop incident to v or v' and hence both vertices are adjacent to at least $k+1$ distinct vertices. But there are only $2k+1$ vertices in Γ_X and so there is a vertex $v''(\neq v, v')$ adjacent to both v and v'. When $r > 1$ any such vertex v'' is distinct from every u_i since Γ_X is independent. In the case that $r = 1$, both v and v' are adjacent to k vertices not equal to u_1, v or v'. Hence it is possible to choose $v'' \neq u_1$. Now either $\{(v, v''), (v', v'')\}$, $\{(v'', v), (v'', v')\}$, $\{(v, v''), (v'', v')\}$ or $\{(v', v''), (v'', v)\}$ is a subset of $E(\Gamma_X)$. In the first case, the path $(v, u_1), (u_1, u_2), \ldots, (u_r, v'), (v', v'')$ does not contain (v, v''). Secondly, $(v'', v), (v, u_1), \ldots, (u_r, v')$ does not contain (v'', v'). Thirdly, $(v, v''), (v'', v'), (v', u_r), (u_r, u_{r-1}), \ldots, (u_2, u_1)$ does not contain (v, u_1). The final case follows by an analogous argument to the third. In each case we have reached a contradiction and the result follows.

The case when $n = 2k$ follows by the same argument. □

In order to prove that there is a loop at every vertex of an independent graph, with maximum number of edges, we require the following result:

Lemma 2.5. *Let Γ_X be an independent digraph with n $(n \geq 4)$ vertices and no loops. Then there is at least one vertex in Γ_X with degree at most $\lfloor n/2 \rfloor + 1$.*

Proof. If Γ_X contains a 2-cycle then the result follows immediately from Lemma 2.4. Therefore we may suppose that Γ_X contains no 2-cycles. It follows, by the assumption that Γ_X has no loops, that the underlying graph of Γ_X is simple. As above assume that $d(v) \geq k+2$ for all vertices v. Let v_0 be an arbitrary vertex in Γ_X and let v_1 be any vertex adjacent to v_0. Since there are at most $k-2$ vertices in Γ_X not adjacent, or equal, to v_0 it follows that there is a vertex $v_2(\neq v_1, v_0)$ adjacent to both v_1 and v_0. Likewise, there are vertices $v_3(\neq v_2, v_1, v_0)$ and $v_4(\neq v_0, v_2, v_3)$ such that v_3 is adjacent to both v_2 and v_0, and v_4 is adjacent to both v_3 and v_0. Now, either $v_4 = v_1$ or $v_4 \neq v_1$.

If the former is true, then the subgraph of Γ_X induced by v_0, v_1, v_2 and v_3 is the complete graph K_4 on 4 vertices. It is easy to verify that any digraph with underlying graph isomorphic to K_4 is dependent, and we obtain a contradiction.

In the latter case, each of the subgraphs of Γ_X with vertex-sets $\{v_0, v_1, v_2\}$, $\{v_0, v_3, v_4\}$ and $\{v_0, v_2, v_3\}$ has underlying graph isomorphic to the complete graph with 3 vertices. Now, either (v_0, v_1) or (v_1, v_0) is an edge in Γ_X. In the first case, since Γ_X is independent $\{(v_0, v_1), (v_1, v_2), (v_2, v_0)\}$, $\{(v_2, v_0), (v_0, v_3), (v_3, v_2)\}$ and $\{(v_0, v_3), (v_3, v_4), (v_4, v_0)\}$ are subsets of $E(\Gamma_X)$. But then $(v_3, v_2) \in E(\Gamma_X)$ and $(v_3, v_4), (v_4, v_0), (v_0, v_1), (v_1, v_2) \in E(\Gamma_X)$, a contradiction. Using a similar argument, we obtain a contradiction in the case that $(v_1, v_0) \in E(\Gamma_X)$.

The result follows by a similar argument when $n = 2k$. □

Next, we show that inequality (1) is sharp for $n > 1$.

Theorem 2.6. *Let X be a maximum independent set in $\mathfrak{B}_n$ $(n > 1)$. Then $|X| = \lfloor n^2/4 \rfloor + n$ and the underlying graph of Γ_X is a complete independent $\lceil n/2 \rceil, \lfloor n/2 \rfloor$-bipartite graph.*

Proof. For $n = 2, 3, 4$ it is easy to verify that the conditions of the theorem are satisfied. We proceed by induction. Assume that $n = 2k$. Now, suppose that every maximum independent set in $\mathfrak{B}_{n-1}$ has the required properties and let X be any maximum independent set in $\mathfrak{B}_n$. By (1) we have $|X| \geq \lfloor n^2/4 \rfloor + n = k^2 + 2k$.

Suppose that Γ_X contains a 2-cycle $(p, q), (q, p)$, for vertices p, q ($p \neq q$). By Lemma 2.4 there exists a vertex v such that $d(v) \leq k+1$. Hence $\Gamma_X \setminus \{v\}$ has at least $k^2 + 2k - k - 1 = k^2 + k - 1$ edges. But $\lfloor (n-1)^2/4 \rfloor + n - 1 = k^2 + k - 1$, and hence by the inductive hypothesis $\Gamma_X \setminus \{v\}$ represents a maximum independent set in $\mathfrak{B}_{n-1}$. It follows that the underlying graph of $\Gamma_X \setminus \{v\}$ contains no 2-cycles, and hence v is either p or q, say p. But then q has no incident loop, a contradiction.

Suppose that Γ_X has no loops. Then by Lemma 2.5 there exists a vertex v such that $d(v) \leq k + 1$. We conclude, as in the previous paragraph, that $\Gamma_X \setminus \{v\}$ represents a maximum independent set in $\mathfrak{B}_{n-1}$. Therefore there is a loop at every vertex of $\Gamma_X \setminus \{v\}$, a contradiction.

It follows that Γ_X contains no 2-cycles, and there exists a vertex v' in Γ_X with a loop. Suppose that every vertex in Γ_X with a loop has degree strictly greater than $\lceil n/2 \rceil + 2 = k + 2$ and every vertex without a loop has degree strictly greater than $\lceil n/2 \rceil + 1 = k + 1$. Let $v''(\neq v')$ be any vertex adjacent to v'. Since $d(v'), d(v'') > k + 1$ it follows that there exists a vertex $v'''(\neq v', v'')$ adjacent to both v' and v''. Since v' has an incident loop it follows that Γ_X is dependent, a contradiction. Therefore, there exists a vertex u in Γ_X with a loop and $d(u) \leq k+2$ or without a loop and $d(u) \leq k + 1$.

We now prove that there is a loop at every vertex of Γ_X. Assume otherwise, then there exists a $v' \in V(\Gamma_X)$ with no incident loop. Since X is a maximum independent set we deduce that v' is contained in a cycle of length at least three consisting of vertices without loops. Hence there is a vertex in $\Gamma_X \setminus \{u\}$ with no incident loop. It follows, by our inductive hypothesis, that $\Gamma_X \setminus \{u\}$ has strictly less than $\lfloor (n-1)^2/4 \rfloor + n - 1 = k^2 + k - 1$ edges. In the case that u has an incident loop we have $|X| = |E(\Gamma_X)| = |E(\Gamma_X \setminus \{u\})| + d(u) - 1 < \lfloor (n-1)^2/4 \rfloor + n - 1 + \lceil n/2 \rceil + 1 = k^2 + 2k = \lfloor n^2/4 \rfloor + n \leq |X|$, a contradiction. Otherwise, $|X| = |E(\Gamma_X)| = |E(\Gamma_X \setminus \{u\})| + d(u) < \lfloor (n-1)^2/4 \rfloor + n - 1 + \lceil n/2 \rceil + 1 \leq |X|$, again a contradiction. We deduce that Γ_X contains a loop at every vertex. Moreover, by the discussion following Lemma 2.3, we deduce that Γ_X is triangle-free, and hence $|X| = \lfloor n^2/4 \rfloor + n$.

The fact that the underlying graph of Γ_X is a complete $\lceil n/2 \rceil, \lfloor n/2 \rfloor$-bipartite graph follows immediately from Exercise 2.23 in [7]. This exercise states that there is only one, up to isomorphism, triangle-free graph with $\lfloor n^2/4 \rfloor + n$ edges.

The result follows by a similar argument in the case that $n = 2k + 1$. □

Corollary 2.7. *Let X be an arbitrary subset of $\mathfrak{B}_n$ $(n > 1)$. Then X is a maximum independent set in $\mathfrak{B}_n$ if and only if the underlying graph of Γ_X is a complete independent $\lceil n/2 \rceil, \lfloor n/2 \rfloor$-bipartite graph.*

Proof. ($\Rightarrow$) This implications follows immediately from Theorem 2.6.

($\Leftarrow$) Every maximum independent set in $\mathfrak{B}_n$ has $\lfloor n^2/4 \rfloor + n$ elements. The graph Γ_X represents an independent set with $\lfloor n^2/4 \rfloor + n$ elements, and the result follows. □

3. Final remarks

We end the paper by giving some bounds for the lower, intermediate and upper ranks of the Brandt semigroup $\mathcal{M}^0[G; n, n; P]$, where G is an arbitrary group, $n \in \mathbb{N}$ and P is the $n \times n$ identity matrix as defined in the introduction. We denote this semigroup by $\mathfrak{B}(G, n)$.

Note that if $n = 1$ then $\mathfrak{B}(G, n)$ is isomorphic to the group G with a zero adjoined. In this case it follows that $\text{rank}_L(\mathfrak{B}(G, n)) = \text{rank}_L(G) + 1$, $\text{rank}_I(\mathfrak{B}(G, n)) = \text{rank}_I(G) + 1$ and $\text{rank}_U(\mathfrak{B}(G, n)) = \text{rank}_U(G) + 1$.

Next, we give the lower rank of $\mathfrak{B}(G, n)$.

Proposition 3.1. *Let G be a finite group and $n > 1$. (We suppose that the trivial group has rank 1.) Then the lower rank of $\mathfrak{B}(G, n)$ is $(n-1) +$ $\text{rank}_L(G)$.*

Proof. Let X be a minimum generating set for G and let 1_G denote the identity of G. Then the set $A = \{\ (i, 1_G, i+1)\ :\ i \in \{2, \ldots, n-1\}\ \} \cup \{(n, 1_G, 1)\} \cup \{\ (1, g, 2)\ :\ g \in X\ \}$ is a generating set for $\mathfrak{B}(G, n)$. That A is a minimum generating set for $\mathfrak{B}(G, n)$ follows immediately from Theorem 3.3 in [6]. Indeed, in this result it is shown that the rank of $\mathfrak{B}(G, n)$, as an inverse semigroup, is $(n-1) + \text{rank}_L(G)$. It follows that

$\text{rank}_L(\mathfrak{B}(G,n)) \geq (n-1) + \text{rank}_L(G)$, and the result follows. □

We give a lower bound for the intermediate rank of $\mathfrak{B}(G,n)$.

Proposition 3.2. *Let G be a finite group and $n > 1$. Then the intermediate rank of $\mathfrak{B}(G,n)$ is not less than $(n-1) + \text{rank}_I(G)$.*

Proof. Let X be a maximum independent generating set for G. Then the set $\{ (i, 1_G, i+1) : i \in \{2, \dots, n-1\} \} \cup \{(n, 1_G, 1)\} \cup \{ (1, g, 2) : g \in X \}$ is an independent generating set for $\mathfrak{B}(G,n)$. □

Finally, we give a lower bound for the upper rank of $\mathfrak{B}(G,n)$.

Proposition 3.3. *Let G be a finite group and $n > 1$. Then the upper rank of $\mathfrak{B}(G,n)$ is not less than $|G|\lfloor n^2/4 \rfloor + n$.*

Proof. Partition the set $N = \{1, 2, \dots, n\}$ into two sets I and J where $|I| = \lceil n/2 \rceil$ and $|J| = \lfloor n/2 \rfloor$. Then the set $\{ (i, g, j) : i \in I,\ j \in J \text{ and } g \in G \} \cup \{ (i, 1_G, i) : i \in N \}$ is independent in $\mathfrak{B}(G,n)$. □

The author verified that the bound in the previous lemma is sharp in the cases that $G = \mathbb{Z}_2$ (the cyclic group of order 2) and $n = 3$ or 4 using GAP. However, it seems unlikely that this bound is sharp in general.

We conclude the paper by posing some open problems.

Open Problem. What are the intermediate and upper rank of a Brandt semigroup $\mathfrak{B}(G,n)$ for an arbitrary group G and arbitrary $n \in \mathbb{N}$?

Open Problem. What are the intermediate and upper rank of a Rees zero matrix semigroup $\mathcal{M}^0[G; I, I; P]$ where G is trivial, I is finite and P is an arbitrary matrix?

There are a great many semigroups for which the maximum cardinalities of independent or independent generating sets are unknown.

Open Problem. What are the intermediate and upper ranks of a free semilattice with an n element generating set?

This question may be restated as follows:

Open Problem. Let $\mathcal{X}$ denote the set of subsets of an n element set. What is the maximum cardinality of a subset $\mathcal{Y}$ of $\mathcal{X}$ with the property that no set in $\mathcal{Y}$ is the union of other sets in $\mathcal{Y}$?

A related problem was considered in [10].

Another example of a variety of semigroups whose finitely generated free objects are finite is that of *bands*. A *free band* is the quotient of a finitely generated free semigroup A^+ by the congruence generated by $\{ (w^2, w) \ : \ w \in A^+ \}$. For more details see [9].

Open Problem. What are the intermediate and upper ranks of a free band with an n element generating set?

The proof that the upper rank of the symmetric group on n elements is $n-1$, in [13], relies on the classification of finite simple groups. We give as an open problem a statement from [14].

Open Problem. Find an elementary proof that the maximum cardinality of an independent generating set and an independent set in the symmetric group on n elements is $n-1$.

As the symmetric group is important in group theory, the full transformation semigroup is important in semigroup theory. As such it is desirable to answer the following question:

Open Problem. Find the maximum cardinality of an independent generating set or an independent set in the semigroup $\mathcal{T}_X$ of all mappings from a finite set X to itself.

Acknowledgments

The author would like to acknowledge the support of FCT and FEDER through 'Financiamento Programático do CAUL'. The author would also like to thank Mario Branco, and the anonymous referee for their careful reading of this paper and helpful comments.

References

1. M. Aigner and G.M. Ziegler, *Proofs from the Book*, Springer-Verlag (2001).
2. A.Ya. Aĭzenshtat, *Defining relations of finite symmetric semigroups* (Russian), Mat. Sbornik **45** (87) (1958) 261-280.

3. B. Bollobás, *Extremal graph theory*, London Mathematical Society, Monographs, **11**, Academic Press (1978).
4. B. Bollobás, *Modern graph theory*, Graduate Texts in Mathematics, **184**, Springer-Verlag (1998).
5. P.J. Cameron and P. Cara, *Independent generating sets and geometries for symmetric groups*, J. Algebra **258** (2002) 641-650.
6. G.M.S. Gomes and J.M. Howie, *On the ranks of certain finite semigroups of transformations*, Math. Proc. Cambridge Philos. Soc. **101** (1987) 395-403.
7. F. Harary, *Graph theory*, Addison-Wesley Publishing Co. (1969).
8. J.M. Howie and M.I.M. Ribeiro, *Rank properties in finite semigroups*, Comm. Algebra **27** (1999) 5333-5347.
9. J.M. Howie, *Fundamentals of semigroup theory*, London Mathematical Society, New Series **12**, Oxford Science Publications, The Clarendon Press, Oxford University Press (1995).
10. D.J. Kleitman, *Extremal properties of collections of subsets containing no two sets and their union*, J. Combinatorial Theory Ser. A **20** (1976) 390-392.
11. N. Ruškuc, *On the rank of completely* 0*-simple semigroups*, Math. Proc. Cambridge Philos. Soc. **116** (1994), no. 2, 325-338.
12. P. Turán, *Eine Extremalaufgabe aus der Graphentheorie* (Hungarian), Mat. Fiz. Lapok **48** (1941) 436-452.
13. J. Whiston, *Maximal independent generating sets of the symmetric group*, J. Algebra **232** (2002) 255-268.
14. J. Whiston and J. Saxl, *On the maximal size of independent generating sets of $PSL_2(q)$*, J. Algebra **232** (2002) 651-657.

ON SEMILATTICES OF ARCHIMEDEAN SEMIGROUPS — A SURVEY

MELANIJA MITROVIĆ*

Faculty of Mechanical Engineering,
University of Niš,
Aleksandra Medvedeva 14,
18000 Niš, Serbia and Montenegro
E-mail: meli@junis.ni.ac.yu

We provide an exposition of the approach to the structure theory of semigroups through semilattice decompositons. This approach was introduced by A. H. Clifford, extended by T. Tamura and others including M. Petrich, and, in particular, by the members of the Niš school of semigroup theory, headed by S. Bogdanović. An emphasis is placed on the author's results some published in various articles, and some appearing for the first time in print in her book "Semilattices of Archimedean Semigroups".

0. Introduction

Among methods used for describing a structure of a certain type of semigroup there is a method of decomposition of a semigroup, based on the partition of a semigroup, describing the structure of each of the components and establishing connections between them. In fact, semilattice decompositions of semigroups play the central role in this paper.

A congruence ξ on a semigroup S is a *semilattice congruence* of S if the factor S/ξ is a semilattice. The partition and the factor determined by such ξ are called *semilattice decomposition* and *semilattice homomorphic image* of S, respectively. A semigroup S is *semilattice-indecomposable* if $S \times S$ is the only semilattice congruence of S. Let us denote a ξ-class of S by S_i, $i \in I$, where I is a semilattice isomorphic to S/ξ, with isomorphism defined by $i \longmapsto S_i$. If $\mathcal{K}$ is a certain class of semigroups, and if $S_i \in \mathcal{K}$, for every $i \in I$, then we say that S is a *semilattice* I *of semigroups* S_i, $i \in I$, *from a class* $\mathcal{K}$.

*This work is partially support by Grant 1379 of the Serbian Ministry of Science, Technology and Development

Semilattice decompositions of semigroups were introduced by A. H. Clifford in 1941(see [25]). Since that time several authors have been working on the topic. Special contributions to the development of the theory of semilattice decompositions were made by T. Tamura, and co-authors N. Kimura and J. Shafer. The series of their papers began in 1954, [84], where semilattice decompositions of commutative semigroups were considered. The existence of the greatest semilattice decomposition of an arbitary semigroup was established in 1955 (see [85,89]). The fundamental result that components in the greatest semilattice decomposition of a semigroup are semilattice-indecomposable, was proved in 1956 (see [79]). Various characterizations of the smallest semilattice congruence on a semigroup can be found also in [80,82]. Among the authors who also made contributions to the theory of semilattice decompositions are M. Petrich (see [61,62]), M. S. Putcha (see [70]), R. Šulka, (see [78]), M. Ćirić and S. Bogdanović (see [28,29,30]).

Here, we are interested, in particular, in the decomposability of a certain type of semigroups into a semilattice of archimedean semigroups. Most of the results of on decompositions of completely π-regular semigroups into semilattices of archimedean semigroups obtained by 1992 were cumulated in [12]. The present survey is however not a mere continuation of [12]. It is structured as follows: Section 1 gives an overview of results concerning different types of regularity , archimedean semigroups and semilattices of archimedean semigroups. Although there are a few overlaps with [12], this section is self-contained. Sections 2-6 are based on author's original results some appearing in [55], some published in various articles ([19,20,57]), some appearing for the first time in print in [56]. The main characterization of the approach given in Sections 2-6 is making connections of semilattice decomposition of a certain type of semigroup into archimedean components, on the one hand, and equalities between (generalized) regular subsets and Green's subsets of such semigroups on the other. Following that point of view, we have the new characterizations of semilattices of nil-extensions of simple semigroups (Theorem 2.1.) in Section 2, which contains some basic results to be used in the other parts of this paper. In Section 3, which is the core of the survay, we consider left quasi-π-regular semigroups (Theorem 3.1.), and give conditions under which we can decompose such semigroups (Theorem 3.5.). The semilattices of left strongly archimedean semigroups, as we call left quasi-π-regular semigroups which are also semilattices of archimedean semigroups, are generalizations of some already known semigroups such as: semilattices of left completely archimedean semigroups

(Section 4), semilattices of nil-extensions of simple and regular semigroups (Section 5) and semilattices of completely archimedean semigroups (Section 6). The proofs of results given in these sections can be found in [56].

Throughout this paper N will denote the set of positive integers. Let S be a semigroup. If S contains an idempotent, then we will denote by $E(S)$ a set of its idempotents. The intersection of all subsemigroups of S containing $E(S)$ is the *idempotent-generated* subsemigroup of S, denoted by $\langle E(S)\rangle$.

If $a \in S$, then the intersections of the ideals (left ideals, right ideals) of S which contain a is *the principal ideal (principal left ideal, principal right ideal) of S generated by a* and is denoted by $J(a)$ $(L(a), R(a))$. By using principal ideals of certain elements of a semigroup S, we can define various very important relations on S. Let $a, b \in S$. *Division relations* on S are defined by: $a|b \Leftrightarrow b \in J(a)$, $a\,{}_l^{|}\,b \Leftrightarrow b \in L(a)$, $a\,{}_r^{|}\,b \Leftrightarrow b \in R(a)$, $a\,{}_t^{|}\,b \Leftrightarrow a\,{}_l^{|}\,b \,\&\, a\,{}_r^{|}\,b$. Relations $\mathcal{J}$, $\mathcal{L}$, $\mathcal{R}$, $\mathcal{D}$ and $\mathcal{H}$ defined on S by $a\,\mathcal{J}\,b \Leftrightarrow J(a) = J(b)$, $a\,\mathcal{L}\,b \Leftrightarrow L(a) = L(b)$, $a\,\mathcal{R}\,b \Leftrightarrow R(a) = R(b)$, $\mathcal{D} = \mathcal{LR} = \mathcal{RL}$, $\mathcal{H} = \mathcal{L} \cap \mathcal{R}$ are well-known *Green's relations* or *Green's equivalences*. For any element a of a semigroup S and $\mathcal{T} \in \{\mathcal{J}, \mathcal{L}, \mathcal{R}, \mathcal{D}, \mathcal{H}\}$, the $\mathcal{T}$-class of S containing a is denoted by T_a. Now, once again, we use the division relations for defining the following relations on S: $a \longrightarrow b \Leftrightarrow (\exists n \in \mathrm{N})\; a|b^n$, $a \overset{l}{\longrightarrow} b \Leftrightarrow (\exists n \in \mathrm{N})\; a\,{}_l^{|}\,b^n$, $a \overset{r}{\longrightarrow} b \Leftrightarrow (\exists n \in \mathrm{N})\; a\,{}_r^{|}\,b^n$, $a \overset{t}{\longrightarrow} b \Leftrightarrow a \overset{l}{\longrightarrow} b \,\&\, a \overset{r}{\longrightarrow} b$.

Given an arbitrary relation ρ on S, then $a\,\tau(\rho)\,b \Leftrightarrow (\exists n \in \mathrm{N})\;\; a^n\,\rho\,b^n$ is a relation on S caled the *τ-radical of ρ*.

Before we proceed we will present tables and diagrams that visualize connections of semigroups which will be studied together with their semilattices in Sections 1 - 6.

Table 1. (Generalized) regular semigroups

notation	class of semigroups	characterization
$\mathcal{SS}$	semisimple	$a \in SaSaS$
$\mathcal{LQR}$	left quasi-regular	$a \in SaSa$
$\mathcal{RQR}$	right quasi-regular	$a \in aSaS$
$\mathcal{CQR}$	completely quasi-regular	$a \in aSaS \cap SaSa$
$\mathcal{IR}$	intra-regular	$a \in Sa^2S$
$\mathcal{LR}$	left regular	$a \in Sa^2$
$\mathcal{RR}$	right regular	$a \in a^2S$
$\mathcal{R}$	regular	$a \in aSa$
$\mathcal{CR} = \mathcal{UG}$	completely regular	$a \in aSa^2$

Table 2. π-regular semigroups

notation	class of semigroups	characterization
$\pi\mathcal{SS}$	π-semisimple	$a^n \in Sa^nSa^nS$
$\mathcal{LQ}\pi\mathcal{R}$	left quasi-π-regular	$a^n \in Sa^nSa^n$
$\mathcal{RQ}\pi\mathcal{R}$	right quasi-π-regular	$a^n \in a^nSa^nS$
$\mathcal{CQ}\pi\mathcal{R}$	completely quasi-π-regular	$a^n \in a^nSa^nS \cap Sa^nSa^n$
$\mathcal{I}\pi\mathcal{R}$	intra-π-regular	$a^n \in Sa^{2n}S$
$\mathcal{L}\pi\mathcal{R}$	left π-regular	$a^n \in Sa^{n+1}$
$\mathcal{R}\pi\mathcal{R}$	right π-regular	$a^n \in a^{n+1}S$
$\pi\mathcal{R}$	π-regular	$a^n \in a^nSa^n$
$\mathcal{C}\pi\mathcal{R}$	completely-π-regular	$a^n \in a^nSa^{n+1}$

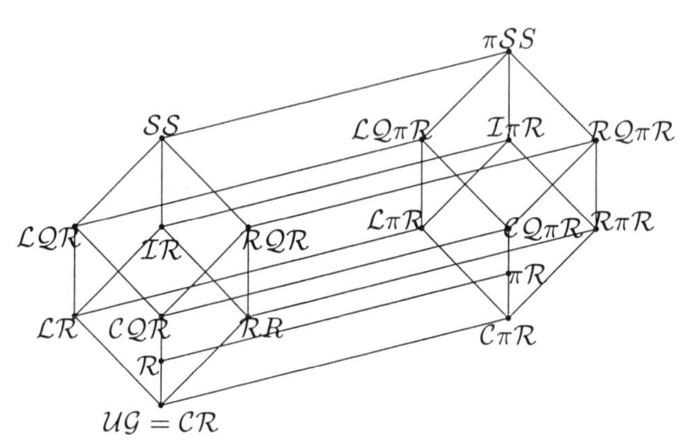

Table 3. Simple semigroups

notation	class of semigroups	characterization
$\mathcal{S}$	simple	$a \in SbS$
$\mathcal{LS}$	left simple	$a \in Sb$
$\mathcal{RS}$	right simple	$a \in bS$
$\mathcal{G}$	group	$a \in bS \cap Sb$
$\mathcal{LSS}$	left strongly simple	$a \in SbSa$
$\mathcal{RSS}$	right strongly simple	$a \in aSbS$
$\mathcal{SSS}$	strongly simple	$a \in aSbS \cap SbSa$
$\mathcal{LCS}$	left completely simple	$a \in Sba$
$\mathcal{RCS}$	right completely simple	$a \in abS$
$\mathcal{SR}$	simple and regular	$a \in aSbSa$
$\mathcal{CS}$	completely simple	$a \in abSa$

Table 4. Archimedean semigroups

notation	class of semigroups	characterization
$\mathcal{A}$	archimedean	$a^n \in SbS$
$\mathcal{LA}$	left archimedean	$a^n \in Sb$
$\mathcal{RA}$	right archimedean	$a^n \in bS$
$\mathcal{TA}$	t-archimedean	$a^n \in bS \cap Sb$
$\mathcal{NES}$	nil-extension of simple	$a^n \in Sb^nS$
$\mathcal{NELS}$	nil-extension of left simple	$a^n \in Sb^n$
$\mathcal{NERS}$	nil-extension of right simple	$a^n \in b^nS$
$\mathcal{NEG}$	nil-extension of group	$a^n \in b^nSb^n$
$\mathcal{LSA}$	left strongly archimedean	$a^n \in SbSa^n$
$\mathcal{RSA}$	right strongly archimedean	$a^n \in a^nSbS$
$\mathcal{SA}$	strongly archimedean	$a^n \in a^nSbS \cap SbSa^n$
$\mathcal{LCA}$	left completely archimedean	$a^n \in Sba^n$
$\mathcal{RCA}$	right completely archimedean	$a^n \in a^nbS$
$\pi\mathcal{A}$	π-regular and archimedean	$a^n \in a^nSbSa^n$
$\mathcal{CA}$	completely archimedean	$a^n \in a^nbSa^n$

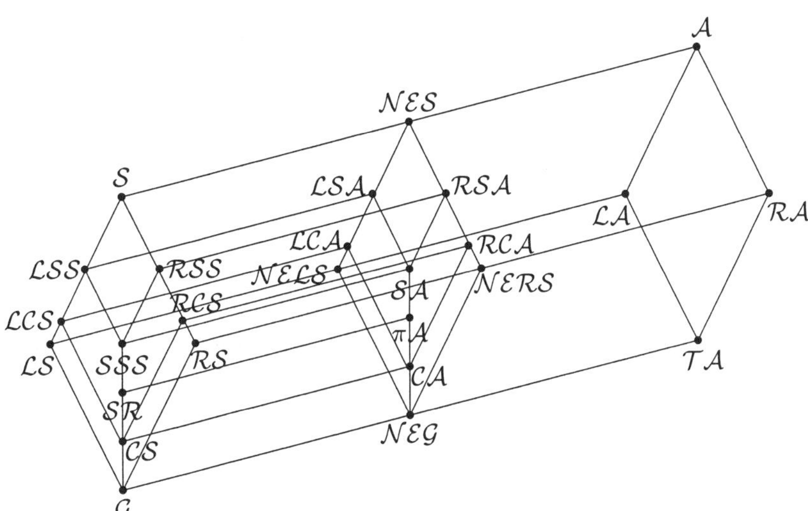

According to the pictures we have to emphasize here that the ascending lines go from a class to a larger class, or equivalently, from a condition to a weaker condition. Thus, we have an ordered set, which might look like a lattice, but, it is not so: for example, $\mathcal{A}$ is not the union of $\mathcal{LA}$ and $\mathcal{RA}$, or $\mathcal{G}$ is not the intersection of $\mathcal{LS}$ and $\mathcal{RSS}$. (This does not in contradict to the fact that sometimes a class, which looks like the intersection of two larger classes on the diagram, is indeed equal to their intersection.)

All the classes in the first diagram (which encompasses regularity and π-regularity conditions) are closed under forming semilattices. In fact, we have:

Proposition 0.1. *Let $\mathcal{K}$ be one of the class of semigroups given in Table 1 or Table 2. Let S be a semilattice Y semigroups S_α, $\alpha \in Y$. Then S is a semigroup from class $\mathcal{K}$ if and only if S_α is in class $\mathcal{K}$, for every $\alpha \in Y$.*

The classes from the second diagram (which encompasses simplicity and archimedeaness) are not closed under forming semilattices (see Theorem 1.2 below). We want to establisg relations between semilattice closures of the classes in the second diagram and the classes in the first diagram.

For undefined notions and notations we refer to [7], [26], [43], [45] or [62].

1. Preliminaries

GENERALIZED REGULAR SEMIGROUPS

As a generalization of the concept of an idempotent element, J. von Neumann [1936] introduced the notion of regularity for rings. The class of regular semigroups and its subclasses are the main subject of many books. The authors of some of them are: A. H. Clifford and G. B. Preston [26], M. Petrich [62,63,64], J. M. Howie [45], P. A. Grillet [41], and some others.

An element a of a semigroup S is *regular* if $a \in aSa$. The set of all regular elements of S is denoted by $Reg(S)$ and callcd *thc rcgular part of a semigroup S*. A semigroup S is *regular* if $S = Reg(S)$. Concept of π-regularity, in its various forms appeared first in ring theory, in paper [53] of McCoy from 1939 (see also [46]). In semigroup theory, this concept attracted great attention both as a generalization of regularity and a generalization of finiteness and periodicity, and was studied under different names, for example, as: quasi-regularity in the papers of M. Putcha, J. L. Galbiati and M. L. Veronesi; power-regularity initially, and then π-regularity in the papers of S. Bogdanović and S. Milić; eventual regularity in the papers of D. Easdown, R. Edwards and P. Higgins. (See, for example, [4,22,33,34,35,37,43,67].)

An element a of a semigroup S is *π-regular* if there exists $n \in \mathrm{N}$ such that $a^n \in a^nSa^n$, i.e. if some power of a is in $Reg(S)$. A semigroup S is *π-regular* if all of its elements are π-regular. The description of π-regular semigroups given below is from [4].

Theorem 1.1. *The following conditons on a semigroup S are equivalent:*

(*i*) *S is π-regular;*
(*ii*) *for every $a \in S$ there exists $n \in \mathrm{N}$ and $e \in E(S)$ such that $R(a^n) = eS$ $(L(a^n) = Se)$;*
(*iii*) *for every $a \in S$ there exists $n \in \mathrm{N}$ such that $R(a^n)$ $(L(a^n))$ has a left (right) identity.*

An element a of a semigroup S is *intra-regular* if $a \in Sa^2S$. The set of all intra-regular elements of a semigroup S is denoted by $Intra(S)$. A semigroup S is *intra-regular* if $S = Intra(S)$. Intra-regular semigroups were introduced by R. Croisot, [27]. The result which follows is from that paper.

Theorem 1.2. *A semigroup S is intra-regular if and only if it is a semilattice of simple semigroups.*

Semilattices of simple semigroups are also described in [1,26,61,62]. Chains of simple semigroups were considered in [61] in the following way:

Theorem 1.3. *A semigroup S is a chain of simple semigroups if and only if $a \in SabS$ or $b \in SabS$, for all $a, b \in S$.*

As a generalization of the previous concept, the notion of an intra-π-regular element was introduced in [67] under the name quasi-intra-regular element. An element a of a semigroup S is *intra-π-regular* if there exists $n \in \mathrm{N}$ such that $a^n \in Sa^{2n}S$, i.e. if $a^n \in Intra(S)$. A semigroup S is *intra-π-regular* if each of its elements is intra-π-regular.

Various concepts of regularity were investigated by R. Croisot in [27], and his study is presented in A. H. Clifford's and G. B. Preston's book, [26], as Croisot's theory. A very important concept in that theory is the notion of left regularity. An element a of a semigroup S is *left* (*right*) *regular* if $a \in Sa^2$ $(a \in a^2S)$. The set of all left (right) regular elements of a semigroup S is denoted by $LReg(S)$ $(RReg(S))$. A semigroup S is *left* (*right*) *regular* if $S = LReg(S)$ $(S = RReg(S))$.

An interesting generalization of the concept of left regularity appears in [2]. So, an element a of a semigroup S is *left* (*right*) *π-regular* if there exists $n \in \mathrm{N}$ such that $a^n \in Sa^{n+1}$ $(a^n \in a^{n+1}S)$. A semigroup S is *left* (*right*) *π-regular* if all its elements are left (right) π-regular. The first description of left π-regular semigroups was given by S. Bogdanović and M. Ćirić (see [16]).

Theorem 1.4. *A semigroup S is left π-regular if and only if it is intra-π-regular and $LReg(S) = Intra(S)$.*

Among various characterizations of left regular semigroups given in [26], there is one about decompositions of these semigroups into a union of left simple semigroups. For decompositions of these semigroups into left simple components in some special cases see [4,61,62,65,66]. Left regular semigroups were described by their left, right and bi-ideals and connections between intersections and products of these ideals in [48,49,50]. As a generalization of completely simple semigroups, we have the following notion: a semigroup S is *left (right) completely simple* if it is simple and left (right) regular (see [16]). Further, using that notion, left regular semigroups were described as semilattices of left completely simple semigroups. In the theorem which follows, equivalences (i)⇔(ii)⇔(iii)⇔(iv) are from [16], while (iv)⇔(v)⇔(vi) is from [4] (for some related results one can, also, see [26 - Theorem 6.36]).

Theorem 1.5. *The following conditions on a semigroup S are equivalent:*

(*i*) *S is left completely simple;*
(*ii*) *S is simple and left π-regular;*
(*iii*) *S is a matrix (right zero band) of left simple semigroups;*
(*iv*) $(\forall a, b \in S)\ a \in Sba$*;*
(*v*) *every principal ideal of S is a left simple subsemigroup of S.*

We now present the result mentioned above about left regular semigroups (see [16]):

Theorem 1.6. *The following conditions on a semigroup S are equivalent:*

(*i*) *S is left regular;*
(*ii*) *S is intra-regular and left π-regular;*
(*iii*) *S is a semilattice of left completely simple semigroups;*
(*iv*) $(\forall a, b \in S)\ a|b \ \Rightarrow\ ab \,\underset{l}{|}\, b$.

An element a of a semigroup S is a *completely regular* or *group element* if there exists $x \in S$ such that $a = axa$ and $ax = xa$. The set of all completely regular elements of a semigroup S is denoted by $Gr(S)$ and is called the *group part of a semigroup S*. Now we have a well known result:

Lemma 1.1. *The following conditions on an element a of a semigroup S are equivalent:*

(*i*) $a \in Gr(S)$;
(*ii*) $a \in LReg(S) \cap RReg(S)$;
(*iii*) *a is contained in a subgroup of* S.

A semigroup S is *completely regular* if $S = Gr(S)$. Part (iii) from the previous lemma explains why completely regular semigroups are often called *unions of groups.*

Now we have the following definition: an element a of a semigroup S is *completely* π*-regular* if there exist $x \in S$ and $n \in \mathrm{N}$ such that $a^n = a^n x a^n$ and $a^n x = x a^n$, i.e. if some power of an element a belongs to $Gr(S)$. A semigroup S is *completely* π*-regular* if each of its elements is completely π-regular.

The notion of a pseudoinverse of an element, which is essential for completely π-regular semigroups, is considered by Drazin in [31]. Considerations of completely π-regular semigroups, i.e. the first investigations of such semigroups, were conducted by W. D. Munn in [60] where they were called *pseudo-invertible* semigroups. Since that time, with completely π-regular semigroups as the main object of interest numerous papers have appeared. Various terms have been used for completely π-regular semigroups in them. Namely, we have: quasi-completely regular semigroups, as well as semigroups in which power of each element lies in a subgroup, in Putsha's papers, [67,69], completely quasi-regular semigroups in the papers of J. L. Galbiati, M. L. Veronesi, [39], groupbound in the papers of B. L. Madison, T. K. Mukherjee and M. K. Sen, [51,52] and P. M. Higgins, [43], quasi periodic at first and then epigroups in the papers of L. N. Shevrin, [75,76], power completely regular in the paper of S. Bogdanović, [4].

Theorem 1.7. *The following conditions on a semigroup* S *are equivalent:*

(*i*) S *is completely* π*-regular;*
(*ii*) *some power of every element of* S *lies in a subgroup of* S*;*
(*iii*) $(\forall a \in S)(\exists n \in \mathrm{N})\ a^n \in a^n S a^{n+1}$ $(a^n \in a^{n+1} S a^n)$;
(*iv*) S *is* π*-regular and left* π*-regular;*
(*v*) S *is left and right* π*-regular;*
(*vi*) *every principal bi-ideal of* S *is* π*-regular;*
(*vii*) *every left (right) ideal of* S *is* π*-regular.*

The idempotent-generated subsemigroup $\langle E(S)\rangle$ of S often belongs to the same class of semigroups as S does.

Lemma 1.2. [32] *If a semigroup* S *is (completely)* π*-regular then* $\langle E(S)\rangle$

is (completely) π*-regular too.*

Theorem 1.8. *The following conditions on a semigroup* S *are equivalent:*

(*i*) S *completely simple;*
(*ii*) S *is simple and completely* π*-regular;*
(*iii*) S *is simple and completely regular;*
(*iv*) S *is a matrix of groups;*
(*v*) $(\forall a, b \in S)\ a \in abSa$.

The equivalence (i)$\Leftrightarrow$(ii) from the theorem above is known as Munn's theorem, (i)$\Leftrightarrow$(iii) is an immediate consequence of Munn's theorem, and (i)$\Leftrightarrow$(iv) is a direct consequence of the definition of Rees matrix semigroup. It is easy to verify that a completely simple semigroup is both left and right completely simple.

A semigroup S isomorphic to the direct product of the group and the left (right) zero band is called *left (right) group.*

Theorem 1.9. *The following conditions on a semigroup* S *are equivalent:*

(*i*) S *is a left group;*
(*ii*) S *is a left zero band of groups;*
(*iii*) $(\forall a, b \in S)\ a \in aSb$;
(*iv*) S *is regular and* $E(S)$ *is a left zero band;*
(*v*) S *is left simple and contains an idempotent.*

Completely regular semigroups or unions of groups were introduced by A. H. Clifford in 1941 (see [25]). The equivalence (i)$\Leftrightarrow$(ii), from the theorem given below, is one of his results.

Theorem 1.10. *The following conditions on a semigroup* S *are equivalent:*

(*i*) S *is completely regular;*
(*ii*) S *is a semilattice of completely simple semigroups;*
(*iii*) $(\forall a \in S)\ a \in aSa^2 \quad (a \in a^2Sa)$;
(*iv*) S *is left and right regular;*
(*v*) S *is regular and left (right) regular;*
(*vi*) *every left ideal of* S *is a regular semigroup.*

Description of chains of completely simple semigroups are from [61].

Theorem 1.11. *A semigroup* S *is a chain of completely simple semigroups if and only if* $a \in abSa$ *or* $b \in baSb$, *for all* $a, b \in S$.

As a consequence of Theorem 1.10., part (i)$\Leftrightarrow$(ii), we have the following result (see [25]).

Theorem 1.12. *A semigroup S is a band if and only if it is a semilattice of matrices (rectangular bands).*

ARCHIMEDEAN SEMIGROUPS

Studying the semilattice decompositions of commutative semigroups T. Tamura and N. Kimura in 1954 (see [84]), and, independently of them, G. Thierrin, [87], in the same year, introduced the notion of archimedean semigroups, i.e. semigroups in which for any two elements one of them divides some power of the other.

A semigroup S is *archimedean* if $a \longrightarrow b$, for all $a, b \in S$. A semigroup S is *left* (*right*) *archimedean* if $a \stackrel{l}{\longrightarrow} b$ ($a \stackrel{r}{\longrightarrow} b$), for all $a, b \in S$. Further, S is *t-archimedean* if it is both left and right archimedean, i.e. if $a \stackrel{t}{\longrightarrow} b$, for all $a, b \in S$. It is clear that an archimedean (left archimedean, right archimedean, t-archimedean) semigroup is a generalization of a simple (left simple, right simple, group) semigroup.

Now, a result which is a consequence of Tamura's characterization of semilattice-indecomposable semigroups (see [82]) will be given.

Proposition 1.1. *An archimedean semigroup is a semilattice-indecomposable.*

The previous proposition gives the reason why the study of archimedean semigroups is important. Archimedean semigroups are the most investigated class of semilattice-indecomposable semigroups. The concept of archimedeaness of a semigroup is in close connection with the notion of *nil-extension* of a semigroup, i.e. with an ideal extension of a semigroup by a nil-semigroup. Namely, the answer of the question: what can we say about archimedean semigroups which belong to the certain classes of semigroups from 1.1. leads to the notion of nil-extensions. The widest such known classes of semigroups are archimedean and intra-π-regular semigroups, i.e. archimedean semigroups with kernels. In the theorem given below, equivalences (i)$\Leftrightarrow$(iii)$\Leftrightarrow$(iv) are from [67], and (i)$\Leftrightarrow$(ii) is from [7].

Theorem 1.13. *The following conditions on a semigroup S are equivalent:*

(*i*) *S is a nil-extension of a simple semigroup;*
(*ii*) $(\forall a, b \in S)(\exists n \in \mathrm{N})\ a^n \in Sb^{2n}S$;
(*iii*) *S is archimedean and intra-π-regular;*

(*iv*) S *is archimedean and* $Intra(S) \neq \emptyset$.

An archimedean semigroup, in general, need not contain an idempotent. But, if an archimedean semigroup does contain an idempotent, then, by the previous theorem, it is a nil-extension of a simple semigroup too.

Theorem 1.14. *A semigroup S is archimedean and contains an idempotent if and only if S is a nil-extension of simple semigroup with an idempotent.*

A semigroup S is *left (right) completely archimedean* if it is archimedean and left (right) π-regular. This class of semigroups was introduced in [17]. The result which follows is also from that paper.

Theorem 1.15. *The following conditions on a semigroup S are equivalent:*

(*i*) S *is left completely archimedean;*
(*ii*) S *is a nil-extension of a left completely simple semigroup;*
(*iii*) $(\forall a, b \in S)(\exists n \in \mathrm{N})\ a^n \in Sba^n$.

The class of left archimedean semigroups which are intra-π-regular coincides with the class of left archimedean semigroups which are left π-regular (see [7]).

Theorem 1.16. *The following conditions on a semigroup S are equivalent:*

(*i*) S *is a nil-extension of a left simple semigroup;*
(*ii*) $(\forall a, b \in S)(\exists n \in \mathrm{N})\ a^n \in Sb^{n+1}$;
(*iii*) S *is left archimedean and left* π*-regular.*

A semigroup S is *completely archimedean* if it is archimedean and contains a primitive idempotent. In the theorem given below the equivalence (i)$\Leftrightarrow$(ii) is from [24], the part (ii)$\Leftrightarrow$(iii)$\Leftrightarrow$(iv) is from [22], and (i)$\Leftrightarrow$(v) is from [40].

Theorem 1.17. *The following conditions on a semigroup S are equivalent:*

(*i*) S *is completely archimedean;*
(*ii*) S *is a nil-extension of a completely simple semigroup;*
(*iii*) S *is* π*-regular and all its idempotents are primitive;*
(*iv*) $(\forall a, b \in S)(\exists n \in \mathrm{N})\ a^n \in a^nSba^n\ (a^n \in a^nbSa^n)$;
(*v*) S *is archimedean and completely* π*-regular.*

It is easy to verify that completely archimedean semigroups are both left and right completely archimedean.

What follows now is a description of maximal subgroups of completely archimedean semigroups taken from [22].

Lemma 1.3. *Let S be a completely archimedean semigroup. Then, maximal subgroups of S are given by $G_e = eSe$, $e \in E(S)$.*

According to Theorem 1.14., an archimedean semigroup with an idempotent need not be completely π-regular, but for a left archimedean semigroup (see [22]), we have the following:

Theorem 1.18. *The following conditions on a semigroup S are equivalent:*

(*i*) *S is left archimedean with an idempotent;*
(*ii*) *S is a nil-extension of the left group;*
(*iii*) *S is π-regular and $E(S)$ is a left zero band;*
(*iv*) $(\forall a, b \in S)(\exists n \in \mathrm{N})\ a^n \in a^n S a^n b \quad (a^n \in b a^n S a^n)$.

A semigroup S is a *π-group* if it is π-regular and contains exactly one idempotent, or equivalently, by T. Tamura's result, if it is archimedean with exactly one idempotent. The remaining characterizations from the theorem given below are from [10,22].

Theorem 1.19. *The following conditions on a semigroup S are equivalent:*

(*i*) *S is a π-group;*
(*ii*) *S is a nil-extension of a group;*
(*iii*) $(\forall a, b \in S)(\exists n \in \mathrm{N})\ a^n \in b^n S b^n$.

Semilattices of Archimedean Semigroups

Investigations of semigroups which can be decomposed into a semilattice with archimedean components began in [84], where it was proved that any commutative semigroup is a semilattice of archimedean semigroups (commutativity means that components are t-archimedean). This result was extended to the class of medial semigroups first in [24], and later to the class of exponential semigroups ([86]). The first complete description of a semigroups which are a semilattice of archimedean semigroups was given by M. S. Putcha ([67]). That is why many authors call such semigroups *Putcha's semigroups.* Some other characterizations of these semigroups can be found in [11,21,28,47,62,71,81].

Theorem 1.20. *The following conditions on a semigroup S are equivalent:*

(*i*) *S is a semilattice of archimedean semigroups;*
(*ii*) *S is a band of archimedean semigroups;*
(*iii*) $(\forall a, b \in S)\ a|b \ \Rightarrow\ a^2 \longrightarrow b;$
(*iv*) $(\forall a, b \in S)\ a^2 \longrightarrow ab;$
(*v*) $(\forall a, b \in S)(\forall n \in \mathrm{N})\ a^n \longrightarrow ab;$
(*vi*) *in every homomorphic image of S containing zero the set of all nilpotent elements is an ideal;*
(*vii*) $(\forall a, b, c \in S)\ a \longrightarrow b \ \&\ b \longrightarrow c \ \Rightarrow \quad a \longrightarrow c;$
(*viii*) $(\forall a, b, c \in S)\ a \longrightarrow c \ \&\ b \longrightarrow c \ \Rightarrow \quad ab \longrightarrow c;$
(*ix*) *every bi-ideal of S is a semilattice of archimedean semigroups;*
(*x*) *every one-sided ideal of S is a semilattice of archimedean semigroups.*

In the previous theorem, equivalences (i)⇔(ii)⇔(iii) are from [67], (ii)⇔(iv)⇔(v) are from [28], (i)⇔(vi) is from [11], (vii) and (viii) are from [81].

Semilattices of left archimedean and t-archimedean semigroups have been studied in many papers. In the theorem given below, the equivalence (i)⇔(ii) is from [72], and (i)⇔(iii) is from [5] (for the results concerning these semigroups see also [62,67,68]).

Theorem 1.21. *The following conditions on a semigroup S are equivalent:*

(*i*) *S is a semilattice of left archimedean semigroups;*
(*ii*) $(\forall a, b \in S)\ a|b \ \Rightarrow\ a \overset{l}{\longrightarrow} b;$
(*iii*) $(\forall a, b \in S)\ a \overset{l}{\longrightarrow} b.$

Chains of archimedean semigroups were introduced in [3]. For some results concerning that type of semigroups see also [14].

Theorem 1.22. *A semigroup S is a chain of archimedean semigroups if and only if* $ab \longrightarrow a$ *or* $ab \longrightarrow b$*, for all* $a, b \in S$.

The rest of this section is devoted to semilattices of archimedean semigroups which are intra-π-regular, left π-regular or completely π-regular. Semilattices of archimedean semigroups which are intra-π-regular were studied for the first time in [67]. In the theorem which follows, the equivalence (i)⇔(ii) is from that paper, (i)⇔(iii)⇔(iv) is from [17], and (i)⇔(v) is from [18,21].

Theorem 1.23. *The following conditions on a semigroup S are equivalent:*

(*i*) *S is a semilattice of nil-extensions of simple semigroups;*
(*ii*) *S is a semilattice of archimedean semigroups and intra-π-regular;*
(*iii*) $(\forall a, b \in S)(\exists n \in \mathrm{N})\ (ab)^n \in S(ba)^n(ab)^n S$;
(*iv*) *S is an intra-π-regular semigroup and every $\mathcal{J}$-class of S which contains an intra-regular element is a subsemigroup;*
(*v*) $\tau(\mathcal{J})$ *is a semillatice (band) congruence on S.*

The semilattices of archimedean semigroups which are left π-regular were introduced and described in [17]. Here is the main result from that paper.

Theorem 1.24. *The following conditions on a semigroup S are equivalent:*

(*i*) *S is a semilattice of left completely archimedean semigroups;*
(*ii*) *S is a semilattice of nil-extensions of left completely simple semigroups;*
(*iii*) *S is a semilattice of archimedean semigroups and left π-regular;*
(*iv*) $(\forall a, b \in S)(\exists n \in \mathrm{N})\ (ab)^n \in Sa(ab)^n$;
(*v*) *S is a left π-regular semigroup and every $\mathcal{L}(\mathcal{J})$-class of S which contains a left regular element is a subsemigroup.*

The semigroups described in the previous theorem can be treated as a generalization of the semilattices of nil-extensions of left simple semigroups (introduced in [67], and also characterized in [7,13]).

Theorem 1.25. *The following conditions on a semigroup S are equivalent:*

(*i*) *S is a semilattice of nil-extensions of left simple semigroups;*
(*ii*) *S is a semilattice of left archimedean semigroups and left π-regular;*
(*iii*) $(\forall a, b \in S)(\exists n \in \mathrm{N})\ (ab)^n \in Sa^{2n+1}$.

A semigroup S is *uniformly* π-regular if it is a semilattice of completely archimedean semigroups. Studies of decompositions of semigroups into semilattices of completely archimedean semigroups began in L. N. Shevrin's papers. The final results of his several year long investigations are given in [75]. Similar results concerning decompositions of completely π-regular semigroups into semilattices of archimedean semigroups were obtained by J. L. Galbiati and M. L. Veronesi in [38], where they started such investigations, and by M. L. Veronesi in paper [88], where it ended. The main result is

Theorem 1.26. **[Shevrin-Veronesi]** *A semigroup S is uniformly π-regular (a semilattice of completely archimedean semigroups) if and only*

if S is π-regular and $Reg(S) = Gr(S)$.

Some other characterizations of semilattices of completely archimedean semigroups, in general and some special cases, can be found in papers of S. Bogdanović, and S. Bogdanović and M. Ćirić (see, for example, [7,13,17]).

Theorem 1.27. *The following conditions on a semigroup S are equivalent:*

(*i*) *S is a uniformly π-regular semigroup;*
(*ii*) *S is π-regular and $Reg(S) \subseteq LReg(S)$;*
(*iii*) *S is a semilattice of archimedean semigroups and completely π-regular;*
(*iv*) $(\forall a, b \in S)(\exists n \in \mathrm{N})\ (ab)^n \in (ab)^n bS(ab)^n$;
(*v*) *S is completely π-regular and*
$(\forall a \in S)(\forall e \in E(S))\ a|e \Rightarrow a^2|e$;
(*vi*) *S is completely π-regular and*
$(\forall a, b \in S)(\forall e \in E(S))\ a|e$ *and* $b|e \Rightarrow ab|e$;
(*vii*) *S is completely π-regular and*
$(\forall e, f, g \in E(S))\ e|g$ *and* $f|g \Rightarrow ef|g$;
(*viii*) *S is a π-regular semigroup and every $\mathcal{L}(\mathcal{J})$ of S which contains an idempotent is a subsemigroup of S;*
(*ix*) *S is a semilattice of left completely archimedean semigroups and π-regular.*

In the previous theorem, equivalences (i)⇔(iii)⇔(v)⇔(viii) are from [67], (i)⇔(iv) is from [8], (i)⇔(vi)⇔(vii) are from [75], and (i)⇔(ii)⇔(ix) are from [17].

Chains of completely archimedean semigroups were introduced and described in [5]. In the theorem given below, the equivalence (i)⇔(ii) is from [5], and (i)⇔(iii) is from [15].

Theorem 1.28. *The following conditions on a semigroup S are equivalent:*

(*i*) *S is a chain of completely archimedean semigroups;*
(*ii*) *S is completely π-regular and $e \in efS$ or $f \in feS$, for all $e, f \in E(S)$;*
(*iii*) *S is π-regular and $Reg(S)$ is a chain of completely simple semigroups.*

The rest of this section is devoted to two very important subclasses of the class of semigroups considered in Theorem 1.27. Semilattices of nil-extensions of left groups were introduced in [67]. Descriptions of these semigroups given here are from [5,6,8,76].

Theorem 1.29. *The following conditions on a semigroup S are equivalent:*

(*i*) *S is a semilattice of nil-extensions of left groups;*
(*ii*) *S is a semilattice of left archimedean semigroups and π-regular;*
(*iii*) $(\forall a, b \in S)(\exists n \in \mathrm{N})\ (ab)^n \in (ab)^n S(ba)^n$;
(*iv*) $(\forall a, b \in S)(\exists n \in \mathrm{N})\ (ab)^n Sa^{2n}$;
(*v*) *S is uniformly π-regular and left π-inverse;*
(*vi*) *S is uniformly π-regular and for all $e, f \in E(S)$ there exists $n \in \mathrm{N}$ such that $(ef)^n = (efe)^n$;*
(*vii*) *S is π-regular and $a = axa$ implies $ax = xa^2x$;*
(*viii*) *S is a completely π-regular semigroup and every $\mathcal{J}$-class which contains an idempotent is a left group.*

Some other characterizations of the semigroups from Theorem 1.28. can be found in [9,15].

The semilattices of π-groups, i.e. the semilattices of nil-extensions of groups were introduced in [67]. The results concerning these semigroups can also be found in [5,8,36,39,73,76,88].

Theorem 1.30. *The following conditions on a semigroup S are equivalent:*

(*i*) *S is a semilattice of nil-extensions of groups;*
(*ii*) *S is uniformly π-regular and π-inverse;*
(*iii*) *S is uniformly π-regular and for all $e, f \in E(S)$ there exists $n \in \mathrm{N}$ such that $(ef)^n = (fe)^n$;*
(*iv*) *S is π-regular and $a = axa$ implies $ax = xa$;*
(*v*) *S is a semilattice of t-archimedean semigroups and π-regular;*
(*vi*) $(\forall a, b \in S)(\exists n \in \mathrm{N})\ (ab)^n \in b^{2n} Sa^{2n}$;
(*vii*) *S is a π-regular semigroup and every $\mathcal{J}$-class which contains an idempotent is a group.*

In the previous theorem, the equivalences (i)⇔(ii)⇔(iii) is from [88], (i)⇔(iv) is from [5], (i)⇔(vi) is from [8], and (i)⇔(vii) is from [76].

2. Semilattices of Nil-extensions of Simple Semigroups

Studying minimal conditions on semigroups, E. Hotzel, [44], introduced weakly periodic, or as we call them, π-semisimple semigroups. It is easy to see that π-semisimple semigroups can be treated as a generalization of semisimple ones. According to the decomposability of π-semisimple semigroups into semilattices of archimedean semigroups, it will be seen that such semigroups coincide with intra-π-regular semigroups which are semilattices of archimedean semigroups. Apart from that result we will give some other criteria in the term of equality of some subsets of such semigroups.

An element a of a semigroup S is *semisimple* if $a \in SaSaS$. A set of all semisimple elements of S will be denoted by $Semis(S)$. A semigroup S is *semisimple* if $S = Semis(S)$. According to the results from [58,74,77], a semigroup S is semisimple if none of its principal factors are null, or, equivalently, if $I = I^2$, for every ideal I of S. The importance of the previous concept can also be seen from [59]. A semigroup S is *π-semisimple*, if every $a \in S$ has a power a^n, $n \in \mathrm{N}$, satisfying $a^n \in Semis(S)$. The next result is obvious:

Lemma 2.1. *A semigroup S is π-semisimple if and only if for every $a \in S$ there exists $n \in \mathrm{N}$ such that $J^2(a^n) = J(a^n)$.*

Now, we will give connections between the existence of semisimple and intra-regular elements.

Lemma 2.2. *For a semigroup S we have $Semis(S) \neq \emptyset$ if and only if $Intra(S) \neq \emptyset$.*

A characterization of semisimple semigroups which follows is to a great extent a consequence of the previous lemma.

Lemma 2.3. *The following conditions on a semigroup S are equivalent:*

(i) *S is semisimple;*
(ii) $(\forall a \in S)\ J^2(a) = J(a)$;
(iii) *$J(a)$ is generated by an intra-regular element for every $a \in S$.*

Now, we will consider archimedean semigroups which are π-semisimple. In fact, we will show that these semigroups coincide with already known nil-extensions of simple semigroups.

Lemma 2.4. *The following conditions on a semigroup S are equivalent:*

(i) *S is a nil-extension of a simple semigroup;*
(ii) *S is archimedean and intra-π-regular;*
(iii) *S is archimedean and is π-semisimple;*
(iv) *S is archimedean and there exists $b \in S$ such that $b \in Sb^kS$, for every $k \in \mathrm{N}$;*

(v) $(\forall a, b \in S)(\exists n \in \mathrm{N})\ a^n \in Sb^nS$.

Let us introduce *Green's subsets* of a semigroup S. Let $\mathcal{T}$ be one of well-known Green's relations on a semigroup S. Then by $\mathcal{U}_{\mathcal{T}}(S)$ we will denote the union of all $\mathcal{T}$-classes of S which are subsemigroups of S, by $L^S_{\mathcal{T}}(S)$ the union of all $\mathcal{T}$-classes of S which are left simple subsemigroups of S, by $L_{\mathcal{T}}(S)$ the union of all $\mathcal{T}$-classes of S which are left groups and by $H_{\mathcal{T}}(S)$ the union of all $\mathcal{T}$-classes of S which are groups.

Semigroups which are semilattices of nil-extensions of simple semigroups, or, intra-π-regular semigroups which are decomposed into archimedean components, admit a great number of different characterizations (see, Theorem 1.23). It is an easy-to-prove consquence of Lemma 2.4 and Proposition 0.1 that these semigroups coincide with semilattices of archimedean semigroups which are π-semisimple. The main result of this section gives us a connections between decomposability of π-semisimple (intra-π-regular) semigroups into archimedean components and certain equalities of (generalized) regular and Green's subsets of such a semigroup.

Theorem 2.1. *The following conditions on a semigroup S are equivalent:*

(i) *S is a semilattice of nil-extensions of simple semigroups;*
(ii) *S is intra π-regular and $Intra(S) = \mathcal{U}_{\mathcal{J}}(S)$;*
(iii) *S is π-semisimple and $Semis(S) = \mathcal{U}_{\mathcal{J}}(S)$.*

Semigroups with a non-empty set of idempotents which are semilattices of semigroups with kernels were studied in [67]. Here, the description of these semigroups is given with the help of Tamura's *cm*-property (see [81]) applied to idempotents.

Theorem 2.2. *Let $E(S) \neq \emptyset$. Then the following conditions on a semigroup S are equivalent:*

(i) $(\forall a \in S)(\forall e \in E(S))\ a|e \Rightarrow a^2|e$;
(ii) $(\forall a, b \in S)(\forall e \in S)\ a|e\ \&\ b|e \Rightarrow ab|e$;
(iii) $(\forall e, f, g \in E(S))\ e|g\ \&\ f|g \Rightarrow ef|g$;
(iv) *S is a semilattice Y of semigroups S_α, $\alpha \in Y$, where S_α has a kernel K_α such that $E(S_\alpha) \subseteq K_\alpha$ or $E(S_\alpha) = \emptyset$, for an arbitary $\alpha \in Y$.*

3. Semilattices of Left Strongly Archimedean Semigroups

Within the class of π-semisimple semigroups, mentioned in the previous section, two interesting subclasses will be taken into consideration. First, we have left quasi-regular semigroups introduced by J. Calais, [23]. Second, as a generalization of these semigroups, we introduce and describe left quasi-π-regular semigroups. In order to describe left quasi-regular and intra-regular semigroups we will introduce the notion of left strongly simple semigroups, we will give various characterizations of these semigroups, and then describe intra-regular and left quasi-regular semigroups as semilattices of left strongly simple semigroups. Analogously, we will introduce the notion of left strongly archimedean semigroup, and, finaly, we will see that left strongly archimedean semigroups are precisely archimedean components in semilattice decompositions of left quasi-π-regular semigroups.

An element a of a semigroup S is a *left (right, completely) quasi-regular* if $a \in SaSa$ ($a \in aSaS$, $a \in aSaS \cap SaSa$), for all $a \in S$. A set of all left (right, completely) quasi-regular elements of S will be denoted by $LQReg(S)$ ($RQReg(S)$, $QGr(S)$). A semigroup S is *left (right, completely) quasi-regular* if $LQReg(S) = S$ ($RQReg(S) = S$, $QGr(S) = S$). Here, as the generalization of the previous concept we have the following notion: a semigroup S is *left (right) quasi-π-regular* if for every $a \in S$ some power of a belongs to $LQReg(S)$ ($RQReg(S)$). A semigroup S is *completely quasi-π-regular* if it is both left and right quasi-π-regular. The structure of left quasi-π-regular semigroups is given in the next theorem.

Theorem 3.1. *The following conditions on a semigroup S are equivalent:*

(i) *S is left quasi-π-regular;*
(ii) $(\forall a \in S)(\exists n \in \mathrm{N})\ L^2(a^n) = L(a^n)$;
(iii) *S is π-semisimple and $Semis(S) = LQReg(S)$.*

Now, connections between the existence of left quasi-regular and left regular elements will be given.

Lemma 3.1. *For a semigroup S we have that $LReg(S) \neq \emptyset$ if and only if $LQReg(S) \neq \emptyset$.*

Using the previous lemma and Theorem 3.1., descriptions of left quasi-regular semigroups can be obtained.

Lemma 3.2. *The following conditions on a semigroup S are equivalent:*

(i) *S is a left quasi-regular semigroups;*
(ii) $(\forall a \in S)\ L^2(a) = L(a)$;
(iii) *$L(a)$ is generated by left regular element, for every $a \in S$.*

As the consequence of the previous lemma and its dual we can describe completely quasi-regular semigroups as the ones with idempotent (principal) one-sided ideals. As a consequence of Theorem 3.1. and its dual we will give the result concerned with semigroups which are the generalization of completely quasi-regular ones.

Corollary 3.1. *The following conditions on a semigroup S are equivalent:*

(*i*) *S is completely quasi-π-regular;*
(*ii*) *S is left quasi-π-regular and $Semis(S) = RQReg(S)$;*
(*iii*) *S is π-semisimple and $Semis(S) = LQReg(S) = RQReg(S)$;*
(*iv*) $(\forall a \in S)(\exists n \in \mathrm{N})\ a^n \in Sa^nSa^n \cap a^nSa^nS$.

Next, let us turn our attention to the simple semigroups and their semilattices, i.e., according to Theorem 1.2., intra-regular semigroups which belong to the classes of semigroups which have just been described. First, it is convenient to introduce the following terminology: a semigroup S is a *left (right) strongly simple semigroup* if S is simple and left (right) quasi-regular. A semigroup S is a *strongly simple semigroup* if S is both left and right strongly simple.

Lemma 3.3. *The following conditions on a semigroup S are equivalent:*

(*i*) *S is left strongly simple;*
(*ii*) $(\forall a, b \in S)\ a \in SbSa$;
(*iii*) *every left ideal of S is a simple semigroup;*
(*iv*) *S is simple and every left ideal of S is idempotent;*
(*v*) *S is simple and every left ideal of S is an intra-regular semigroup;*
(*vi*) *S is simple and left quasi-π-regular semigroup.*

The equivalence (i)$\Leftrightarrow$(vi) from Lemma 3.3. can be viewed as generalization of the well known Munn's theorem (see [26, Theorem 2.55]).

From the previous lemma and its dual we can easily deduce that strongly simple semigroups coincide with simple and completely quasi-(π-)regular semigroups. The next characterizations of strongly simple semigroups, given in terms of their ideals and elements, are consequences of the previous lemma, too.

Corollary 3.2. *The following conditions on a semigroup S are equivalent:*

(*i*) *S is strongly simple;*
(*ii*) $(\forall a, b \in S)\ a \in SbSa \cap aSbS$;

(*iii*) *every one-sided ideal of S is a simple semigroup;*
(*iv*) *S is simple and every one-sided ideal of S is idempotent;*
(*v*) *S is simple and every one-sided ideal of S is intra-regular.*

Now, we will describe left quasi-regular and intra-regular semigroups from the perspective of the results given in Lemma 3.3. and Proposition 0.1.

Theorem 3.2. *The following conditions on a semigroup S are equivalent:*

(*i*) *S is a semilattice of left strongly simple semigroups;*
(*ii*) $(\forall a \in S)\ a \in Sa^2Sa$;
(*iii*) *every left ideal of S is an intra-regular subsemigroup of S;*
(*iv*) *every left ideal of S is a semisimple subsemigroup of S.*

We can give a description of a semilattice of strongly simple semigroups, i.e. intra-regular and completely quasi-(π-)regular semigroups, as an easy consequence of Theorem 3.2. and its dual.

Corollary 3.3. *The following conditions on a semigroup S are equivalent:*

(*i*) *S is a semilattice of strongly simple semigroups;*
(*ii*) $(\forall a \in S)\ a \in Sa^2Sa \cap aSa^2S$;
(*iii*) *every one-sided ideal of S is an intra-regular subsemigroup of S;*
(*iv*) *every one sided ideal of S is a semisimple subsemigroup of S.*

Within the class of semigroups considered in Theorem 3.2. there is a subclass of semigroups which is a generalization of those from [61].

Theorem 3.3. *A semigroup S is a chain of left strongly simple semigroups if and only if $a \in SabSa$ or $b \in SabSb$, for all $a, b \in S$.*

At the end of this section we will give results concerning archimedean semigroups which are left quasi π-regular and their semilattices. For that purpose we introduce the following notion: a semigroup S is *left* (*right*) *strongly archimedean* if S is archimedean and left (right) quasi-π-regular. A semigroup S is *strongly archimedean* if S is both left and right strongly archimedean.

Theorem 3.4. *The following conditions on a semigroup S are equivalent:*

(*i*) *S is left strongly archimedean;*
(*ii*) *S is a nil-extension of a left strongly simple semigroup;*
(*iii*) $(\forall a, b \in S)(\exists n \in \mathrm{N})\ a^n \in SbSa^n$;

(iv) $(\forall a, b \in S)(\exists n \in \mathrm{N})\ a^n \in Sb^nSa^n$.

Using Theorem 3.4. we can describe left quasi-π-regular semigroups which are decomposable into archimedean components.

Theorem 3.5. *The following conditions on a semigroup S are equivalent:*

(i) *S is a semilattice of left strongly archimedean semigroups;*
(ii) *S is a semilattice of archimedean semigroups and left quasi-π-regular semigroup;*
(iii) $(\forall a, b \in S)(\exists n \in \mathrm{N})\ (ab)^n \in Sa^2S(ab)^n$;
(iv) *S is left quasi-π-regular and $Semis(S) = \mathcal{U}_{\mathcal{J}}(S)$;*
(v) *S is intra π-regular and $LQReg(S) = Intra(S) = \mathcal{U}_{\mathcal{J}}(S)$;*
(vi) *S is intra π-regular and every $\mathcal{J}$-class of S containing an intra-regular element is a left strongly simple semigroup.*

Because of Theorem 3.4. and its dual, strongly archimedean semigroups are nil-extensions of strongly simple semigroups. That fact as well as the previous theorem and its dual are used for characterizing semilattices of strongly archimedean semigroups.

Corollary 3.4. *The following conditions on a semigroup S are equivalent:*

(i) *S is a semilattice of strongly archimedean semigroups;*
(ii) $(\forall a, b \in S)(\exists n \in \mathrm{N})\ (ab)^n \in Sa^2S(ab)^n \cap (ab)^nSa^2S$;
(iii) *S is completely quasi-π-regular and $Semis(S) = \mathcal{U}_{\mathcal{J}}(S)$;*
(iv) *S is intra π-regular and $QGr(S) = Intra(S) = \mathcal{U}_{\mathcal{J}}(S)$;*
(v) *S is intra-π-regular and every $\mathcal{J}$-class of S containing an intra-regular element is a strongly simple semigroup.*

4. Semilattices of Left Completely Archimedean Semigroups

Left π-regular, left completely simple, left regular, left completely archimedean semigroups and their semilattices are already known types of semigroups (see Section 1). Here, in this section, we consider them from the perspective of the results from Sections 2 and 3. First, some connections of left π-regular semigroups with left quasi-π-regular ones will be given.

Theorem 4.1. *The following conditions on a semigroup S are equivalent:*

(i) *every bi-ideal of S is left quasi-π-regular;*
(ii) *every left ideal of S is left quasi-π-regular;*

(*iii*) S *is left* π*-regular.*

Concerning connections of left completely simple semigroups with the left strongly simple ones from the previous section we have the next:

Lemma 4.1. *A semigroup* S *is left completely simple if and only if every left ideal of* S *is a left strongly simple subsemigroup of* S*.*

Left regular semigroups are described by their left ideals through membership in certain classes of semigroups, i.e. by their connection with left quasi-regular semigroups.

Lemma 4.2. *A semigroup* S *is left regular if and only if every left ideal of* S *is left quasi-regular.*

It is known that left regular semigroups coincide with semilattices of left completely simple semigroups (see Theorem 1.6.). Here, we distinguish chains of left completely simple semigroups.

Theorem 4.2. *The following conditions on a semigroup* S *are equivalent:*

(*i*) S *is a chain of left completely simple semigroups;*
(*ii*) $(\forall a, b \in S)\ a \in Sba$ *or* $b \in Sab^2$*;*
(*iii*) $(\forall a, b \in S)\ a \in SabSa^2$ *or* $b \in SabSb^2$*.*

Archimedean semigroups which are left π-regular, i.e. left completely archimedean semigroups and their connection with left strongly archimedean semigroups are given in the following lemma.

Lemma 4.3. *A semigroup* S *is left completely archimedean if and only if every left ideal of* S *is a left strongly archimedean semigroup.*

Now, at the end of this section, we turn our attention to the left π-regular semigroups decomposable into archimedean components (see Theorem 1.24.). Namely, we have left π-regular semigroups with certain equalities between their (generalized) regular, on the one hand, and Green's subsets on the other one.

Theorem 4.3. *The following conditions on a semigroup* S *are equivalent:*

(*i*) S *is a semilattice of left completely archimedean semigroups;*
(*ii*) S *is left* π*-regular and* $LReg(S) = \mathcal{U}_{\mathcal{L}}(S) = \mathcal{U}_{\mathcal{J}}(S)$*;*
(*iii*) S *is left* π*-regular and* $Semis(S) = \mathcal{U}_{\mathcal{J}}(S)$*;*

(*iv*) *every left ideal of* S *is a semilattice of left strongly archimedean semigroups.*

According to Theorem 1.16, we have that the class of left archimedean semigroups which are intra-π-regular coincides with the class of left archimedean semigroups which are left π-regular, i.e. we have the class of nil-extensions of left simple semigroups. Taking into account the results from Section 2 and Section 3 we end this section with some other characteristisc of nil-extensions of left simple semigroups and their semilattices.

Lemma 4.4. *The following conditions on a semigroup* S *are equivalent*

(*i*) S *is a nil-extension of a left simple semigroup;*
(*ii*) $(\forall a, b \in S)(\exists n \in \mathrm{N})\ a^n \in Sb^n$*;*
(*iii*) S *is left archimedean and left quasi* π*-regular.*

Theorem 4.4. *The following conditions on a semigroup* S *are equivalent:*

(*i*) S *is a semilattice of nil-extensions of left simple semigroups;*
(*ii*) $(\forall a, b \in S)(\exists n \in \mathrm{N})\ (ab)^n \in Sa^{2n}$*;*
(*iii*) S *is* π*-semisimple and* $Semis(S) = L^S_{\mathcal{J}}(S)$*;*
(*iv*) S *is a left quasi-*π*-regular semigroup and* $LQReg(S) = L^S_{\mathcal{J}}(S)$*.*

5. Semilattices of Nil-exstensions of Simple and Regular Semigroups

The concept of regularity as well as its generalization, the concept of π-regularity, in its various forms appeared first in the ring theory. These concepts have awakened enormous attention among the specialists in semigroup theory, as evidenced by a number of monographs and papers (see, for example [25,26,41,42,45,49,62,63] for regular, and [4,7,13,38] for π-regular semigroups). Using results from the previous sections we will give some other results concerning π-regular, regular, regular and simple, regular and intra-regular, semilattices of nil-extensions of simple and regular semigroups.

Lemma 5.1. *The following conditions on a semigroup* S *are equivalent:*

(*i*) S *is* π*-regular;*
(*ii*) S *is left quasi-*π*-regular and* $Reg(S) = LQReg(S)$*;*
(*iii*) S *is* π*-semisimple and* $Reg(S) = Semis(S)$*.*

Lemma 5.2. *The following conditions on a semigroup* S *are equivalent:*

(*i*) *S is left (right) quasi-regular and* π*-regular;*
(*ii*) *S is semisimple and* π*-regular;*
(*iii*) *S is regular.*

Talking about regular and simple semigroups, we will see that π-regularity on simple semigroups goes down to the "ordinary" regularity.

Lemma 5.3. *The following conditions on a semigroup S are equivalent:*

(*i*) *S is regular and simple;*
(*ii*) *S is* π*-regular and simple;*
(*iii*) $(\forall a, b \in S) a \in aSbSa$*;*
(*iv*) *S is regular and left (right) strongly simple;*
(*v*) *every bi-ideal of S is a simple semigroup.*

Regular and intra-regular semigroups were studied in many papers (for example, see [49,50,67]). Characterizations of these semigroups given here are consequences of the previous results.

Theorem 5.1. *The following conditions on a semigroup S are equivalent:*

(*i*) *S is a semilattice of simple and regular semigroups;*
(*ii*) $(\forall a \in S)\ a \in aSa^2Sa$*;*
(*iii*) *every bi-ideal of S is an intra-regular subsemigroup of S;*
(*iv*) *every bi-ideal of S is a semisimple subsemigroup of S;*
(*v*) *every left ideal of S is right quasi-regular;*
(*vi*) *S is* π*-regular and intra-regular.*

Theorem 5.2. *A semigroup S is a chain of simple and regular semigroups if and only if* $a \in aSabSa$ *or* $b \in bSabSb$*, for every* $a, b \in S$.

Archimedean and π-regular semigroups, as it is shown below, are in close connection with the structure of simple and regular semigroups.

Theorem 5.3. *The following conditions on a semigroup S are equivalent:*

(*i*) *S is* π*-regular and archimedean;*
(*ii*) *S is a nil-extension of a simple and regular semigroup;*
(*iii*) $(\forall a, b \in S)(\exists n \in \mathrm{N})\ a^n \in a^nSbSa^n$.

Now, we can give the main result of this section, i.e. we can describe π-regular semigroups which are decomposable into semilattices of archimedean semigroups.

Theorem 5.4. *The following conditions on a semigroup S are equivalent:*

(*i*) *S is a semilattice of nil-extensions of simple and regular semigroups;*
(*ii*) *S is π-regular and a semilattice of archimedean semigroups;*
(*iii*) $(\forall a,b \in S)(\exists n \in \mathrm{N})\ (ab)^n \in (ab)^n S a^2 S (ab)^n$;
(*iv*) *S is π-regular and*
$$(\forall a \in S)(\forall e \in E(S))\ a|e \Rightarrow a^2|e;$$
(*v*) *S is π-regular and*
$$(\forall a,b \in S)(\forall e \in E(S))\ a|e \text{ and } b|e \Rightarrow ab|e;$$
(*vi*) *S is π-regular and*
$$(\forall e,f,g \in E(S))\quad e|g \text{ and } f|g \Rightarrow ef|g;$$
(*vii*) *S is intra-π-regular and each $\mathcal{J}$-class of S containing an intra-regular element is a regular subsemigroup of S;*
(*viii*) *S is π-regular and each $\mathcal{J}$-class of S containing an idempotent is a subsemigroup of S;*
(*ix*) *S is π-regular and $\tau(\mathcal{J})$ is a semilattice (or a band) congruence on S;*
(*x*) *S is a semilattice of nil-extensions of simple semigroups and $Intra(S) = Reg(S)$;*
(*xi*) *S is π-regular and $Reg(S) = Intra(S) = \mathcal{U}_{\mathcal{J}}(S)$.*

Within the class of semigroups mentioned in the previous theorem, chains of archimedean and π-regular semigroups will be considered.

Theorem 5.5. *The following conditions on a semigroup S are equivalent:*

(*i*) *S is a chain of nil-extensions of simple and regular semigroups;*
(*ii*) $(\forall a,b \in S)(\exists n \in \mathrm{N})\ a^n \in a^n SabSa^n$ *or* $b^n \in b^n SabSb^n$;
(*iii*) *S is π-regular and $(\forall e,f \in E(S))\ ef|e$ or $ef|f$;*
(*iv*) *S is π-regular and $Reg(S)$ is a chain of simple and regular semigroups.*

6. Semilattices of Completely Archimedean Semigroups

The class of completely π-regular semigroups and its subclasses, completely simple, completely regular, completely archimedean semigroups or semilattices of completely archimedean semigroups are the main subject of this section. Of course, they are described from the perspective of the results from Sections 2-5.

Theorem 6.1. *The following conditions on a semigroup S are equivalent:*

(*i*) *S is completely π-regular;*

(*ii*) *every bi-ideal of a semigroup S is completely quasi-π-regular;*
(*iii*) *S is left π-regular and right quasi-π-regular.*

Lemma 6.1. *The following conditions on a semigroup S are equivalent:*

(*i*) *S is completely simple;*
(*ii*) *S is left (right) completely simple and has an idempotent;*
(*iii*) *S is simple, left regular and right quasi-regular;*
(*iv*) *S is simple, left regular and right quasi-π-regular.*

Completely regular smigroups (or union of groups) are considered by using results from Theorem 6.1. and Lemma 6.1.

Theorem 6.2. *The following conditions on a semigroup S are equivalent:*

(*i*) *S is a completely regular semigroup;*
(*ii*) *every left ideal of S is right regular;*
(*iii*) *every left ideal of S is completely quasi-regular;*
(*iv*) *every bi-ideal of S is left quasi-regular;*
(*v*) *S is left regular and right quasi-regular;*
(*vi*) *S is left regular and right quasi-π-regular.*

Archimedean semigroups which are completely π-regular, i.e. completely archimedean semigroups are the most described subclass of the class of archimedean semigroups (see, for example, Theorem 1.17.). Here we have the following result.

Theorem 6.3. *The following conditions on a semigroup S are equivalent:*

(*i*) *S is completely archimedean;*
(*ii*) *S is π-regular and eSe is a unipotent monoid for every $e \in E(S)$;*
(*iii*) *S is π-regular and $E(eS)$ is a semigroup of right zeros for every $e \in E(S)$;*
(*iv*) *S is completely π-regular and $\langle E(S) \rangle$ is a (completely) simple semigroup;*
(*v*) *S is left completely archimedean and contains an idempotent.*

There are many results concerning completely π-regular semigroups which are decomposable into semilattices of archimedean components (see Theorem 1.27). We will describe semilattices of completely archimedean semigroups by certain equalities between the group part and Green's subsets of completely π-regular semigroups.

Theorem 6.4. *The following conditions on a semigroup S are equivalent:*

(*i*) *S is a semilattice of completely archimedean semigroups (uniformly π-regular);*
(*ii*) *S is completely π-regular, $Reg(\langle E(S)\rangle) = Gr(\langle E(S)\rangle)$ and for all $e, f \in E(S)$, $f|e$ in S implies $f|e$ in $\langle E(S)\rangle$;*
(*iii*) *S is completely π-regular and $Gr(S) = \mathcal{U}_{\mathcal{J}}(S) = \mathcal{U}_{\mathcal{L}}(S)$;*
(*iv*) *every left ideal of S is a semilattice of nil-extensions of simple and regular semigroups;*
(*v*) *every bi-ideal of S is a semilattice of strongly archimedean semigroups.*

Chains of completely archimedean semigroups were studied in [14,15]. Here characterizations of these semigoups are given in terms of their idempotents.

Theorem 6.5. *The following conditions on a semigroup S are equivalent:*

(*i*) *S is a chain of completely archimedean semigroups;*
(*ii*) *S is completely π-regular and $e \in ef\langle E(S)\rangle fe$ or $f \in fe\langle E(S)\rangle ef$, for all $e, f \in E(S)$;*
(*iii*) *S is completely π-regular and $e \in ef\langle E(S)\rangle$ or $f \in \langle E(S)\rangle ef$, for all $e, f \in E(S)$;*
(*iv*) *S is a completely π-regular semigroup and $\langle E(S)\rangle$ is a chain of completely simple semigrups.*

Corollary 6.1. *A semigroup S is a semilattice of nil-extensions of left groups if and only if S is completely π-regular and $Gr(S) = L_{\mathcal{J}}(S)$.*

Corollary 6.2. *A semigroup S is a semilattice of nil-extensions of groups if and only if S is completely π-regular and $Gr(S) = H_{\mathcal{J}}(S)$.*

References

1. O. Anderson *Ein Bericht uber Structur abstrakter Halbgruppen*, Thesis, Hamburg (1952)
2. R. Arens and I. Kaplansky, Topological representation of algebras, *Trans. Amer. Math. Soc.* **63** (1948), 457–481
3. S. Bogdanović, A note on strongly reversible semiprimary semigroups , *Publ. Inst. Math.* **28(42)** (1980), 19–23
4. S. Bogdanović, Power regular semigroups, *Zb. rad. PMF Novi Sad* **12** (1982), 417–428
5. S. Bogdanović, Semigroups of Galbiati-Veronesi, *Proceedings of the conference "Algebra and Logic"*, Zagreb 1984, Novi Sad 1984, 9–20
6. S. Bogdanović, Right π-inverse semigroups, *Zbornik radova PMF Novi Sad* **14(2)** (1984), 187–195

7. S. Bogdanović, *Semigroups with a system of subsemigroups*, Inst. of Math., Novi Sad (1985)
8. Bogdanović, Semigroups of Galbiati-Veronesi II, *Facta Univ. Niš, Ser. Math. Inform.* **2** (1987), 61–66
9. S. Bogdanović and M. Ćirić, Semigroups of Galbiati-Veronesi III (Semilattice of nil-extensions of left and right groups), *Facta Univ. Niš, Ser. Math. Inform.* **4** (1989), 1–14
10. S. Bogdanović and M. Ćirić, A nil-extension of a regular semigroup, *Glasnik matematički* **25(2)** (1991), 3–23
11. S. Bogdanović and M. Ćirić, Semigroups in which the radical of every ideal is a subsemigroup, *Zbornik rad. Fil. Fak. Niš, Ser. Mat.* **6** (1992), 129–135
12. S. Bogdanović and M. Ćirić, Semilattices of Archimedean semigroups and (*completely*) π-regular semigroups I (*A survey*), *Filomat (Niš)* **7** (1993), 1–40
13. S. Bogdanović and M. Ćirić, *Semigroups*, Prosveta, Niš (1993), (in Serbian)
14. S. Bogdanović and M. Ćirić, Chains of Archimedean semigroups (*Semiprimary semigroups*), *Indian J. Pure Appl. Math.* **25(3)** (1994), 331–336
15. S. Bogdanović and M. Ćirić, Semilattices of nil-extensions of rectangular groups, *Publ. Math. Debrecen* **47/3-4** (1995), 229–235
16. S. Bogdanović and M. Ćirić, A note on left regular semigroups, *Publ. Math. Debrecen* **48/3-4** (1996), 285–291
17. S. Bogdanović and M. Ćirić, Semilattices of left completely Archimedean semigroups, *Math. Moravica* **1** (1997), 11–16
18. S. Bogdanović and M. Ćirić, Radicals of Green's relations, *Czechoslovak Math. J.*, (to appear)
19. S. Bogdanović, M. Ćirić and M. Mitrović, Semilattices of nil-extensions of simple regular semigroups, *Algebra Colloquium* **10:3** (2003), 81–90
20. S. Bogdanović, M. Ćirić i M. Mitrović, Semigroups Satisfying Certain Regularity Conditions, *"Advances in Algebra - Proceedings of the ICM Satelite Conference in Algebra and Related Topics"*, World Scientific
21. S. Bogdanović, M. Ćirić and Ž. Popović, Semilattice decomposition of semigroups revistated, *Semigroup Forum* **61** (2000), 263–276
22. S. Bogdanović and S. Milić, A nil-extension of a completely simple semigroup, *Publ. Inst. Math.* **36(50)** (1984), 45–50
23. J. Calais, Demi-groupes quasi-inversifs, *C. R. Acad. Sci. Paris* **252** (1961), 2357–2359
24. J. L. Chrislock, On a medial semigroups, *J. Algebra* **12** (1969), 1–9
25. A. H. Clifford, Semigroups admiting relative inverses, *Ann. of Math.* **42** (1941), 1037–1042
26. A.H. Clifford and G.B.Preston, *The algebraic theory of semigroups*, Amer. Math. Soc. Vol I, (1961), Vol II, (1967)
27. R. Croisot, Demi-groupes inversifs et demi-groupes réunions de demi groupes simples, *Ann. Sci. Ecole Norm. Sup. (3)* **70** (1953), 361–379
28. M. Ćirić and S. Bogdanović, Decompositions of semigroups induced by identities, *Semigroup Forum* **46** (1993), 329–346
29. M. Ćirić and S. Bogdanović, Semilattice decompositions of semigroups,

Semigroup Forum **52** (1996), 119–132

30. M. Ćirić and S. Bogdanović, Theory of greatest semilattice decompositions of semigroups, *Filomat (Niš)* **46** (1993), 329–346
31. M. P. Drazin, Pseudoinverses in associative rings and semigroups, *Amer. Math. Mon.* **65** (1958), 506–514
32. D. Easdown, Biordered sets of eventually regular semigroups, *Proc. Lond. Math. Soc.* **(3)49** (1984), 483–503
33. P. Edwards, Eventually regular semigroups, *Bull. Austral. Math. Soc.* **28** (1983), 23–28
34. P. Edwards, On lattice of congruences on an eventually regular semigroups, *J. Austral. Math. Soc.* **A 38** (1985), 281–286
35. P. Edwards, Eventually regular semigroups that are group-bound, *Bull. Austral. Math. Soc.* **34** (1986), 127–132
36. J. L. Galbiati, Some semilattices of semigroups each having one idempotent, *Semigroup Forum* **55** (1997), 206–214
37. J. L. Galbiati and M. L. Veronesi, Sui semigruppi quasi regolari, *Instituto Lombardo (Rend. Sc.)* **(A) 116** (1982), 1–11
38. J. L. Galbiati and M. L. Veronesi, Semigruppi quasi regolari, *Atti del convegno: Teoria dei semigruppi, Siena (ed. F. Migliorini(*, (1982), 91–95
39. J. L. Galbiati and M. L. Veronesi, Sui semigruppi quasi completamente inversi, *Rend. Inst. Lombardo Cl. Sc.* **(A) 118** (1984), 37–51
40. J. L. Galbiati and M. L. Veronesi, On quasi completely regular semigroups, *Semigroup Forum* **29** (1984), 271–275
41. P. A. Grillet, *Semigroups*, Marcel Dekker, Inc (1995)
42. T. E. Hall, On regular semigroups, *J. Algebra* **24** (1973), 1–24
43. P. M. Higgins, *Techniques of semigroup theory*, Oxford Univ. Press. (1992)
44. E. Hotzel, On semigroups with maximal conditions, *Semigroup Forum* **11** (1975/76), 337–362
45. J. M. Howie, *Fundamentals of Semigroup Theory*, London Mathematical Society Monographs. New Series, Oxford: Clarendon Press (1995)
46. I. Kaplansky, Topological representation of algebras, *Trans. Amer. Math. Soc.* **68** (1950), 62–75
47. F. Kmeť, Radicals and their left ideal analogues in a semigroup, *Math. Slovaca* **38** (1988), 139–145
48. S. Lajos, Generalized ideals in semigroups, *Acta Sci. Math.* **22** (1961), 217–222
49. S. Lajos, Bi-ideals in semigroups, I, A survey, *PU. M. A.* **Ser. A 2** (1991) no 3-4, 215–237
50. S. Lajos and I. Szász, Generalized regularity in semigroups, *Depth. of Math. Karl Marx Univ. of Econ. Budapest* **7** (1975), 1–23
51. B. L. Madison, T. K. Mukherjee and M. K. Sen, Periodic properties of groupbound semigroups, *Semigroup Forum* **22** (1981), 225–234
52. B. L. Madison, T. K. Mukherjee and M. K. Sen, Periodic properties of groupbound semigroups II, *Semigroup Forum* **26** (1983), 229–236
53. N. McCoy, Generalized regular rings, *Bull. Amer. Math. Soc.* **45** (1939), 175–178

54. M. Mitrović, *Semilaittice Decompositions of Semigroups*, M. Sc. (1992), University of Niš, (in Serbian)
55. M. Mitrović, *Regular subsets and Semilaittice Decompositions of Semigroups*, PhD Thesis, University of Niš (1999), (in Serbian)
56. M. Mitrović, *Semilattices of Archimedean Semigroups*, University of Niš - Faculty of Mechanical Engineering (2003)
57. M. Mitrović, S. Bogdanović and M. Ćirić, Locally uniformly π-regular semigroups, *In Proceedings of the International Conference of Semigroups, Braga, Portugal, 18-23 Jun* (1999), World Scientific, 106–113
58. W. D. Munn, On a semigroup algebras, *Proc. Cambridge Phil. Soc.* **51** (1955), 1–15
59. W. D. Munn, Matrix representation of semigroups, *Proc. Cambridge Phil. Soc.* **53** (1957), 5–12
60. W. D. Munn, Pseudoinverses in semigroups, *Proc. Cambridge Phil. Soc.* **57** (1961), 247–250
61. M.Petrich, The maximal semilattice decomposition of a semigroup, *Math. Zeitschr.* **85** (1964), 68–82
62. M. Petrich, *Introduction to semigroups*, Merill, Ohio (1973)
63. M. Petrich, *Structure of regular semigroups*, Cahiers Math. Montpellier (1977)
64. M. Petrich, *Inverse semigroups*, J. Wiley & Sons, New York (1984)
65. P. Protić, The band and the semilattice decompositions of some semigroups, *Pure Math. and Appl. Ser. A no.* **1-2** (1991), 141–146
66. P. Protić, On some band decompositions of semigroups, *Publ. Math. Debrecen* **45/1-2** (1994), 205–211
67. M. S. Putcha, Semilattice decompositions of semigroups, *Semigroup Forum* **6** (1973), 12–34
68. M. S. Putcha, Bands of t-Archimedean semigroups, *Semigroup Forum* **6** (1973), 232–239
69. M. S. Putcha, Semigroups in which a power of each element lies in a subgroup, *Semigroup Forum* **5** (1973), 354–361
70. M. S. Putcha, Minimal sequences in semigroups, *Trans. Amer. Math. Soc.* **189** (1974), 93–106
71. M. S. Putcha, Paths in Graphs and minimal π-sequences in semigroups, *Discrete Mathematics* **11** (1975), 173–185
72. M. S. Putcha, Rings which are semilattices of Archimedean semigroups, *Semigroup Forum* **23** (1981), 1–5
73. M. S. Putcha and J. Weissglass, A semilattice decomposition into semigroups with at most one idempotent, *Pacific J. Math.* **39** (1971), 225–228
74. B. M. Schein, Homomorphisms and subdirect product decompositions of semigroups, *Pacific J. Math.* **17** (1966), 81–94
75. L. N. Shevrin, Theory of epigroups I, *Mat. Sb.* **185 no 8** (1994), 129–160, (in Russian)
76. L. N. Shevrin, Theory of epigroups II, *Mat. Sb.* **T. 185 no 9** (1994), 133–154, (in Russian)
77. G. Szász, Semigroups with idempotent ideals, *Publ. Math. Debrecen* **21/1-2**

(1974), 115–117

78. R. Šulka, The maximal semilattice decomposition of semigroup, radicals and nilpotency, *Mat. časopis* **20** (1970), 172–180
79. T. Tamura, The theory of construction of finite semigroups I, *Osaka Math. J.* **8** (1956), 243–261
80. T. Tamura, Another proof of a theorem concerning the greatest semilattice decomposition of a semigroup, *Proc. Japan Acad.* **40** (1964), 777–780
81. T. Tamura, On Putcha's theorem concerning semilattice of Archimedean semigroups, *Semigroup Forum* **4** (1972), 83–86
82. T. Tamura, Note on the greatest semilattice decomposition of semigroups, *Semigroup Forum* **4** (1972), 255–261
83. T. Tamura, Quasi-orders, generalized Archimedeaness, semilattice decompositions, *Math. Nachr.* **68** (1975), 201–220
84. T. Tamura and N. Kimura, On decomposition of a commutative semigroup, *Kodai Math. Sem. Rep.* **4** (1954), 109–112
85. T. Tamura and N. Kimura, Existence of greatest decomposition of semigroup, *Kodai Math. Sem. Rep.* **7** (1955), 83–84
86. T. Tamura and J. Shafer, On exponential semigroups I, *Proc. Japan Acad.* **48** (1972), 474–478
87. G. Thierrin, Sur queiques propriétiés de certaines classes de demi-groupes, *C. R. Acad. Sci. Paris* **239** (1954), 33–34
88. M. L. Veronesi, Sui semigruppi quasi fortemente regolari, *Riv. Mat. Univ. Parma* **(4) 10** (1984), 319–329
89. M. Yamada, On the greatest semilattice decomposition of a semigroup, *Kodai Mat. Sem. Rep.* **7** (1955), 59–62

RELATIVE REWRITING SYSTEMS

S. J. PRIDE
Department of Mathematics,
University of Glasgow,
Glasgow G12 8QW, U.K.
E-mail: sjp@maths.gla.ac.uk

JING WANG
Department of Mathematics and Statistics,
University of Windsor,
Windsor N9B 3P4, Canada
E-mail: jingwang_ca@yahoo.com

1. Rewriting Systems and Ideal Pairs

A *rewriting system* is a pair $\mathcal{P} = [\mathbf{x}; \mathbf{r}]$ where $\mathbf{x}$ is a set (the alphabet) and $\mathbf{r}$ is a set of ordered pairs $R = (R_{+1}, R_{-1})$ of words on $\mathbf{x}$ (rewriting rules). We usually write the ordered pairs in the form $R : R_{+1} = R_{-1}$, or simply $R_{+1} = R_{-1}$.

Let $\mathbf{x}^*$ denote the free monoid on $\mathbf{x}$. We define elementary transformations of words $W \in \mathbf{x}^*$ as follows: if $W = UR_\varepsilon V$ $(U, V \in \mathbf{x}^*, R \in \mathbf{r}, \varepsilon = \pm 1)$ then replace R_ε by $R_{-\varepsilon}$ to obtain the word $W' = UR_{-\varepsilon}V$. Two words are said to be equivalent if one can be transformed to the other by a finite number of elementary transformations. This is a congruence on $\mathbf{x}^*$, and the quotient monoid is called the *monoid defined by* $\mathcal{P}$, denoted $S = S(\mathcal{P})$. We write $[W]_\mathcal{P}$, or simply $[W]$, for the congruence class of $W \in \mathbf{x}^*$. If S' is a monoid with $S' \cong S(\mathcal{P})$ then we say that $\mathcal{P}$ is a *rewriting system for* S'.

A subset $\mathbf{x}_0 \subseteq \mathbf{x}$ will be called *separating* if it has the following property:

$$\text{for each } R \in \mathbf{r} \text{ either } R_{+1}, R_{-1} \in \mathbf{x}_0^* \text{ or } R_{+1}, R_{-1} \in \mathbf{x}^* \setminus \mathbf{x}_0^*.$$

The number of such subsets will be called the *separating number* of $\mathcal{P}$, denoted $\sigma(\mathcal{P})$.

Example 1. *Let* $\mathcal{P}_1 = [x, y, z;\ xy = yx,\ yz = zy,\ zx = xz]$. *Then all* *subsets are separating, so* $\sigma(\mathcal{P}_1) = 8$.

Example 2. *Let* $\mathcal{P}_2 = [x, y, z;\ xy = y^3,\ yz = zy,\ zx = xz]$. *Then all subsets except* $\{y\}$, $\{y, z\}$ *are separating, so* $\sigma(\mathcal{P}_2) = 6$.

Example 3. *Let* $\mathcal{P}_3 = [x, y, z;\ x^2 = y^2,\ y^2 = z^2]$. *Only the subsets* $\emptyset$ *and* $\{x, y, z\}$ *are separating, so* $\sigma(\mathcal{P}_2) = 2$.

Example 4. *Let* $\mathcal{P}_4 = [x, y, z;\ x^2y^3z^4 = 1]$. *Only the subset* $\{x, y, z\}$ *is separating, so* $\sigma(\mathcal{P}_2) = 1$.

It turns out that the separating number is a monoid invariant, that is, if $\mathcal{P}$, $\mathcal{P}'$ are two rewriting systems for a monoid S, then $\sigma(\mathcal{P}) = \sigma(\mathcal{P}')$. This can be proved using Tietze transformation, or alternatively, it follows from our discussion below. Thus we may speak unambiguously of the *separating number* $\sigma(S)$ *of a monoid* S. Of course $\sigma(S)$ could be infinite if S is not finitely generated.

A pair of monoids (S, S_0) will be called an *ideal pair* if S_0 is a submonoid of S, and $S \setminus S_0$ is an ideal of S (that is, if s is *not* in S_0 then us and su are *not* in S_0 for all $u \in S$).

The following result was proved in [15].

Theorem 1 [15]. *Let* S *be given by a rewriting system* $\mathcal{P} = [\mathbf{x}; \mathbf{r}]$ *and let* S_0 *be a submonoid of* S. *Then* (S, S_0) *is an ideal pair if and only if*

$$\mathbf{x}_0 = \{x \in \mathbf{x}; [x] \in S_0\}$$

is a separating subset and $\{[x] : x \in \mathbf{x}_0\}$ *generates* S_0. *In this case, if* $\mathbf{r}_0$ *consists of all rules* $R \in \mathbf{r}$ *with* $R_{+1}, R_{-1} \in \mathbf{x}_0^*$, *then*

$$\mathcal{P}_0 = [\mathbf{x}_0; \mathbf{r}_0]$$

is a rewriting system for S_0.

Thus $\sigma(\mathcal{P})$ is equal to the number of submonoids S_0 of S such that (S, S_0) is an ideal pair.

It turns out that if (S, S_0) is an ideal pair, then many finiteness properties of S are inherited by S_0.

Theorem 2 [15]. *Let* (S, S_0) *be an ideal pair. If* S *has any of the following properties then so does* S_0:

(a) finitely generated;

(b) finitely presented;
(c) having a finite complete rewriting system;
(d) finite derivation type (FDT);
(e) finite homological type (FHT).

(For definitions of FDT and FHT see below.)

(a), (b) are clear from Theorem 1, and (c), (d), (e) are proved in [15] (for (c), see also [9]).

It can also be shown that the properties FDT_2, FHT_2 ([10]) are inherited, and also the properties bi-FP_n $(n \geq 1)$ ([7]).

A useful way to think about ideal pairs is via *relative rewriting systems.* A relative rewriting system over a monoid M is a triple

$$\mathcal{R} = [M, \mathbf{y}; \mathbf{u}]$$

where $\mathbf{y}$ is a set, and each element $U \in \mathbf{u}$ is an ordered pair (U_{+1}, U_{-1}) of words on $M \cup \mathbf{y}$, *both involving at least one symbol from* $\mathbf{y}$. We write $U : U_{+1} = U_{-1}$, and define $U^{-1} : U_{-1} = U_{+1}$. We say that $\mathcal{R}$ is *finite* if $\mathbf{y}$ and $\mathbf{u}$ are finite.

The monoid $S = S(\mathcal{R})$ defined by $\mathcal{R}$ is the quotient of the free product $M * \mathbf{y}^*$ by the congruence generated by $\mathbf{u}$. More concretely, we can define operations on words W on $M \cup \mathbf{y}$ as follows:

(I) If W contains two successive terms $m, m' \in M$ then replace them by a single term which is their product in M.

(II) If W contains a term which is the identity of M then delete that term.

(III) If W contains a subword U_ε $(U : U_{+1} = U_{-1} \in \mathbf{u}, \varepsilon = \pm 1)$ then replace it by $U_{-\varepsilon}$.

Then two words are *equivalent* (relative to $\mathcal{R}$) if one can be obtained from the other by a finite number of operations (I), $(\mathrm{I})^{-1}$, (II), $(\mathrm{II})^{-1}$, (III). The monoid S then consists of the equivalence classes $[W]$, with (well-defined) multiplication $[W_1][W_2] = [W_1 W_2]$. There is a natural homomorphism

$$M \longrightarrow S, \qquad m \mapsto [m].$$

which is injective, and we identify M with its image in S.

Clearly if $\mathcal{R}$ is a relative rewriting system over M then $(S(\mathcal{R}), M)$ is an ideal pair.

Conversely, suppose (S, S_0) is an ideal pair. Let $\mathcal{P} = [\mathbf{x}; \mathbf{r}]$ be a rewriting system for S, let $\mathcal{P}_0 = [\mathbf{x}_0; \mathbf{r}_0]$ be the subpresentation as in Theorem 1 corresponding to S_0, and let $\mathbf{y} = \mathbf{x} \setminus \mathbf{x}_0$. We have the monoid epimorphism

$$\theta: \quad \mathbf{x}_0^* * \mathbf{y}^* \longrightarrow S_0 * \mathbf{y}^*, \qquad x_0 \mapsto [x_0], \quad y \mapsto y$$

where $x_0 \in \mathbf{x}_0, y \in \mathbf{y}$. Then

$$\mathcal{R} = [S_0, \mathbf{y};\ \theta(R_{+1}) = \theta(R_{-1}) \quad (R \in \mathbf{r} \setminus \mathbf{r}_0)]$$

is a relative rewriting system, and there is an induced isomorphism

$$\hat{\theta}: \quad S = S(\mathcal{P}) \longrightarrow S(\mathcal{R})$$

which fixes S_0.

Example 2 (continued). *If S_0 is the submonoid of $S(\mathcal{P}_2)$ corresponding to the separating subset $\{x, z\}$, then the associated relative rewriting system is*

$$\mathcal{R} = [\ S_0, y;\ [x]y = y^3, y[z] = [z]y\ ].$$

Lemma 1. *Let (S, S_0) be an ideal pair. Then S is finitely presented if and only if S_0 is finitely presented and S has a finite relative rewriting system over S_0.*

Proof. The "only if" part follows from our discussion above (taking $\mathcal{P}$ to be finite).

On the other hand if $\mathcal{R} = [S_0, \mathbf{y}; \mathbf{u}]$ is a finite relative rewriting system for S over S_0, and if S_0 is finitely presented, then we may "lift" $\mathcal{R}$ to an ordinary finite rewriting system for S (by essentially reversing the procedure in our discussion above).

2. Finite Derivation and Homological Types (*FDT* and *FHT*)

Given a rewriting system $\mathcal{P} = [\mathbf{x}; \mathbf{r}]$, the 2-complex $\mathcal{D}(\mathcal{P})$ associated with $\mathcal{P}$ is defined as follows. The underlying graph Γ has vertex set $\mathbf{x}^*$, and the edges are all *atomic monoid picture* $e = (U, R, \varepsilon, V)$ $(U, V \in \mathbf{x}^*, R \in \mathbf{r}, \varepsilon = \pm 1)$ (see Figure 1). The initial, terminal and inversion functions $\iota, \tau, ^{-1}$ are defined by

$$\iota(e) = UR_\varepsilon V,\ \tau(e) = UR_{-\varepsilon}V,\ e^{-1} = (U, R, -\varepsilon, V).$$

An edge is called *positive* if $\varepsilon = +1$. There is a left action of $\mathbf{x}^*$ on Γ: if $Z \in \mathbf{x}^*$ then for any vertex W of $\Gamma, Z \cdot W = ZW$ (product in $\mathbf{x}^*$), and for any edge e as above $Z \cdot e = (ZU, R, \varepsilon, V)$. There is a similar right action of $\mathbf{x}^*$ on Γ. Paths in Γ are called (monoid) *pictures*, and closed paths are called *spherical* (monoid) pictures. We now attach some 2-cells to Γ as follows. Let e_1, e_2 be atomic monoid pictures. Then

$$(e_1 \cdot \iota(e_2))\ (\tau(e_1) \cdot e_2)\ (e_1^{-1} \cdot \tau(e_2))\ (\iota(e_1) \cdot e_2^{-1})$$

is a closed path in Γ, which we denote by $[e_1, e_2]$. Then $\mathcal{D} = \mathcal{D}(\mathcal{P})$ is the 2-complex obtained from Γ by attaching 2-cells to Γ along the paths $[e_1, e_2]$ for all pairs e_1, e_2. Notice that the left and right actions of $\mathbf{x}^*$ on Γ can be extended to actions on $\mathcal{D}$. For further information on $\mathcal{D}$ see [11] (and also [4], [12], [15]).

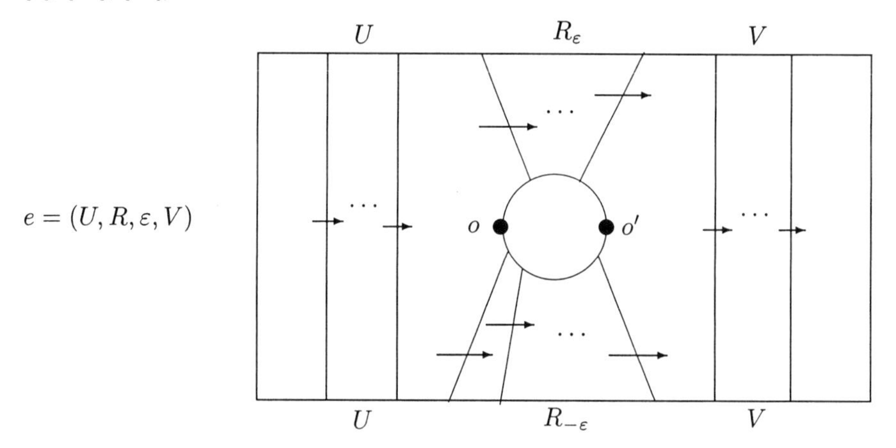

Figure 1

If $\mathbf{X}$ is a set of closed paths in $\mathcal{D}$, then the 2-complex obtained from $\mathcal{D}$ by adjoining additional 2-cells with boundaries $W \cdot p \cdot V, (W, V$ words on $\mathbf{x}, p \in \mathbf{X})$ will be denoted by $\mathcal{D}^{\mathbf{X}}$. We will say that $\mathbf{X}$ *trivializes* $\mathcal{D}$ if $\mathcal{D}^{\mathbf{X}}$ has trivial fundamental groups. (The term *collapses* was used in [12].)

A finite rewriting system $\mathcal{P} = [\mathbf{x}; \mathbf{r}]$ is said to be of *finite derivation type* (*FDT* for short) if there is a *finite* set of closed paths in $\mathcal{D}$ which trivializes $\mathcal{D}$.

We also have the chain complex

$$0 \longrightarrow C_2(\mathcal{D}) \xrightarrow{\partial_2} C_1(\mathcal{D}) \xrightarrow{\partial_1} C_0(\mathcal{D}) \longrightarrow 0$$

of $\mathcal{D}$. Here $C_0(\mathcal{D})$ is the free abelian group with basis $\mathbf{x}^*$, $C_1(\mathcal{D})$ is the free abelian group with basis the set of positive edges in $\mathcal{D}$, and $C_2(\mathcal{D})$ is the free abelian group on the set of defining paths $[e_1, e_2]$. Notice that these chain groups are $(\mathbb{Z}\mathbf{x}^*, \mathbb{Z}\mathbf{x}^*)$-bimodules via the two-sided action of $\mathbf{x}^*$ on the bases of the chain groups. The boundary maps are given by:

$$\partial_2[e_1^{\varepsilon_1}, e_2^{\varepsilon_2}] = \varepsilon_1\varepsilon_2 e_1 \cdot (\iota(e_2) - \tau(e_2)) - \varepsilon_1\varepsilon_2(\iota(e_1) - \tau(e_1)) \cdot e_2,$$

$$\partial_1(e) = \iota(e) - \tau(e),$$

where e_1, e_2 and e are positive edges in $\mathcal{D}$, $\varepsilon_1, \varepsilon_2 = \pm 1$.

It was shown in [11] that the first homology group $Ker\partial_1/Im\partial_2$ is a $(\mathbb{Z}S(\mathcal{P}), \mathbb{Z}S(\mathcal{P}))$-bimodule under the action induced from the action of $\mathbb{Z}\mathbf{x}^*$. As in [15], [18], we will denote this bimodule by $\pi_2^{(b)}(\mathcal{P})$.

For each closed path p in $\mathcal{D}$, say $p = e_1^{\varepsilon_1} \cdots e_k^{\varepsilon_k}$ (e_i a positive edge, $\varepsilon_i = \pm 1, i = 1, \cdots, k$), we have the cycle

$$z_p = \sum_{i=1}^{k} \varepsilon_i e_i \in Ker\partial_1.$$

We will say that $\mathcal{P}$ is of *finite homological type* (*FHT* for short) if $\mathcal{P}$ is finite and there is a *finite* set $\mathbf{X}$ of closed paths in $\mathcal{D}$ such that the elements $z_p + Im\partial_2$ ($p \in \mathbf{X}$) generate the bimodule $\pi_2^{(b)}(\mathcal{P})$.

It can be shown that if S is a finitely presented monoid and if one finite rewriting system for S is *FDT* or *FHT*, then any other finite rewriting system for S has the same property ([17], [18]). Thus we may speak of *FDT*, *FHT monoids.* Clearly, *FDT* implies *FHT*. The converse is false in general [14]. However, for finitely presented *groups* the two properties *are* equivalent and coincide with the homological property FP_3 ([3], see also [13]).

Let $\mathcal{P}_0 = [\mathbf{x}_0; \mathbf{r}_0]$ be a subpresentation of $\mathcal{P} = [\mathbf{x}; \mathbf{r}]$. We will say that $\mathcal{P}$ is *aspherical over* $\mathcal{P}_0$ if the set of all spherical monoid pictures over $\mathcal{P}_0$ trivializes $\mathcal{D}(\mathcal{P})$. Note that this is equivalent to asserting that if $\mathbf{X}_0$ is a set of spherical pictures over $\mathcal{P}_0$ which trivializes $\mathcal{D}(\mathcal{P}_0)$, then $\mathbf{X}_0$ trivializes

$\mathcal{D}(\mathcal{P})$. Then we have the following lemma.

Lemma 2. *Suppose $\mathcal{P}$ is aspherical over $\mathcal{P}_0$.*
(i) If $\mathcal{P}_0$ is FDT then so is $\mathcal{P}$.
(ii) If $\mathcal{P}_0$ is FHT then so is $\mathcal{P}$.

Proof. (i) It is clear.

(ii) Let $\mathcal{D}_0 = \mathcal{D}(\mathcal{P}_0)$ and $\mathcal{D} = \mathcal{D}(\mathcal{P})$. The inclusion map from $\mathcal{P}_0$ to $\mathcal{P}$ induces a mapping of 2-complexes $\mu : \mathcal{D}_0 \longrightarrow \mathcal{D}$. Then the mapping μ induces a chain map from the chain complex

$$0 \longrightarrow C_2(\mathcal{D}_0) \xrightarrow{\partial_2^0} C_1(\mathcal{D}_0) \xrightarrow{\partial_1^0} C_0(\mathcal{D}_0) \longrightarrow 0$$

of $\mathcal{D}_0$ to the chain complex

$$0 \longrightarrow C_2(\mathcal{D}) \xrightarrow{\partial_2} C_1(\mathcal{D}) \xrightarrow{\partial_1} C_0(\mathcal{D}) \longrightarrow 0$$

of $\mathcal{D}$, and this chain map then induces a homomorphism

$$\mu^* : \pi_2^{(b)}(\mathcal{P}_0) \longrightarrow \pi_2^{(b)}(\mathcal{P}),\ z_p + Im\partial_2^0 \longmapsto z_p + Im\partial_2,$$

where p is a closed path in $\mathcal{D}_0$.

Let $\mathbf{X_0}$ be a finite set of spherical monoid pictures over $\mathcal{P}_0$ such that $z_p + Im\partial_2^0$ ($p \in \mathbf{X_0}$) generate the bimodule $\pi_2^{(b)}(\mathcal{P}_0)$. We want to show that $z_p + Im\partial_2$ ($p \in \mathbf{X_0}$) generate the bimodule $\pi_2^{(b)}(\mathcal{P})$.

If q is a closed path in $\mathcal{D}$, then since $\mathcal{P}$ is aspherical over $\mathcal{P}_0$, q is homotopic in $\mathcal{D}$ to a path of the form

$$p_1(U_1 \cdot q_1 \cdot V_1)p_1^{-1}p_2(U_2 \cdot q_2 \cdot V_2)p_2^{-1} \cdots p_k(U_k \cdot q_k \cdot V_k)p_k^{-1},$$

where q_i is a closed path in $\mathcal{D}_0$, p_i is a path in $\mathcal{D}$, $U_i, V_i \in \mathbf{x}^*$ $(1 \le i \le k)$. So

$$\begin{aligned} z_q + Im\partial_2 &= \sum_{i=1}^{k} U_i z_{q_i} V_i + Im\partial_2 \\ &= \sum_{i=1}^{k} [U_i](z_{q_i} + Im\partial_2)[V_i]. \end{aligned}$$

Since $\pi_2^{(b)}(\mathcal{P}_0)$ is generated by $z_p + Im\partial_2^0$ ($p \in \mathbf{X_0}$), we have, for each q_i,

$$z_{q_i} + Im\partial_2^0 = \sum_{j=1}^{m_i} [W_{ij}]_0(z_{p_{ij}} + Im\partial_2^0)[W'_{ij}]_0$$

for some $p_{ij} \in \mathbf{X_0}$ and $W_{ij}, W'_{ij} \in \mathbf{x}_0^*$. Applying μ^* we get

$$z_{q_i} + Im\partial_2 = \sum_{j=1}^{m_i} [W_{ij}](z_{p_{ij}} + Im\partial_2)[W'_{ij}].$$

Thus

$$z_q + Im\partial_2 = \sum_{i=1}^{k} \sum_{j=1}^{m_i} [U_i][W_{ij}](z_{p_{ij}} + Im\partial_2)[W'_{ij}][V_i].$$

So $z_p + Im\partial_2$ $(p \in \mathbf{X_0})$ generate the bimodule $\pi_2^{(b)}(\mathcal{P})$, as required.

3. Relative Rewriting Systems over Groups

We consider relative rewriting systems over a *group* H.

Following Kilgour [5], [6], we associate with such a system $\mathcal{R} = [H, \mathbf{y}; \mathbf{u}]$ two labelled graphs, the *left graph* $LG(\mathcal{R})$ and the *right graph* $RG(\mathcal{R})$.

A *labelled graph* over H is a graph Γ (in the sense of Serre [16]) with vertex set $\mathbf{v}$, edge set $\mathbf{e}$, and initial, terminal and inversion function $\iota, \tau, -1$, say, together with a *labelling function*

$$\phi : \mathbf{e} \longrightarrow H,$$

where $\phi(e^{-1}) = \phi(e)^{-1}$ for all $e \in \mathbf{e}$. The label $\phi(\alpha)$ on a path α is the product in H of the labels on the edges making up the path. A *cycle* in the labelled graph is a non-empty reduced closed path whose label is 1. A labelled graph is *cycle-free* if it has no cycles. This is equivalent to the assertion that the induced group homomorphism

$$\pi_1(\Gamma, *) \longrightarrow H, \quad \langle\alpha\rangle \longmapsto \phi(\langle\alpha\rangle)$$

is injective for any basepoint $*$, where $\langle\alpha\rangle$ denotes the homotopy class of a closed path α at $*$.

Both $LG(\mathcal{R})$ and $RG(\mathcal{R})$ have vertex set $\mathbf{y}$, $LG(\mathcal{R})$ has edge set $\{e_U, e_U^{-1} : U \in \mathbf{u}\}$, and $RG(\mathcal{R})$ has edge set $\{f_U, f_U^{-1} : U \in \mathbf{u}\}$. The initial, terminal and labelling functions are as follows. Let

$$U : h_1 y_1 \ldots h_{r-1} y_{r-1} h_r = h'_1 y'_1 \ldots h'_{s-1} y'_{s-1} h'_s$$

be a rule in $\mathbf{u}$, where the h's belong to H and the y's belong to $\mathbf{y}$.

$$LG(\mathcal{R}) : \quad \iota(e_U) = y_1,\ \tau(e_U) = y'_1,\ \phi(e_U) = h_1^{-1} h'_1$$
$$RG(\mathcal{R}) : \iota(f_U) = y_{r-1},\ \tau(f_U) = y'_{s-1},\ \phi(f_U) = h_r h_s'^{-1}.$$

These graphs generalize the left and right graphs of ordinary rewriting systems introduced by Adian [1], [2]. As in [5], [6] we will say that $\mathcal{R}$ is *left* (respectively, *right*) *cycle-free* if $LG(\mathcal{R})$ (respectively, $RG(\mathcal{R})$) is cycle-free.

Example 5. Let $\mathcal{R}_1 = [H, x;\ xk = hx] \qquad (k, h \in H)$.

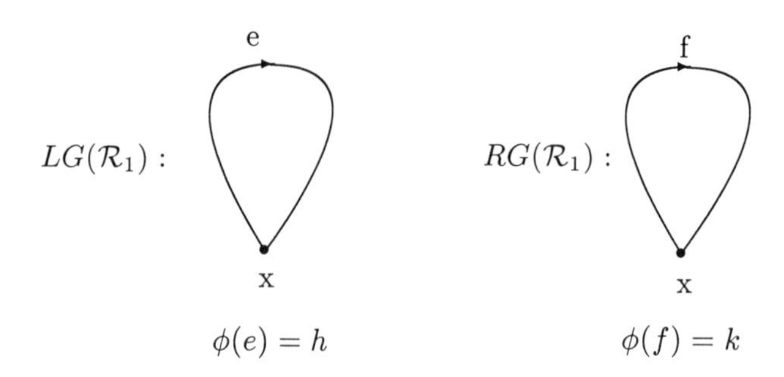

Then $\mathcal{R}_1$ is left cycle-free if and only if h has infinite order, and is right cycle-free if and only if k has infinite order.

Example 6. Let $\mathcal{R}_2 = [H, x, y;\ x^2yk_1 = h_1y^2,\ xyaxk_2 = h_2x^2,\ h_3xbxk_3 = h_4yx,\ y^2xk_4 = h_5y^2cydx^2k_5] \qquad (a, b, c, d, h_i, k_i\ \in H, 1 \le i \le 5)$.

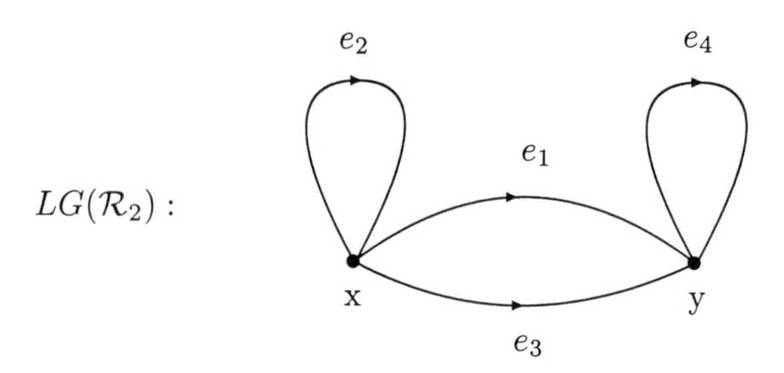

$$\phi(e_1) = h_1,\ \phi(e_2) = h_2,\ \phi(e_3) = h_3^{-1}h_4,\ \phi(e_4) = h_5.$$

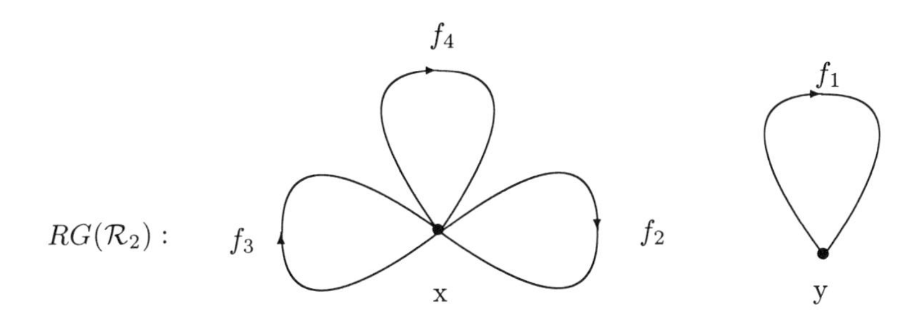

$$\phi(f_1) = k_1,\ \phi(f_2) = k_2,\ \phi(f_3) = k_3,\ \phi(f_4) = k_4 k_5^{-1}.$$

Then $\mathcal{R}_2$ is left cycle-free if and only if $h_2, h_1 h_5 h_1^{-1}, h_3^{-1} h_4 h_1^{-1}$ freely generate a subgroup of H, and is right cycle-free if and only if k_1 is of infinite order and $k_2, k_3, k_4 k_5^{-1}$ freely generate a subgroup of H.

Using geometric methods, Kilgour [5], [6] proved the following results.

Theorem 3 ([5], [6]). (a) *If $\mathcal{R}$ is left (respectively, right) cycle-free then $S = S(\mathcal{R})$ is left (respectively right) cancellative.*
(b) *If $\mathcal{R}$ is both left and right cycle-free then S is embeddable in a group.*

This generalizes results of Adian [1], [2] which deal with the case $H = 1$. Our aim here is to prove the following

Theorem 4. *Suppose $\mathcal{R} = [H, \mathbf{y}; \mathbf{u}]$ is finite, and is either left or right cycle-free. Then the following are equivalent:*
(a) $S(\mathcal{R})$ is FDT;
(b) $S(\mathcal{R})$ is FHT;
(c) H is finitely presented and of homological type FP_3.

Clearly (a) implies (b). Also by Theorem 2 (e), (b) implies that H is FHT which (as previously remarked) is equivalent to (c). So it remains to prove that (c) implies (a).

Let $< \mathbf{a}; \mathbf{t} >$ be a finite *group* presentation for H. Then we have the finite rewriting system

$$\mathcal{P}_0 = \left[\mathbf{a}, \mathbf{a}^{-1}; T = 1 (T \in \mathbf{t}),\ a^{\varepsilon} a^{-\varepsilon} = 1\ (a \in \mathbf{a}, \varepsilon = \pm 1)\right]$$

for H, and then

$$\mathcal{P} = \left[\mathbf{a}, \mathbf{a}^{-1}, \mathbf{y}; T = 1 (T \in \mathbf{t}), a^{\varepsilon} a^{-\varepsilon} = 1 (a \in \mathbf{a}, \varepsilon = \pm 1), \mathbf{u}\right]$$

is a finite rewriting system for S (where the elements of H appearing in the rules in $\mathbf{u}$ are expressed as words on $\mathbf{a} \cup \mathbf{a}^{-1}$). We will prove in the next section that

$$\mathcal{P} \text{ is aspherical over } \mathcal{P}_0. \qquad (*)$$

Then, since H is a group, (c) is equivalent to H being FDT. Thus $\mathcal{P}_0$ is FDT, and hence so is $\mathcal{P}$ by Lemma 2.

Remark: The left or right cycle-free condition in the above theorem cannot be dispensed with in general, as the examples in [14] show.

4. Proof of Theorem 4

It will be convenient to introduce some terminology concerning monoid pictures. If e is as in Figure 1, then we will say that the *label* on the disc Δ is R^{ε}. We will define the *upper and lower boundaries* $\partial^+\Delta, \partial^-\Delta$ of Δ to be the segments of $\partial\Delta$ reading from o to o' in the clockwise, anticlockwise directions respectively. Also, the upper boundary of e, denoted $\partial^+ e$, will be the horizontal part of ∂e lying along the top of e, and the lower boundary $\partial^- e$ of e will be the horizontal part of e lying along the bottom of e. If $p = e_1 e_2 \ldots e_n$ is a picture then we define $\partial^+ p = \partial^+ e_1$, $\partial^- p = \partial^- p_n$.

For concreteness, we will prove $(*)$ under the assumption that $\mathcal{R}$ is left cycle-free. We let $\mathbf{X}$ denote the set of all spherical monoid pictures over $\mathcal{P}_0$.

A *partial (left) dipole* in a picture p consists of two discs Δ, Δ', one labelled by U and the other by U^{-1} for some $U \in \mathbf{u}$, and such that the first $\mathbf{y}$-arc meeting $\partial^-\Delta$ (reading left to right) coincides with the first $\mathbf{y}$-arc meeting $\partial^+\Delta'$. A partial dipole will be called a *dipole* if *all* $\mathbf{y}$-arcs meeting $\partial^-\Delta$ also meet $\partial^+\Delta'$, and will be called a *complete dipole* if *all the arcs* meeting $\partial^-\Delta$ also meet $\partial^+\Delta'$. See Figure 2.

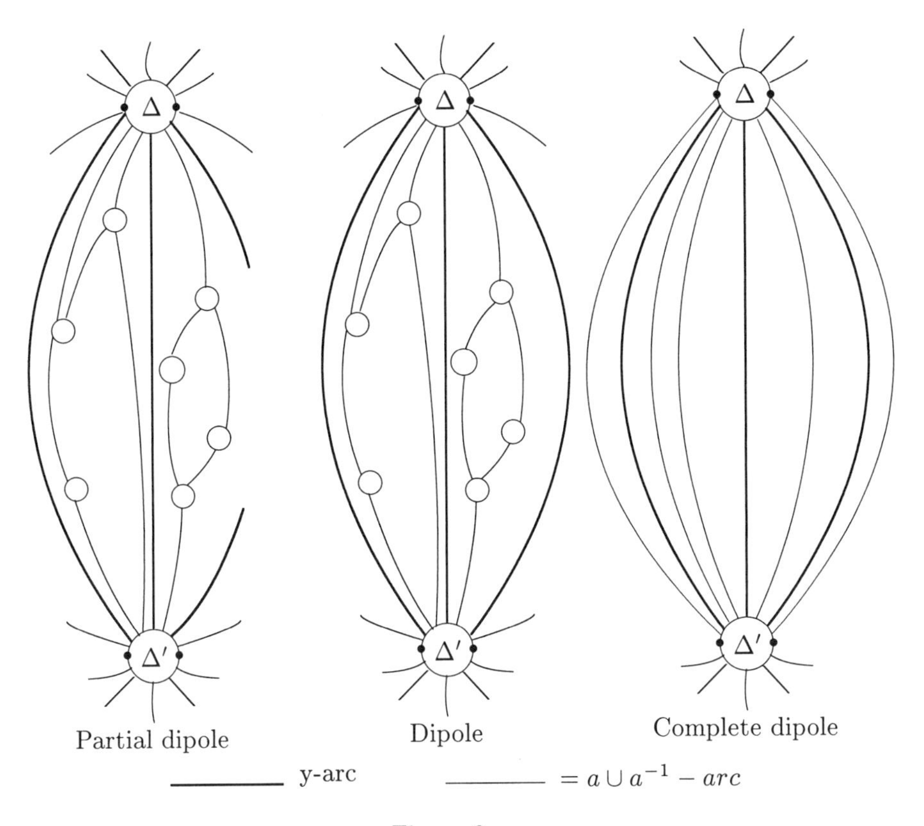

Figure 2

It is clear that if p contains a complete dipole then by swapping levels of discs if necessary (which is permissible up to homotopy in $\mathcal{D}(\mathcal{P})$) we can convert the complete dipole into the discs of a cancelling pair ee^{-1} of edges in p which can be removed giving a picture p_1 homotopic to p with less **u**-discs.

Now suppose that p contains a dipole. Let us show that p is homotopic in $\mathcal{D}(\mathcal{P})^{\mathbf{X}}$ to a picture with the same number of **u**-discs as p but where the dipole has now been converted into a complete dipole.

Let Δ, Δ' be a dipole in p as above and let $\alpha_1, \alpha_2, \dots \alpha_m$ be the **y**-arcs meeting $\partial^- \Delta \cup \partial^+ \Delta'$ (taken in order from left to right). Then the part of p between α_i and α_{i+1} $(1 \leq i < m)$ will be a spherical picture p_i over $\mathcal{P}_0$. After swapping levels of disjoint discs in p if necessary we can suppose that for certain words W_i, V_i, $W_i \cdot p_i \cdot V_i$ is a subpicture of p which, working in $\mathcal{D}(\mathcal{P})^{\mathbf{X}}$, we can remove. In this way we can convert p to a homotopic picture in $\mathcal{D}(\mathcal{P})^{\mathbf{X}}$, in which all the arcs ($\mathbf{a}\cup\mathbf{a}^{-1}$-arcs and $\mathbf{y}$-arcs) meeting $\partial^{-}\Delta$ from α_1 to α_m (reading left to right) also meet $\partial^{+}\Delta'$.

Now consider the $\mathbf{a}\cup\mathbf{a}^{-1}$-arcs meeting $\partial^{-}\Delta$ which lie to the left of α_1. We can connect these to the corresponding $\mathbf{a}\cup\mathbf{a}^{-1}$-arcs meeting $\partial^{+}\Delta'$ to the left of α_1 by using the operation as depicted in Figure 3. This process does not change the homotopy type of p in $\mathcal{D}(\mathcal{P})^{\mathbf{X}}$. A similar process can be used to connect all the $\mathbf{a}\cup\mathbf{a}^{-1}$-arcs meeting $\partial^{-}\Delta$ to the right of α_m, to the corresponding arcs meeting $\partial^{+}\Delta'$ to the right of α_m.

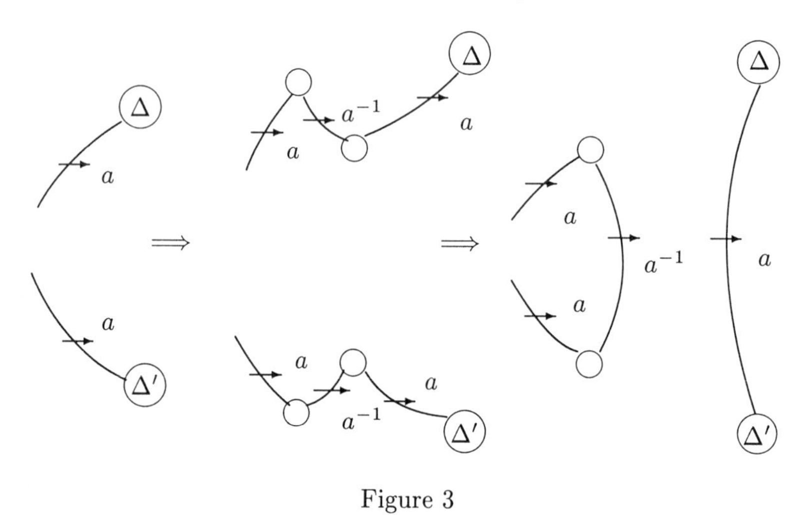

Figure 3

To prove $(*)$, we now see that it suffices to prove the following:

Lemma 3. *Every spherical picture containing a* $\mathbf{u}$*-disc contains a dipole.*

For by the above discussion it will then follow inductively that every spherical picture is homotopic in $\mathcal{D}(\mathcal{P})^{\mathbf{X}}$ to a spherical picture without $\mathbf{u}$-discs; clearly such a picture is homotopic to an empty path in $\mathcal{D}(\mathcal{P})^{\mathbf{X}}$.

To prove Lemma 3 we must consider pictures p_1 satisfying

$(**)$ $\iota(p_1) = AyW, \tau(p_1) = AyV$ where A is a word on $\mathbf{a}\cup\mathbf{a}^{-1}, W, V$ are words on $\mathbf{a}\cup\mathbf{a}^{-1}\cup\mathbf{y}$, $y\in\mathbf{y}$.

Following Kilgour [5], [6] we define the *left* $\mathbf{y}$*-circle* C of such a picture

p_1. This is a sequence $\beta_0, \Delta_1, \beta_1, \Delta_2, \beta_2, \ldots, \Delta_n, \beta_n$ of **y**-arcs β_i and **u**-discs Δ_i, obtained as follows. Let β_0 be the left-most arc meeting $\partial^+ p_1$, labelled y. If β_0 also meets $\partial^- p_1$ then β_0 is C. Otherwise β_0 meets $\partial^+ \Delta_1$ for some disc Δ_1 (and is, moreover, the left-most **y**-arc meeting $\partial^+ \Delta_1$). Let β_1 be the left-most **y**-arc meeting $\partial^- \Delta_1$. If β_1 meets $\partial^- p_1$ then $\beta_0, \Delta_1, \beta_1$ is C. Otherwise, β_1 is the left-most **y**-arc meeting $\partial^+ \Delta_2$ for some disc Δ_2. Let β_2 be the left-most **y**-arc meeting $\partial^- \Delta_2$. And so on. Continuing this process we eventually obtain C. We say that C is non-trivial if $n > 0$.

Lemma 4. *Suppose that p_1 satisfies (**) and that its left **y**-circle is non-trivial. Then p_1 contains a dipole.*

Proof. The proof is by induction on the number of **u**-discs in p_1, and is an adaption of the proof of Lemma 8.1 of [6].

Consider the left **y**-circle C of p_1 as above, and suppose Δ_i is labelled by

$$U_i^{\varepsilon_i} : A_i y_i \ldots = A_i' y_i' \ldots$$

where A_i, A_i' are words on $\mathbf{a} \cup \mathbf{a}^{-1}$ and $y_i, y_i' \in \mathbf{y}$. Note that $y_1 = y_n' = y$. We thus have the closed path $\gamma = e_{U_1}^{\varepsilon_1} \ldots e_{U_n}^{\varepsilon_n}$ at y in $LG(\mathcal{R})$. This path is labelled by the element of H represented by the word

$$A_0 = A_1^{-1} A_1' A_2^{-1} A_2' \ldots A_n^{-1} A_n'.$$

Now the part of p_1 to the left of C can be converted to a group picture over the group presentation $< \mathbf{a}; \mathbf{t} >$ of H by contracting all discs labelled $a^{\pm 1} a^{\mp 1}$ ($a \in \mathbf{a}$) to points, and re-orienting all arcs labelled a^{-1} and relabelling them by a ($a \in \mathbf{a}$). This picture has boundary label $A A_0 A^{-1}$. By the van Kampen lemma (see for example [8]) the boundary label of a picture over a group presentation defines the identity of the group presented, so $A A_0 A^{-1}$ (and hence A_0) represents the identity of H. Thus the label on γ is 1, so γ must not be reduced, as $LG(\mathcal{R})$ has no cycles. We conclude that for some i, $e_{U_i}^{\varepsilon_i} = e_{U_{i+1}}^{-\varepsilon_{i+1}}$. Thus the discs Δ_i, Δ_{i+1} constitute a partial dipole. If this partial dipole is a dipole then we are finished. Otherwise, we can single out a part of p_1 to the right of $\Delta_i, \beta_i, \Delta_{i+1}$ satisfying the assumptions of our lemma but with fewer **u**-discs (see Figure 4). We can then use the inductive hypothesis.

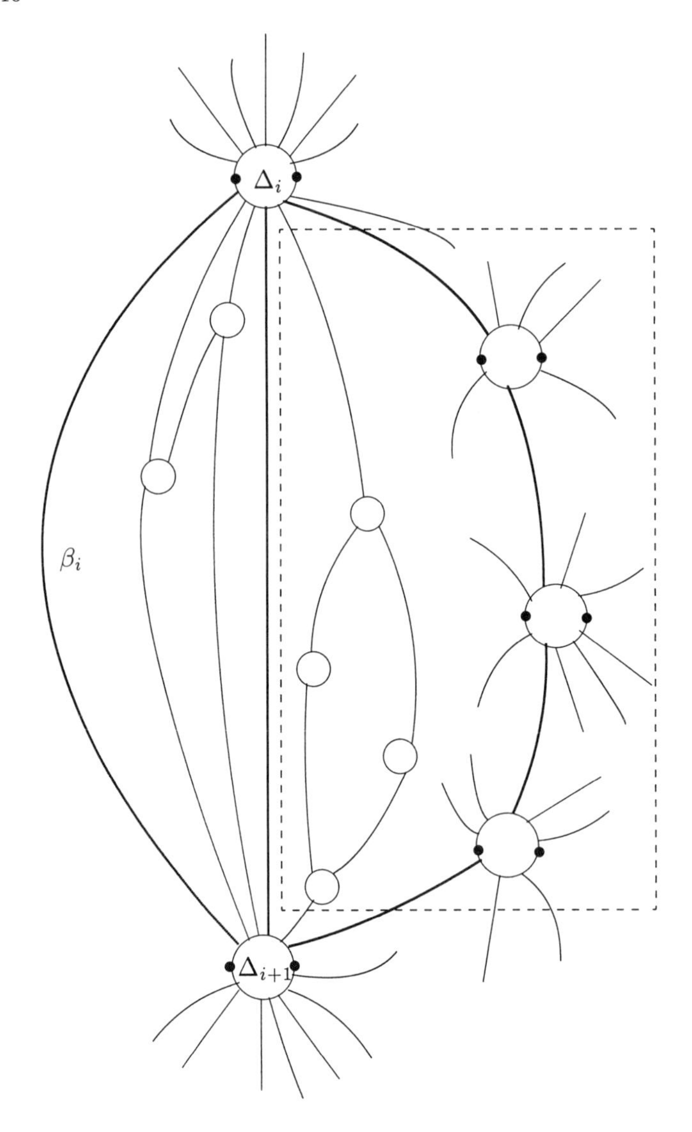

Figure 4

Proof of Lemma 3. If p is a spherical picture over $\mathcal{P}$ containing **u**-discs, then p will be homotopic in $\mathcal{D}(\mathcal{P})$ to a picture $(p_0 \cdot \iota(p_1))(\tau(p_0) \cdot p_1)$ where p_0 contains no **u**-discs and p_1 satisfies the assumptions of Lemma 4. Thus, by Lemma 4, p_1 (and hence p) contains a dipole.

References

1. S.I. Adian, *On the embeddability of semigroups into groups*, Doklady Akad. Nauk SSSR **133** (1960), 255-257; English translation: Soviet Mathematics **1** (1960), Amer. Math. Soc.
2. S.I. Adian, *Defining relations and algorithmic problems for groups and semigroups*, Trudy Mat. Inst. Steklov **85** (1966); Translations Amer. Math. Soc. **152** (1967).
3. R. Cremanns and F. Otto, *For groups the property of having finite derivation type is equivalent to the homological finiteness condition* FP_3, J. Symb. Comput. **22** (1996), 155-177.
4. V.S. Guba and M. Sapir, *Diagram groups*, Memoirs of the American Mathematical Society **130** (1997), 1-117.
5. C.W. Kilgour, *Using pictures in combinatorial group and semigroup theory*, PhD thesis, University of Glasgow, 1995.
6. C.W. Kilgour, *Relative monoid presentations*, Glasgow University Preprint Series **96/44**, 1996.
7. Y. Kobayashi and F. Otto, *For finitely presented monoids the homological finiteness conditions* FHT *and bi-*FP_3 *coincide*, J. Algebra **264** (2003), 327-341.
8. R.C. Lyndon and P.E. Schupp, *Combinatorial Group Theory*, Springer-Verlag, New York, 1977.
9. A. Malheiro, *Complete presentations for submonoids*, preprint, Universidade Nova de Lisboa, 2002.
10. S. McGlashan and S.J. Pride, *Finiteness conditions for rewriting systems*, Int. J. Algebra and Comput, to appear.
11. S.J. Pride, *Low-dimensional homotopy theory for monoids*, Int. J. Algebra and Comput. **5** (1995), 631-649.
12. S.J. Pride, *Geometric methods in combinatorial semigroup theory*, in: Proceedings of International Conference on Semigroups, Formal Languages and Groups, J. Fountain (ed.), Kluwer Publishers, 1995, 215-232.
13. S.J. Pride, *Low dimensional homotopy theory for monoids II: groups*, Glasgow Math. J. **41** (1999), 1-11.
14. S.J. Pride and F. Otto, *For rewriting systems the topological finiteness conditions FDT, FHT are not equivalent*, J. London Math. Soc, to appear.
15. S.J. Pride and Jing Wang, *Rewriting systems, finiteness conditions and associated functions*, in Algorithmic Problems in Groups and Semigroups (ed. J.C. Birget, S. Margolis, J. Meakin and M. Sapir), Birkhauser, Boston, 2000, 195-216.
16. J.-P. Serre, *Trees*, Springer-Verlag, 1980.
17. C.C. Squier, *A finiteness condition for rewriting systems*, revision by F. Otto and Y. Kobayashi, Theoretical Comp. Sci. **131** (1994), 271-294.
18. X. Wang and S.J. Pride, *Second order Dehn functions of groups and monoids*, Int. J. Algebra and Comput. **10** (2000), 425-456.

INVERSE SEMIGROUPS ACTING ON GRAPHS

JAMES RENSHAW
Faculty of Mathematical Studies,
University of Southampton,
Southampton, SO17 1BJ, U.K.
E-mail: j.h.renshaw@maths.soton.ac.uk

There has been much work done recently on the action of semigroups on sets with some important applications to, for example, the theory and structure of semigroup amalgams. It seems natural to consider the actions of semigroups on sets 'with structure' and in particular on graphs and trees. The theory of group actions has proved a powerful tool in combinatorial group theory and it is reasonable to expect that useful techniques in semigroup theory may be obtained by trying to 'port' the Bass-Serre theory to a semigroup context. Given the importance of *transitivity* in the group case, we believe that this can only reasonably be achieved by restricting our attention to the class of inverse semigroups. However, it very soon becomes apparent that there are some fundamental differences with inverse semigroup actions and even such basic notions such as *free actions* have to be treated carefully. We make a start on this topic in this paper by first of all recasting some of Schein's work on representations by partial homomorphisms in terms of actions and then trying to 'mimic' some of the basic ideas from the group theory case. We hope to expand on this in a future paper [5].

1. Introduction

The Bass-Serre theory of group actions on graphs has proved a powerful tool for combinatorial group theorists and the aim of this paper is to consider whether there is any 'mileage' in trying to use these techniques in the context of semigroup theory as well. In particular we aim to develop the theory of (partial) actions of inverse semigroups on graphs and trees and highlight some of the connections with the case for group actions.
In section 2 we give a brief account of the classical theory for group actions on graphs and trees before introducing the basic concept of a partial inverse semigroup action in section 3. We include in this section a brief account of Schein's $\omega-$cosets and partial congruences together with a brief account of a concept of a free action. In section 4, $S-$Graphs are introduced and we present a few examples to illustrate the underlying concepts. We hope to

extend this work in a future publication [5].

2. The Group Case

We 'set the scene' in this section by giving a very brief outline of the main players in the Bass-Serre theory of groups. The notation and terminology is mainly that of [2] and we refer the reader to that text for more details. Let G be a group. By a (left) $G-$set we mean a non-empty set, X, on which G acts by permutations, in the sense that there is a group homomorphism $\rho : G \to \mathrm{Sym}X$, where Sym X is the symmetric group on X. As usual, we denote $\rho(g)(x)$ by gx. If X and Y are $G-$sets then a function $f : X \to Y$ is called a $G-$map if for all $x \in X, g \in G, f(gx) = gf(x)$. Let X be a $G-$set. By the *$G-$stabilizer* of an element $x \in X$ we mean the set of elements of G that 'fix' x, i.e. $G_x = \{g \in G : gx = x\}$. It is easy to see that G_x is a subgroup of G and that for any $g \in G, G_{gx} \simeq gG_xg^{-1}$. A group G is said to act *freely* on X if $G_x = 1$ for all $x \in X$. The *$G-$orbit* of an element x is the set $Gx = \{gx : g \in G\}$ which is a $G-$subset of X and it is easy to prove that Gx is $G-$isomorphic to the $G-$set of cosets of G_x in G, denoted G/G_x. The *quotient set* for the $G-$set X is the set of $G-$orbits, $G\backslash X = \{Gx : x \in X\}$ which clearly has a natural map $X \to G\backslash X, x \mapsto Gx$. A *$G-$transversal in X* is a subset Y of X which contains exactly one element of each $G-$orbit of X. Hence the composite $Y \subseteq X \to G\backslash X$ is a bijection.

A $G-$graph (X, V, E, ι, τ) is a non-empty $G-$set X with disjoint non-empty $G-$subsets V and E such that $X = V \cup E$ and two $G-$maps $\iota, \tau : E \to V$. If Y is a $G-$subset of X then we write $VY = V \cap Y, EY = E \cap Y$. If Y is non-empty and both ιe and τe belong to VY for all e in Y then we say that Y is a $G-$subgraph of X.

By the *quotient graph*, $G\backslash X$, we mean the graph $(G\backslash X, G\backslash V, G\backslash E, \bar{\iota}, \bar{\tau})$ where $\bar{\iota}(Ge) = G\iota e, \bar{\tau}(Ge) = G\tau e$ for all $Ge \in G\backslash E$.

If $G\backslash X$ is connected then it can be shown (see [2]) that there exist subsets $Y_0 \subseteq Y \subseteq X$ such that Y is a $G-$transversal in X, Y_0 is a subtree of X with $VY_0 = VY$ and for each $e \in EY, \iota(e) \in VY$. In this case, Y is called a *fundamental transversal* in X.

The *Cayley graph* of G with respect to a subset T of G is the $G-$graph, $X(G, T)$, with vertex set $V = G$, edge set $E = G \times T$ and incidence function $\iota(g, t) = g, \tau(g, t) = gt$ for all $(g, t) \in E$.

For example, consider the cyclic group $C_4 = \langle s : s^4 \rangle$ and $T = \{s\}$ then the Cayley graph can be represented as

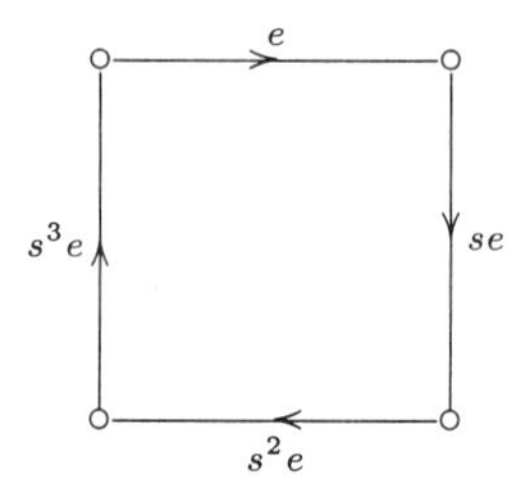

where $e = (1, s) \in G \times T$. The quotient graph is

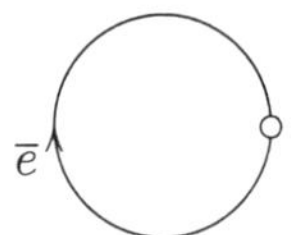

and a corresponding fundamental $G-$transversal is

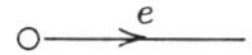

A *graph of groups* $(G(-), Y)$, is a connected graph $(Y, V, E, \overline{\iota}, \overline{\tau})$ together with a function $G(-)$ which assigns to each $v \in V$ a group $G(v)$ and to each edge $e \in E$ a subgroup $G(e)$ of $G(\overline{\iota}e)$ and a group monomorphism $t_e : G(e) \to G(\overline{\tau}e)$.
The standard example is given as follows. We start with a $G-$graph X such that $G\backslash X$ is connected and choose a fundamental transversal Y with subtree Y_0. For each edge e in EY, there are unique vertices $\overline{\iota}e, \overline{\tau}e$ in VY which belong to the same $G-$orbit as $\iota e, \tau e$ respectively. From the way that Y is defined, we see that in fact $\overline{\iota}e = \iota e$. By using the incidence functions $\overline{\iota}, \overline{\tau} : EY \to VY$ we can thus make Y into a graph.
For each e in EY, τe and $\overline{\tau}e$ belong to the same $G-$orbit and so there exists t_e in G such that $t_e\overline{\tau}e = \tau e$ - if $\overline{\tau}e = \tau e$ then we can take $t_e = 1$. Notice then that $G_{\tau e} = t_e G_{\overline{\tau}e} t_e^{-1}$. Now clearly, $G_e \subseteq G_{\iota e}, G_e \subseteq G_{\tau e}$ and so there is an embedding $G_e \to G_{\overline{\tau}e}$ given by $g \mapsto t_e^{-1} g t_e$.
Hence we have constructed a graph of groups *associated to* X.
For our previous example, we have $G_e = \{1\} = G_{\iota e}$ and the connecting element $t_e = s$.

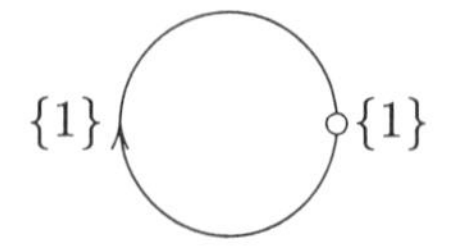

Let $(G(-), Y)$ be a graph of groups. Choose a spanning subtree Y_0 of Y. It follows that $VY_0 = VY$. The *fundamental group* $\pi(G(-), Y, Y_0)$ is the group with generating set $\{t_e : e \in E\} \cup \bigcup_{v \in V} G(v)$ and relations : the relations for $G(v)$, for each $v \in VY$; $t_e^{-1} g t_e = t_e(g)$ for all e in EY; $t_e = 1$ for all $e \in EY_0$.

From the fundamental group, $G = \pi(G(-), Y, Y_0)$,we can construct a *standard* G–graph as follows: Let T be the G–set generated by Y and relations saying that for each y in Y, $G(y)$ stabilizes y. Then T has G–subsets $VT = GV$ and $ET = GE$. Define $\iota, \tau : ET \to VT$ by $\iota(ge) = g\bar{\iota}(e), \tau(ge) = g t_e \overline{\tau}(e)$. Then T is a G–graph with fundamental transversal Y.
It is straightforward to verify that the graph of groups associated to this G–graph is isomorphic to the original graph of groups. Conversely, we can show that given a group, G, acting on a *tree* we can form the graph of groups and the fundamental group, π, is then isomorphic to G and the standard graph is isomorphic to the original G–tree. See [2,§3.4 & Theorem 4.1] for details.

The two classic examples of fundamental groups arise from the following two graphs of groups:

$$A \circ \xrightarrow{\;C\;} \circ B$$

and

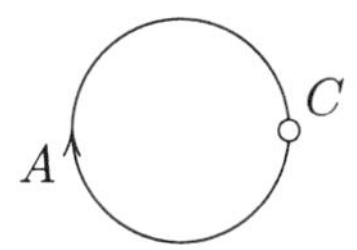

In the former case, the fundamental group is the amalgamated free product $A *_C B$ while in the later case it is the HNN-extension $A *_C t$.

3. Partial Actions of Inverse Semigroups

3.1. *Introduction*

Throughout the rest of this paper, S will denote an inverse semigroup. We refer the reader to [3] or [4] for basic concepts relating to semigroups and inverse semigroups. By a (left) partial S–act, X, we mean a partial action of S on the set X such that $(st)x$ exists if and only if $s(tx)$ exists and then

$$(st)x = s(tx).$$

In addition, we require that whenever $sx = sy$ then $x = y$. The *domain* of an element $s \in S$ is the set

$$D_s^X = \{x \in X : sx \in X\}.$$

We shall denote D_s^X as simply D_s when the context is clear. Notice then that $x \in D_{st}$ if and only if $x \in D_t$ and $tx \in D_s$. Right partial $S-$acts are defined dually. By a *partial biact* X we mean a set which is both a left partial $S-$act and a right partial $S-$act and which satisfies the additional condition that for all $s, t \in S, x \in X$, if sx and xt exist then $s(xt)$ and $(sx)t$ exist and $s(xt) = (sx)t$. To signify that X is an $S-$biact we shall sometimes denote it by ${}_SX_S$.
Note that if X is a left $S-$act then it is also a right $S-$act with multiplication given by $xs = s^{-1}x$. However, it may not be an $S-$biact.

Example 3.1. Let S be an inverse semigroup and let $s \in S$. Define $D_s = s^{-1}sS$ and for $x \in D_s$ define $s \cdot x = sx$. Then $(S, \cdot)$ is a left partial $S-$act. This is the partial act induced by the Preston-Wagner representation of S. In a similar way, S is also a right partial $S-$act and in fact a partial $S-$biact.

Example 3.2. Let S be an inverse semigroup and X a non-empty set. Define a left partial action of S on X by $sx = x$ for all $x \in X, s \in S$. This is called the *trivial left partial action of S on X*. There is clearly a similar construction for trivial right partial actions.

Notice that if X is a left partial $S-$act, then we can consider it a partial $S-$biact with the trivial right partial action.

Lemma 3.1. *Let X be a partial $S-$act and let $s \in S, e \in E(S), x, y \in X$.*

(1) If $sx = y$ then $x = s^{-1}y$;
(2) if $x \in D_s$ then $s^{-1}sx = x$;
(3) if $x \in D_e$ then $ex = x$.

Proof.

(1) $sx = ss^{-1}sx = ss^{-1}y$ and so $x = s^{-1}y$.
(2) $sx = ss^{-1}sx$ and so $x = s^{-1}sx$.
(3) $e(ex) = ex$ and so $ex = x$. □

The set of all elements of S that act on x will be denoted $D^x = \{s \in S : x \in D_s\}$. In addition we shall use the notation $D^X = \bigcup_{x \in X} D^x$.

Let H be a subset of an inverse semigroup S. Denote by $H\omega$ the set $\{s \in S : s \geq h, \text{for some } h \in H\}$. This is called the *closure* of H and we say that H is *closed* if $H\omega = H$. Notice that $H \subseteq H\omega$ and that $(H\omega)\omega = H\omega$.

Lemma 3.2. *Let S be an inverse semigroup and X a left partial $S-$act. Let $x \in X$.*

(1) D^x is a union of $\mathcal{L}-$classes;
(2) D^x is a closed subset of S;
(3) D^X is a union of $\mathcal{D}-$classes;
(4) D^X is closed under the taking of inverses.

Proof.

(1) If $s \in D^x$ and $t \,\mathcal{L}\, s$ then $t^{-1}t = s^{-1}s$ and so $t \in D^x$.
(2) If $s \in D^x$ and $t \geq s$ then there exists $e \in E(S)$ with $s = et$ and so $t \in D^x$.
(3) If $s \in D^x$ and $s \,\mathcal{D}\, t$ then $s \,\mathcal{L}\, u \,\mathcal{R}\, t$ for some $u \in S$. Hence $u \in D^x$ by part (1) and $uu^{-1} = tt^{-1}$ and so $t \in D^{t^{-1}ux}$.
(4) If $s \in D^x$ then $s^{-1} \in D^{sx}$. □

A element x of X is said to be *effective* if $D^x \neq \emptyset$. A partial $S-$act X is *effective* if all its elements are effective. A partial $S-$act is *transitive* if for all $x, y \in X$, there exists $s \in S$ with $y = sx$.

Example 3.3. Let S and T be (disjoint) inverse semigroups and consider the $0-$direct union $R = S \dot{\cup} T \dot{\cup} \{0\}$ which is of course also an inverse semigroup. Then S is a left partial $R-$act if we define $r \cdot s = rs$ whenever $s = r^{-1}rs$. Notice that in this case, we must have $D^s \subseteq S$. Notice also that S is an effective left partial $R-$act since for all $s \in S, \{s^{-1}\} \in D^s$. In a similar way, we can consider T as an effective left partial $R-$act and for all $t \in T, D^t \subseteq T$. Now $S \times T$ is a left partial $R-$act with the induced action $r \cdot (s,t) = (r \cdot s, r \cdot t)$ and it then follows that for all $(s,t) \in S \times T, D^{(s,t)} = \emptyset$.

If X is a left partial $S-$act and Y is a subset of X then we shall say that Y is a partial $S-$*subact* of X if for all $s \in S, y \in D_s^X \cap Y \Rightarrow sy \in Y$. Notice that this makes Y a left partial $S-$act with the partial action that induced from X and $D_s^Y = D_s^X \cap Y$ for all $s \in S$.

Example 3.4. Let U be an inverse subsemigroup of an inverse semigroup S. Then S is a partial $U-$biact with action defined by $u \cdot s = us$ for

$s = u^{-1}us$ and $s \cdot u = su$ for $s = suu^{-1}$. If follows that U (with partial action given by Example 3.1) is then a $U-$subact of S.

Let $x \in X$ and define the $S-$*orbit* of x as $S^1x = \{sx : s \in S\} \cup \{x\}$. We use S^1 instead of S to take account of those elements x which are not effective. Notice that S^1x is a left partial $S-$subact of X (the *partial subact generated by* x) and that the action is such that, for all $tx \in S^1x$ and all $s \in S$, $tx \in D_s^{S^1x}$ if and only if $x \in D_{st}^X$ and in which case $s(tx) = (st)x$. Then we have

Lemma 3.3. *For all $x \in X$ the $S-$orbit S^1x, is a transitive left partial $S-$act. If x is effective then so is S^1x.*
Conversely, if a left partial $S-$act is effective and transitive then it has only one $S-$orbit.

Proof. If $y = s_1x$ and $z = s_2x$ then put $t = s_1s_2^{-1}$ to get $y = tz$.
Suppose that x is effective. Then let $sx \in S^1x$ and notice that $ss^{-1}(sx) = sx \in S^1x$ and so S^1x is effective.
The converse is easy. □

Notice that $S^1x = S^1y$ if and only if $y \in S^1x$.

Example 3.5. Consider S as a left partial $S-$act with partial action given as in Example 3.1. Then for all $s \in S, S^1s = L_s$ the $\mathcal{L}-$class containing s.

This is easy to establish : if $t \in S^1s$ then $t = us$ and $s = u^{-1}us$ and so $u^{-1}t = u^{-1}us = s$ and hence $t \in L_s$. Conversely, if $t \in L_s$ then $t^{-1}t = s^{-1}s$ and so $t = tt^{-1}t = (ts^{-1})s$ with $(ts^{-1})^{-1}(ts^{-1})s = st^{-1}ts^{-1}s = ss^{-1}s = s$ as required.

Let X and Y be two partial $S-$acts. A function $f : X \to Y$ is called an $S-$*map* if for all $s \in S$, $x \in D_s^X$ *if and only if* $f(x) \in D_s^Y$ and then $f(sx) = sf(x)$.
For example, if Y is an $S-$subact of an left partial $S-$act X, then the inclusion map $\iota : Y \to X$ is an $S-$map.
The use of "if and only if" as opposed to "only if" in the definition of $S-$map may seem unnecessary but note that both conditions are needed in the proof of Theorem 3.2 below.

Lemma 3.4. *Let S be an inverse semigroup and let $s, t \in S$ then the following are equivalent:*

(1) $L_s \cong L_t$ *as left partial* $S-$*acts,*
(2) for all $u \in S$, $s \in D_u^{L_s}$ *if and only if* $t \in D_u^{L_t}$,
(3) $s\mathcal{D}t$ *in* S.

Proof.
(1) $\Rightarrow$ (2). Suppose that $L_s \cong L_t$. Then there is a left $S-$ isomorphism $\phi : L_s \to L_t$ which maps s to t' say. But $L_{t'} = L_t$ and so we may as well assume that $t' = t$. Now since ϕ is an $S-$map then for all $u \in S$, $s \in D_u^{L_s}$ if and only if $t \in D_u^{L_t}$.
(2) $\Rightarrow$ (3). Since $s \in D_{s^{-1}}^{L_s}$ then $t \in D_{s^{-1}}^{L_t}$ and so $s^{-1}t \in L_t$. In other words $t = ss^{-1}t$ and therefore $t^{-1} = t^{-1}ss^{-1}$. In a similar way, $s^{-1} = s^{-1}tt^{-1}$. Now $t\,\mathcal{L}\,s^{-1}t$ and so $tt^{-1}\,\mathcal{L}\,s^{-1}tt^{-1} = s^{-1}$. But $t\,\mathcal{R}\,tt^{-1}$ and so $t\,\mathcal{D}\,s^{-1}\,\mathcal{D}\,s$ as required.
(3) $\Rightarrow$ (1). Let $(s,t) \in \mathcal{D}$. If $s = t$ then clearly $L_s \cong L_t$ and so we can assume that $s \neq t$. Then there exists $u \in S$ such that $\rho_u : L_s \to L_t, \rho_{u^{-1}} : L_t \to L_s$ given by $\rho_u(x) = xu, \rho_{u^{-1}}(y) = yu^{-1}$ are mutually inverse $\mathcal{R}-$class preserving bijections. This clearly means that ρ_u and $\rho_{u^{-1}}$ are left $S-$maps and so $L_s \cong L_t$ as required. □

Notice that the superscript in the expression "$s \in D_u^{L_s}$" is superfluous and we have simply inserted it the previous lemma for the sake of clarity. We shall in future denote $D_u^{L_s}$ as simply D_u.

Notice that the orbits of an effective partial $S-$act X, partition X. In this case the *quotient set* of X is defined as the set $S \setminus X = \{S^1x : x \in X\}$.

Let X be a left partial $S-$act and ρ an equivalence on X. If ρ has the additional property that whenever $s \in S, (x,y) \in \rho$ are such that $x \in D_s$ if and only if $y \in D_s$, then $(sx, sy) \in \rho$ whenever $x \in D_s$, then we shall call ρ a *left* $S-$*congruence* on X. In other words, whenever $(x,y) \in \rho$ then either both $x, y \in D_s$ or neither are, and when both are then $(sx, sy) \in \rho$. *Right* $S-$*congruences* are defined dually. It is then easy to check that the quotient X/ρ is a left partial $S-$act with action given by $s(x\rho) = (sx)\rho$ whenever $x \in D_s$.

Example 3.6. Let X be a partial $S-$biact and define a relation $\mathcal{L}^X$ on X by $\mathcal{L}^X = \{(x,y) : S^1x = S^1y\}$. Then it is easy to show that $\mathcal{L}^X$ is a right $S-$congruence on X and so the quotient $X/\mathcal{L}^X \cong S \setminus X$ is a right partial $S-$act. If X is only a left partial $S-$act then $\mathcal{L}^X$ is an equivalence on X.

It is clear (see Example 3.5) that if $X = S$ with partial action given by the Preston-Wagner action, then $\mathcal{L}^X$ is Green's $\mathcal{L}-$relation.

Dually, we define $\mathcal{R}^X = \{(x,y) : xS^1 = yS^1\}$ and note that $\mathcal{R}^X$ is a left $S-$congruence on X and that $\mathcal{R}^S = \mathcal{R}$, Green's $\mathcal{R}-$relation.

The intersection of the equivalences $\mathcal{L}^X$ and $\mathcal{R}^X$ will be denoted $\mathcal{H}^X$ and the join $\mathcal{L}^X \vee \mathcal{R}^X$ by $\mathcal{D}^X$.

Denote the $\mathcal{R}^X-$class of x by R_x^X. Similar meaning is attached to L_x^X, H_x^X, D_x^X. The following results are then easy to prove.

Theorem 3.1. *Let S be an inverse semigroup and X a partial $S-$biact. Then*

(1) $\mathcal{L}^X \circ \mathcal{R}^X = \mathcal{R}^X \circ \mathcal{L}^X = \mathcal{D}^X$.

(2) If $(x,y) \in \mathcal{L}^X$ then there exists $s \in S^1$ such that $x = sy$ and the functions $\lambda_s : R_y^X \to R_x^X, \lambda_{s^{-1}} : R_x^X \to R_y^X$ defined by $\lambda_s(z) = sz, \lambda_{s^{-1}}(z) = s^{-1}z$ are mutually inverse $\mathcal{L}^X-$class preserving bijections.

It is clear that the intersection of two left $S-$congruences is again a left $S-$congruence. Let σ be a relation on a left partial $S-$act X and suppose there exists a left $S-$congruence ρ with $\sigma \subseteq \rho$. Then the smallest left $S-$congruence containing σ will be referred to as the left $S-$congruence *generated by* σ and denoted $\sigma^\sharp$. It is clearly equal to the intersection of all left $S-$congruences which contain σ. However, it is important to realise that not every relation generates an $S-$congruence.
For example, let S be an inverse semigroup which is *not* bisimple. Let $s, t \in S$ be such that $(s,t) \notin \mathcal{D}$ and let $\sigma = \{(s,t)\}$. Then the left $S-$congruence generated by σ on the left partial $S-$act S does not exist. If it did then there would exist a left $S-$congruence on S, ρ say, such that $(s,t) \in \sigma \subseteq \rho$. But then for all $u \in S, s \in D_u$ if and only if $t \in D_u$ which is impossible by Lemma 3.4.

Also, if ρ is an $S-$congruence on X then the natural map $\rho^\natural : X \to X/\rho$ is an $S-$map.

Theorem 3.2. *If $f : X \to Y$ is an $S-$map then* $\ker(f)$ *is an $S-$congruence on X.*
Moreover there exists a unique $S-$monomorphism $\phi : X/\ker(f) \to Y$ such

that

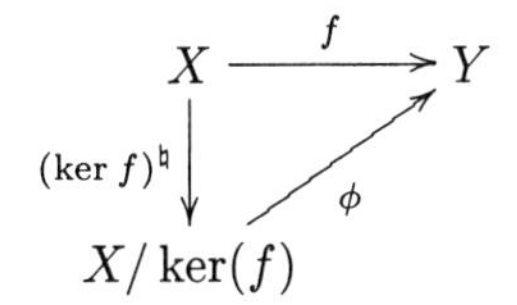

commutes.

Proof. It is clear that $\ker(f)$ is an equivalence and if $(x, x') \in \ker(f), s \in S$ then if $x \in D_s^X$, it follows that $f(x') = f(x) \in D_s^Y$ since f is an $S-$map and so $x' \in D_s^X$. It is then clear that $(sx, sx') \in \ker(f)$.
Define $\phi : X/\ker(f) \to Y$ by $\phi(x\ker(f)) = f(x)$. That this is well-defined and one to one and that the diagram commutes follows in the usual way. Moreover, $x\ker(f) \in D_s^{X/\ker(f)}$ if and only if $x \in D_s$ if and only if $f(x) \in D_s^Y$ and then $sf(x) = f(sx)$ which means that $\phi(s(x\ker(f))) = s\phi(x\ker(f))$. It is clear that ϕ is unique. □

Let X and Y be left partial $S-$acts and consider the disjoint union $X \dot{\cup} Y$. This is clearly a left partial $S-$act and it is easy to check that it satisfies the properties of a *coproduct* of the acts X and Y. In more detail, we have $S-$maps $\iota_X : X \to X \dot{\cup} Y$ and $\iota_Y : Y \to X \dot{\cup} Y$ and if there is a left partial $S-$act Z and $S-$maps $\alpha_X : X \to Z, \alpha_Y : Y \to Z$ then we can define an $S-$map $\phi : X \dot{\cup} Y \to Z$ by $\phi(x) = \alpha_X(x), \phi(y) = \alpha_Y(y)$ which makes the diagram

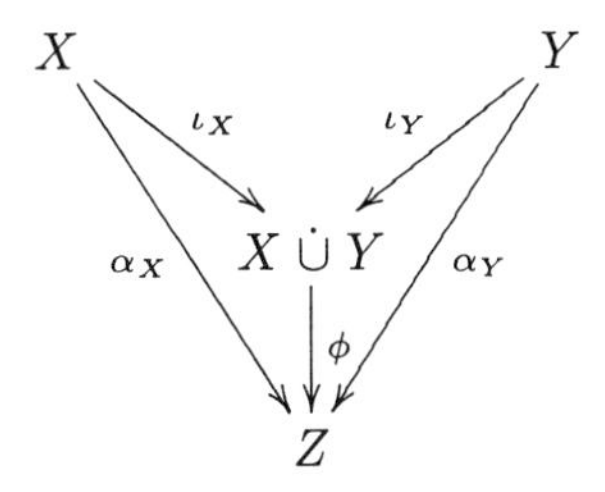

commute.
In a similar way, we can define an obvious left $S-$action on the cartesian product $X \times Y$ by $s(x, y) = (sx, sy)$ whenever both $x, y \in D_s$. This makes $X \times Y$ into a *product* with projections $\pi_X : X \times Y \to X, \pi_Y : X \times Y \to Y$ such that given any left partial $S-$act Z and maps $\alpha_X : Z \to X, \alpha_Y : Z \to Y$ there is a unique $S-$map $\phi : Z \to X \times Y$ (given by $z \mapsto (\alpha_X(x), \alpha_Y(y))$)

such that the diagram

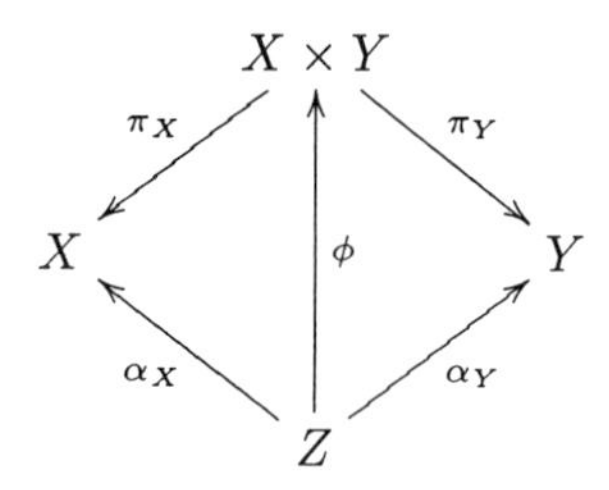

commutes.

Notice that if X is a left partial $S-$act and Y is a partial right $S-$act then $X \times Y$ is a partial $S-$biact with obvious action $s(x,y) = (sx,y), (x,y)t = (x,yt)$.

We shall say that a left partial $S-$act X is *complete* if it satisfies the property that for all $s \in S$ and for all $x, y \in X$, $x \in D_s$ if and only if $y \in D_s$. This means that an element $s \in S$ either acts on all elements of X or on none. Notice that X is complete if and only if the universal equivalence $\nabla_X = X \times X$ is an $S-$congruence on X.
If X is a complete partial $S-$subact of a left partial $S-$act Y, then we can define a left $S-$congruence $\sigma_X = \nabla_X \cup 1_Y$ on Y. We shall denote the quotient Y/σ_X as Y/X and refer to it as the *Rees quotient* of Y by X.

3.2. *Stabilisers and $\omega-$cosets*

Lemma 3.5. *Let H be an inverse subsemigroup of an inverse semigroup S. Then H is closed in S if and only if H is unitary in S.*

Proof. Suppose that H is closed in S and that $hs = h'$ for some $h, h' \in H, s \in S$. Then $h^{-1}hs = h^{-1}h'$ and so $s \geq h^{-1}h' \in H$ and therefore $s \in H$ and H is left unitary in S. Right unitary follows in a similar manner.
Conversely, if H is unitary in S and if $s \geq h$ for some $h \in H, s \in S$ then $hh^{-1} = sh^{-1}$ and so $s \in H$ as required. □

Let T and R be inverse subsemigroups of an inverse semigroup S and suppose that R is closed. If $T \subseteq R$ then $T\omega \subseteq R$. To see this notice that if $h \in T\omega$ then $h \geq t$ for $t \in T \subseteq R$. Hence $h \in R\omega = R$.

We briefly review Schein's theory of partial congruences (see [1] or [3] for more details).

Let $T \subseteq S$ be sets and suppose that ρ is an equivalence on T. Then we say that ρ is a *partial equivalence* on S with domain T. It is easy to establish that ρ is a partial equivalence on S if and only if it is symmetric and transitive. If now T is an inverse subsemigroup of an inverse semigroup S and if ρ is left compatible with the multiplication on S (in the sense that for all $s \in S, (u,v) \in \rho$ either $su, sv \in T$ or $su, sv \in S \setminus T$ and $(su, sv) \in \rho$ in the former case) then ρ is called a *left partial congruence* on S and the set T/ρ of $\rho-$classes will often be denoted by S/ρ.

Theorem 3.3. ([1,Theorem 7.10]) *Let H be a closed inverse subsemigroup of an inverse semigroup S. Define*

$$\pi_H = \{(s,t) \in S \times S : s^{-1}t \in H\}.$$

Then π_H is a left partial congruence on S and the domain of π_H is the set $D_H = \{s \in S : s^{-1}s \in H\}$.
The (partial) equivalence classes are the sets $(sH)\omega$ for $s \in D_H$. The set $(sH)\omega$ is the equivalence class that contains s and in particular H is one of the π_H-classes.

The sets $(sH)\omega$, for $s \in D_H$, are called the *left $\omega-$cosets* of H in S. The set of all left $\omega-$cosets is denoted by S/H. Notice that if $(sH)\omega$ is a left $\omega-$coset then $s^{-1}s \in H$.

Proposition 3.1. ([3, Proposition 5.8.3]) *Let H be an inverse subsemigroup of an inverse semigroup S, and let $(aH)\omega, (bH)\omega$ be left $\omega-$cosets of H. Then the following statements are equivalent:*

(1) $(aH)\omega = (bH)\omega$;
(2) $b^{-1}a \in H\omega$;
(3) $a \in (bH)\omega$;
(4) $b \in (aH)\omega$.

Lemma 3.6. ([1,Lemma 7.11]) *With H and S as in Theorem 3.3,*

(1) precisely one left $\omega-$coset, namely H, contains idempotents,
(2) each left $\omega-$coset is closed,
(3) π_H is left cancellative i.e. $((xa)H)\omega = ((xb)H)\omega$ implies that $(aH)\omega = (bH)\omega$.

Let X be a partial $S-$act and let $x \in X$. Denote by S_x the set of elements of S that fix x, i.e. $S_x = \{s \in S : sx = x\}$. We shall call S_x the *$S-$stabiliser* of x. Notice that $s^{-1}t \in S_x$ if and only if $sx = tx$.

For example, if X is a trivial partial $S-$act, then $S_x = S$ for all $x \in X$, while if S is a group (considered as a partial $S-$act) then $S_s = \{1\}$ for all $s \in S$.

Theorem 3.4. (cf. [1,Lemma 7.18]) *For all $x \in X$, S_x is either empty or a closed inverse subsemigroup of S.*

Proof. Assume that $S_x \neq \emptyset$. If $s, t \in S_x$ then $x = sx = s(tx) = (st)x$ and so S_x is a subsemigroup. Also $sx = x$ implies that $x = s^{-1}x$ and so S_x is an inverse subsemigroup of S. Let $h \geq s$ with $s \in S_x$. Then $s^{-1}h = s^{-1}s$ and so $s^{-1}sx = s^{-1}hx$ which means that $sx = hx$ and so $h \in S_x$. □

From Lemma 3.6 we can easily deduce the following important result.

Theorem 3.5. (cf. [1,Lemma 7.19]) *If H is a closed inverse subsemigroup of an inverse semigroup S then S/H is a left partial $S-$act with action given by $s \cdot X = (sX)\omega$ whenever $X, sX \in S/H$. Moreover, it is easy to establish that $S_{H\omega} = H$*

Lemma 3.7. ([1,Lemma 7.26]) *Let K be a closed inverse subsemigroup of an inverse semigroup S. Let $s^{-1}s \in K$ so that $(sK)\omega$ is a left $\omega-$coset of K. Then $sKs^{-1} \subseteq S_{(sK)\omega}$.*

If H and K are two closed inverse subsemigroups of S then we say that H and K are *conjugate* if $S/H \cong S/K$ (as partial $S-$acts).

Theorem 3.6. ([1,Theorem 7.27]) *Let H and K be closed inverse subsemigroups of an inverse semigroup S. Then H and K are conjugate if and only if there is an element $s \in S$ such that*

$$s^{-1}Hs \subseteq K \text{ and } sKs^{-1} \subseteq H.$$

Moreover, any such element s necessarily satisfies $ss^{-1} \in H, s^{-1}s \in K$.

Theorem 3.7. *Let H and K be closed inverse subsemigroups of an inverse semigroup S. Then H and K are conjugate if and only if there is an element $s \in S$ such that*

$$(s^{-1}Hs)\omega = K \text{ and } (sKs^{-1})\omega = H.$$

Proof. From Theorem 3.6, if H and K are conjugate, then there is an element $s \in S$ such that

$$s^{-1}Hs \subseteq K \text{ and } sKs^{-1} \subseteq H.$$

Now it is clear that $(s^{-1}HS)\omega \subseteq K$ so let $k \in K$ and let $l = sks^{-1} \in H$. Now put $m = s^{-1}ls = s^{-1}sks^{-1}s \in s^{-1}Hs$ and notice that $m \leq k$ and so $k \in (s^{-1}Hs)\omega$ as required. □

Notice that if $ss^{-1} \in H$ then $s^{-1}Hs$ is an inverse subsemigroup of H. To see this note that $s^{-1}hss^{-1}h^{-1}ss^{-1}hs = s^{-1}h(ss^{-1})(h^{-1}ss^{-1}h)s = s^{-1}h(h^{-1}ss^{-1}h)(ss^{-1})s = s^{-1}ss^{-1}hh^{-1}hs = s^{-1}hs$.

Suppose that H is a closed inverse subsemigroup of S and that there exists $s \in S$ such that $ss^{-1} \in H$ is the identity of H. The map $\vartheta : H \to s^{-1}Hs$ given by $\vartheta(t) = s^{-1}ts$ is an embedding, since $\vartheta(tr) = s^{-1}trs = s^{-1}tss^{-1}rs = \vartheta(t)\vartheta(r)$ and if $s^{-1}ts = s^{-1}rs$ then $t = r$.

Conversely, if $ss^{-1} \in H$ is such that $\vartheta : H \to s^{-1}Hs$ given by $\vartheta(t) = s^{-1}ts$ is an embedding, then $\vartheta(tss^{-1}) = s^{-1}tss^{-1}s = s^{-1}ts = \vartheta(t)$ and so $tss^{-1} = t$. Similarly, $ss^{-1}t = t$ and so ss^{-1} is the identity of H.

In fact,

Theorem 3.8. *Let S be an inverse semigroup, $s \in S$ and suppose that H is a closed inverse subsemigroup with $ss^{-1} \in H$. Let $K = (s^{-1}Hs)\omega$. Then the following are equivalent*

(1) $H \to s^{-1}Hs$ given by $h \mapsto s^{-1}hs$ is an embedding of inverse semigroups,
(2) there exists an inverse semigroup embedding $\phi' : H \to s^{-1}Hs$ with $\phi'(ss^{-1}) = s^{-1}s$,
(3) there exists an inverse semigroup embedding $\phi' : H \to s^{-1}Hs$ and ss^{-1} is the identity of H,
(4) ss^{-1} is the identity of H,
(5) $H = sKs^{-1}$,
(6) $H \subseteq (ss^{-1})S(ss^{-1})$,

Proof. (1) ⇒ (2) This is clear.
(2) ⇒ (3) Let $t \in H$ and suppose that $\phi'(t) = s^{-1}t's$ for some $t' \in H$. Then we have $\phi'(tss^{-1}) = \phi'(t)\phi'(ss^{-1}) = s^{-1}t'ss^{-1}s = s^{-1}t's = \phi'(t)$ and so $tss^{-1} = t$. In a similar way, $ss^{-1}t = t$ and so ss^{-1} is the identity of H.
(3) ⇒ (4) Obvious.

(4) $\Rightarrow$ (5) For any $h \in H, h = ss^{-1}hss^{-1} \in sKs^{-1}$. Conversely, let $k \in K$ so that $k \geq s^{-1}hs$ for some $h \in H$. So $ke = s^{-1}hs$ for some $e \in E(S)$. Hence $skes^{-1} = ss^{-1}hss^{-1} = h$ and so $h = (sks^{-1})(ses^{-1})$. Hence $sks^{-1} \geq h$ and so $sKs^{-1} \subseteq H$ since H is closed.
(5) $\Rightarrow$ (6) Clear.
(6) $\Rightarrow$ (1) Let $\theta : H \to s^{-1}Ss$ be given by $\theta(h) = s^{-1}hs$. From the remarks preceding this theorem, it is clear that θ is a morphism. If $\theta(h) = \theta(h')$ then $s^{-1}ss^{-1}tss^{-1}s = s^{-1}ss^{-1}t'ss^{-1}s$ and hence $h = ss^{-1}tss^{-1} = ss^{-1}t'ss^{-1} = h'$ as required. $\square$

Note that under the conditions of the preceding theorem, $E(s^{-1}Hs) = s^{-1}E(H)s$. Moreover, we will see in Example 4.1, that $s^{-1}s$ need not be the identity in K.

Theorem 3.9. *Let X be an effective left partial $S-$act and let $x \in X$. Then $S^1x \cong S/S_x$.*

Proof. Consider the map $\alpha : S^1x \to S/S_x$ given by $\alpha(sx) = (sS_x)\omega$. Notice that if $sx = tx$ then $s^{-1}t \in S_x$ and so $ss^{-1}t \in sS_x$, which means that $t \in (sS_x)\omega$. Similarly, $s \in (tS_x)\omega$ and so α is well-defined. It is clear that α is an $S-$map that is onto. Suppose then that $(sS_x)\omega = (tS_x)\omega$. Then $s^{-1}t \in S_x$ and so $sx = tx$ as required. $\square$

Lemma 3.8. *Let S be an inverse semigroup and X a left partial $S-$act. Let $s \in S$ and $x \in D_s$. Then sS_xs^{-1} is an inverse subsemigroup of S.*

Proof. This follows from Theorem 3.4 and the remarks preceding Theorem 3.8. $\square$

Theorem 3.10. *Let S be an inverse semigroup and X a partial $S-$act. Let $s \in S$ and $x \in D_s$. Then S_x and S_{sx} are conjugate.*

Proof. Since $S^1x = Ssx$ the result follows from Theorem 3.9. In fact, we have that $(sS_xs^{-1})\omega = S_{sx}$. $\square$

3.3. *$\mathcal{L}-$classes and free partial acts*

Let X be a left partial $S-$act, $x \in X$ and consider the $S-$orbit S^1x. In general, not every element of S will act on x so recall that we put $D^x = \{s \in S : x \in D_s\}$. Then we have

Lemma 3.9. *For every $x \in X$, $D^x = \{s \in S : s^{-1}s \in S_x\} = \bigcup_{s \in D^x} L_s$.*

Proof. If $s^{-1}s \in S_x$ then $x \in D_s$ and so $s \in D^x$. Conversely, if $x \in D_s$ then $s^{-1}s \in S_x$ from Lemma 3.1. The last equation follows from Lemma 3.2. □

Let X be a left partial $S-$act and let $x \in D_s$. Notice then that if $y = sx$ then $s^{-1}sx = s^{-1}y = x$ so that $s^{-1}s \in S_x$ and $S_x \cap E(S) \neq \emptyset$. Following the definition for group actions, we shall say that an inverse semigroup S acts *freely* on a partial $S-$act X if for all $x \in X$ there exists $e \in E(S)$ such that $S_x = e\omega(= \{e\}\omega)$. Notice that if $S_x = e\omega$ then $S_{sx} = (ses^{-1})\omega$ for all $sx \in S^1x$. To see this note that if $t \in S_{sx}$ then $tsx = sx$ and so $s^{-1}ts \in S_x$. Hence $s^{-1}ts \geq e$ and so $t \geq ses^{-1}$.

Lemma 3.10. *Let S be an inverse semigroup, $s \in S$ and consider S as a left partial $S-$act. Then $S_s = \{s^{-1}s\}\omega$.*

Proof. If $t \in S_s$ then $s = t^{-1}ts = t^{-1}s$ and so $ss^{-1} = t^{-1}ss^{-1}$ and so $t^{-1} \in (ss^{-1})\omega$ and hence $t \in (ss^{-1})\omega$. Moreover, if $t \in (ss^{-1})\omega$ then $t^{-1} \in (ss^{-1})\omega$ and so $ss^{-1} = t^{-1}ss^{-1}$. But then $s = t^{-1}s$ and so $t \in D_s$ and $ts = tt^{-1}s = s$. Hence $S_s = (ss^{-1})\omega$. □

Theorem 3.11. *The partial $S-$act X is free, effective and transitive if and only if it is isomorphic to an orbit of the left partial $S-$act S (i.e. an $\mathcal{L}-$class of S).*

Proof. Let $s \in S$ and consider S as a left partial $S-$act. By Lemma 3.10 we see that the $S-$orbit S^1s is a free partial $S-$act.
Conversely, let X be a free, effective and transitive partial $S-$act. Then X has only 1 orbit and so choose any $x \in X$ so that $X = S^1x$. By Theorem 3.9 and Lemma 3.10 $X \cong S/S_x = S/e\omega = S/S_e \cong S^1e$ for some $e \in E(S)$. □

Notice that each $\mathcal{L}-$class of S is then a free left partial $S-$subact of S. Moreover, If $(s,t) \in \mathcal{D}$ then $L_s \cong L_t$. Hence each $\mathcal{D}-$class contains, up to isomorphism, a unique free left partial $S-$subact. So in general, there may be more than one non-isomorphic free left partial $S-$act of rank 1.

The following is clear

Lemma 3.11. *Coproducts of free partial $S-$acts are free. Hence S is a free partial $S-$act.*

Theorem 3.12. *An inverse semigroup S is $E-$unitary if and only if for all free left partial $S-$acts X and all $x \in X$, $S_x \subseteq E(S)$.*

Proof. Suppose S is $E-$unitary and that X is a free left partial $S-$act. Then if $x \in X$ and $s \in S_x$ it follows since X is free that there exists $e \in E(S)$ such that $s \in e\omega$ and so there exists $f \in E(S)$ with $sf = e$. Hence $s \in E(S)$ as required.
Conversely, suppose that $sf = e$ for some $s \in S, e, f \in E(S)$. Then clearly $se = e$ and so $e = se = ss^{-1}se = ses^{-1}s = es^{-1}s = s^{-1}se$. Hence in the left partial $S-$act L_e we see that $e \in D_s$ and so $s \in S_e$. Consequently $s \in E(S)$ and S is $E-$unitary as required. □

3.4. *Generators and relations*

When dealing with actions over a group G, it is easy to establish that every $G-$set X is a quotient of a free $G-$set. In particular we can construct the (free) $G-$set $G \times X$ with action defined by $g(h, x) = (gh, x)$ and construct a $G-$map $G \times X \to X$ with $(g, x) \mapsto gx$ which is clearly onto. A similar construction for inverse semigroups and partial actions however fails. While we can construct a left partial $S-$act $S \times X$ with action induced by the left $S-$action of S on S (which is free being a coproduct of $|X|$ copies of S), the corresponding map, $S \times X \to X$ given by $(s, x) \mapsto sx$ is not in general an $S-$map. To see this notice that for $t \in S$, $(s, x) \in D_t^{S \times X}$ if and only if $s \in D_t^S$ (in other words $s = t^{-1}ts$ in S) whereas $sx \in D_t^X$ if and only if $x \in D_{ts}^X$. As an example, let X be any set with the trivial $S-$action, $sx = x$ for all $s \in S, x \in X$.

We wish to give a useful meaning to the notation $\langle x|R \rangle$ where x is a symbol and R is a set of equations of the form $sx = tx$ for $s, t \in S$.
First notice that this equation is equivalent to $t^{-1}s \in S_x$ and that S_x is a closed inverse subsemigroup of S. Suppose then that $R = \{E_i : i \in I\}$ where E_i is the equation $s_i x = t_i x$ for $s_i, t_i \in S$. Let $P = \{t_i^{-1}s_i, s_i^{-1}t_i : i \in I\} \subseteq S$, let K be the smallest inverse subsemigroup of S containing P and let $H = K\omega$. Then $\langle x|R \rangle$ shall denote the left partial $S-$act S/H. It is clear that in practice we do not need to include both elements $t_i^{-1}s_i$ and $s_i^{-1}t_i$ but it disposes of the need for a "well-defined" argument. Notice also that H is the smallest closed inverse subsemigroup of S containing P.
If a partial $S-$act contains x and satisfies an equation $s_i x = t_i x$ then there are a number of other consequences for the partial act. In particular note that $s_i^{-1}s_i, t_i^{-1}t_i \in S_x$. Also, if the element s_i can be factorised as

$s_i = p_i q_i$ then $x \in D_{q_i}$ and so $q_i^{-1} q_i \in S_x$ and if there exists $r_i \in S$ with $x \in D_{r_i s_i} \cap D_{r_i t_i}$ then $r_i s_i x = r_i t_i x$ and so $(r_i t_i)^{-1}(r_i s_i) \in S_x$.
With the notation above we have:

Lemma 3.12.

(1) If $s \in H$ and $s = se$ for $e \in E(S)$ then $e \in H$.
(2) For all $i \in I$, $s_i^{-1} s_i, t_i^{-1} t_i \in H$.
(3) If for some $i \in I$, $s_i = p_i q_i$ then $q_i^{-1} q_i \in H$.
(4) If for some $i \in I$, there exists $r_i \in S$ with $s_i, t_i \in D_{r_i}^S$ then $(r_i t_i)^{-1}(r_i s_i) \in H$.
(5) For each $i \in I$, $(s_i H)\omega = (t_i H)\omega$.

Proof.

(1) This follows since H is closed in S.
(2) Since $t_i^{-1} s_i = t_i^{-1} s_i s_i^{-1} s_i$ and $t_i^{-1} s_i \in H$ then $s_i^{-1} s_i \in H$.
(3) Notice that $t_i^{-1} s_i = t_i^{-1} s_i q_i^{-1} q_i$ and so the result follows from (1).
(4) Notice that $r_i t_i \in L_{t_i}$ and so $t_i^{-1} r_i^{-1} r_i t_i = t_i^{-1} t_i$. Hence $t_i^{-1} r_i^{-1} r_i s_i = t_i^{-1} r_i^{-1} r_i t_i t_i^{-1} s_i = t_i^{-1} s_i \in H$.
(5) Notice that $(s_i H)\omega = (t_i H)\omega$ if and only if $t_i^{-1} s_i \in H$. □

Given any transitive, effective left partial S−act X we can construct a presentation associated with X in an obvious way. Let $X = S^1 y$ for some $y \in X$ and consider the presentation $\langle x | sx = x, s \in S_y \rangle$.

Theorem 3.13. *Let $X = S^1 y$ be an effective, transitive left partial S−act. Then the left partial S−act associated with the presentation $\langle x | sx = x, s \in S_y \rangle$ is isomorphic to X.*

Proof. This follows almost immediately from Theorem 3.9. □

4. Graphs

4.1. *S−graphs*

We define S−graphs for an inverse semigroup S in a similar way as we do for groups. Specifically, an S−graph (X, V, E, ι, τ) is a non-empty S−set X with disjoint non-empty S−subsets V and E such that $X = V \cup E$ and two maps $\iota, \tau : E \to V$ with the property that for all $e \in E, s \in S$ if $s \in D^e$ then $s \in D^{\iota e} \cap D^{\tau e}$ and $s\iota e = \iota(se), s\tau e = \tau(se)$.

If we have the stronger property that ι and τ are $S-$maps then we shall refer to X as a *complete $S-$graph.* Notice that in this case, since ι and τ are $S-$maps then for any edge $e \in E$ and any $s \in S$, $\iota e \in D_s$ if and only if $\tau e \in D_s$. Hence it follows that for all $s \in S$, either s acts on all vertices in a connected component of X or it acts on none. It also follows that if Y is a connected subgraph of a complete $S-$graph X and $s \in S$ acts on an element (and hence all) of Y then sY is a connected subgraph of X.

A *path* p in X is a finite sequence

$$v_0, e_1^{\epsilon_1}, v_1, \ldots, v_{n-1}, e_n^{\epsilon_n}, v_n$$

where $n \geq 0, v_i \in VX$ for each $1 \leq i \leq n$, $e_i^{\epsilon_i} \in EX^{\pm 1}, \iota(e_i^{\epsilon_i}) = v_{i-1}, \tau(e_i^{\epsilon_i}) = v_i$ for each $1 \leq i \leq n$.
We extend the incidence functions to paths by defining $\iota(p) = v_0, \tau(p) = v_n$ and define the *length* of p to be n. The inverse path p^{-1} is defined to be the path $v_n, e_n^{-\epsilon_n}, v_{n-1}, \ldots, v_1, e_1^{-\epsilon_1}, v_0$. If for each $1 \leq i \leq n-1, e_{i+1}^{\epsilon_{i+1}} \neq e_i^{-\epsilon_i}$ then p is said to be *reduced.*
Let $u, v \in V$ and consider $P = \{p : p \text{ is a reduced path with } \iota(p) = u, \tau(p) = v\}$. Any path $p \in P$ of minimal length is called a *geodesic* from u to v. A path p of length $n \geq 1$ is called a *closed path* or *cycle* if $\iota(p) = \tau(p)$. A connected $S-$graph with no cycles will be called an $S-$tree. It is clear that there is a unique path connecting any two vertices in an $S-$tree and so this will be a geodesic.
If (X, V, E, ι, τ) is a tree and $W \subseteq V$ then by the *subtree generated by W* we mean the subgraph of X consisting of all edges and vertices which occur in the geodesics joining all pairs of vertices in W. This is essentially the "smallest" subtree of X containing W.

Suppose now that X is a connected complete $S-$graph. Then for $x \in X$, it follows that D^x is a closed inverse subsemigroup of S which is a union of $\mathcal{D}-$classes. To see this first notice that $D^x = D^y = D^X$ for all $x, y \in X$. Now suppose that $s, t \in D^x$. Then $t \in D^x$ implies $t \in D^{sx}$ and so $ts \in D^x$. Hence by Lemma 3.2, D^X is a closed inverse subsemigroup of S and is a union of $\mathcal{D}-$classes of S.

Let X be a left partial $S-$act. We shall say that S *stabilizes an element* $x \in X$ if $s \in S_x$ if and only if $x \in D_s$.

Lemma 4.1. (cf. [2, Proposition 4.7]) *Let S be an inverse semigroup, T an effective, complete $S-$tree and v a vertex of T. Then S stabilizes a vertex of T if and only if there is an integer n such that the distance from v to each element of $S^1 v$ is at most n.*

Proof. Suppose that S stabilizes a vertex v_0 and that the geodesic, p, from v to v_0 has length N. Then for each $s \in D^T$, there is a path, p, sp^{-1} of length $2N$ from v to sv.
Conversely, let T' be the subtree generated by S^1v. Then it follows that T' is an $S-$subtree of T and no reduced path has length greater than $2n$. If T' has at most one edge then every element of T' is $S-$stable and so in particular v is $S-$stable. So we can assume that T' has at least two edges and therefore at least one vertex of T' has degree at least two. Now remove from T' all the leaves of the tree and their incident edges to leave an $S-$subtree, T'', (to see that T'' is an $S-$subtree notice that given an edge e of T'', then $\deg(\iota e), \deg(\tau e) \geq 2$ in T' and so for any $s \in D^T$, $\deg(\iota se), \deg(\tau se) \geq 2$ in T' and hence $se \in T''$) in which no reduced path has length greater than $2n-2$. Hence by induction, $S-$stabilizes a vertex$\square$

Corollary 4.1. *If there is a finite $S-$orbit in VT then S stabilizes a vertex of T.*

Corollary 4.2. *A finite inverse semigroup acting completely on a tree must stabilize a vertex.*

If X is a partial $S-$biact, the (left) *Schützenberger* graph of X with respect to a subset T of S, will be denoted $\Gamma = \Gamma(X,T)$, and is the (left) $S-$graph with vertex set $V = X$, edge set $E = \{(x,t) \in X \times T : xt \text{ exists and } xt \neq x\}$ and incidence functions $\iota(x,t) = x, \tau(x,t) = xt$ for all $(x,t) \in E$. The partial action is that induced by the left action of S on X.

$$x \xrightarrow{t} xt$$

Notice that $\Gamma(X,T)$ is a complete $S-$graph.
In particular, we are interested in the case $X = {}_SS_S$. In the following theorem, we shall denote the element S^1x of the quotient $S \setminus \Gamma(X,T)$ by $\overline{x}$.

Theorem 4.1. *Let X be an $S-$biact and let T be a generating set for S. Then*

(1) $x\,\mathcal{R}^X\,y$ in X if and only if there exists a finite path in $\Gamma(X,T)$ connecting x and y (we include here the null path of length 0),
(2) $x\,\mathcal{L}^X\,y$ in X if and only if $\overline{x} = \overline{y}$ in $S \setminus \Gamma(X,T)$,
(3) $x\,\mathcal{D}^X\,y$ in X if and only if there exists a finite path in $S \setminus \Gamma(X,T)$ connecting $\overline{x}$ to $\overline{y}$.

Proof. (1) If $x\,\mathcal{R}^X\,y$ then $y = x \cdot s$ for some $s \in S$ and if $s = t_1 \dots t_n$ for $t_i \in T$ then there are edges in $\Gamma(X,T)$

$$x \xrightarrow{t_1} xt_1 \xrightarrow{t_2} xt_1t_2 \qquad \cdots \qquad \xrightarrow{t_n} xt_1 \dots t_n = y$$

The converse is clear since for any edge e in $\Gamma(X,T)$, $(\iota(e), \tau(e)) \in \mathcal{R}^X$. The other results follow easily. □

Corollary 4.3. *Let S be an inverse semigroup with generating set T. Then S is bisimple if and only if $S \setminus \Gamma = S \setminus \Gamma(S,T)$ is a connected graph.*

Lemma 4.2. *If in $\Gamma(X,T)$, the vertex x is* isolated *(i.e. has no edges incident on it) then $x \cdot t$ does not exists for any $t \in T$. Consequently, if T generates S then there are no isolated vertices in $\Gamma(S,T)$.*

As an illustrative example of some of these constructions, let us consider the inverse subsemigroup S of $\mathcal{J}_{\{1,2,3,4,5\}}$ generated by

$$\alpha = \begin{pmatrix} 1\,2\,3\,4 \\ 2\,3\,1\,5 \end{pmatrix}.$$

It is easy to calculate that $S = \{\alpha, \alpha^{-1}, \alpha^2, \alpha^{-2}, \alpha\alpha^{-1}, \alpha^{-1}\alpha, \alpha^2\alpha^{-2}\}$ and that $E(S) = \{\alpha\alpha^{-1}, \alpha^{-1}\alpha, \alpha^2\alpha^{-2}\}$.
In fact S has multiplication table given by

	α	α^{-1}	α^2	α^{-2}	$\alpha\alpha^{-1}$	$\alpha^{-1}\alpha$	$\alpha^2\alpha^{-2}$
α	α^2	$\alpha\alpha^{-1}$	$\alpha^2\alpha^{-2}$	α^2	α^{-2}	α	α^{-2}
α^{-1}	$\alpha^{-1}\alpha$	α^{-2}	α^{-2}	$\alpha^2\alpha^{-2}$	α^{-1}	α^2	α^2
α^2	$\alpha^2\alpha^{-2}$	α^{-2}	α^{-2}	$\alpha^2\alpha^{-2}$	α^2	α^2	α^2
α^{-2}	α^2	$\alpha^2\alpha^{-2}$	$\alpha^2\alpha^{-2}$	α^2	α^{-2}	α^{-2}	α^{-2}
$\alpha\alpha^{-1}$	α	α^2	α^2	α^{-2}	$\alpha\alpha^{-1}$	$\alpha^2\alpha^{-2}$	$\alpha^2\alpha^{-2}$
$\alpha^{-1}\alpha$	α^{-2}	α^{-1}	α^2	α^{-2}	$\alpha^2\alpha^{-2}$	$\alpha^{-1}\alpha$	$\alpha^2\alpha^{-2}$
$\alpha^2\alpha^{-2}$	α^{-2}	α^2	α^2	α^{-2}	$\alpha^2\alpha^{-2}$	$\alpha^2\alpha^{-2}$	$\alpha^2\alpha^{-2}$

and so the $\mathcal{R}$–classes are $\{\alpha, \alpha\alpha^{-1}\}, \{\alpha^{-1}, \alpha^{-1}\alpha\}, \{\alpha^2, \alpha^{-2}, \alpha^2\alpha^{-2}\}$.
Let $T = \{\alpha\}$. Then the Schützenberger graph, $\Gamma = \Gamma(S,T)$, of S with respect to T is

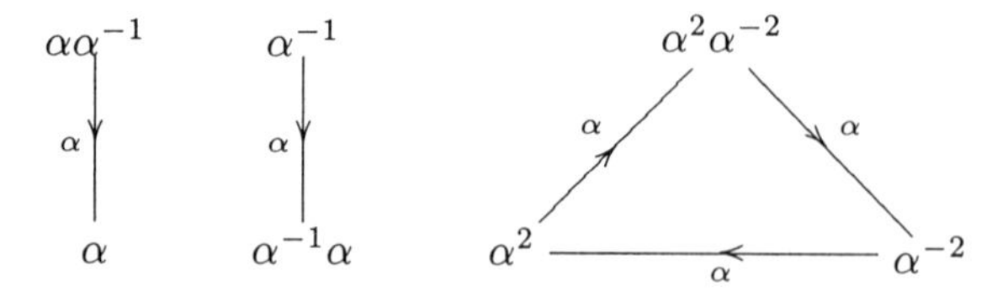

The action of S (induced by the Preston-Wagner representation) on the graph is as follows

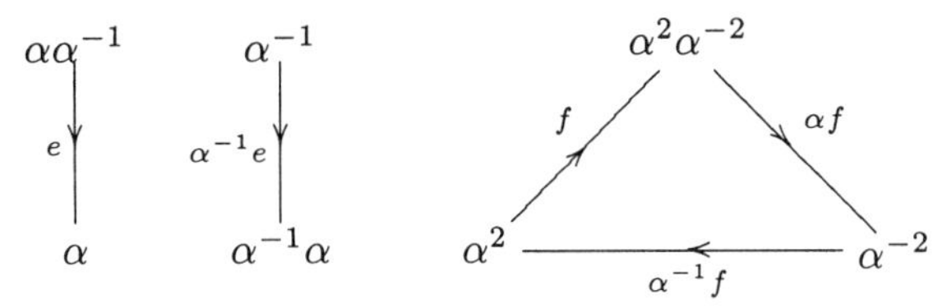

Also, we can calculate the orbits of Γ; the edge orbits are

$$S^1 e = \{e, \alpha^{-1}e\}, S^1 f = \{f, \alpha f, \alpha^{-1} f\}$$

while the vertex orbits are

$S^1 \cdot \alpha = \{\alpha, \alpha^{-1}\alpha\}$, $S^1 \cdot \alpha^{-1} = \{\alpha^{-1}, \alpha\alpha^{-1}\}$ and $S^1 \cdot \alpha^2 = \{\alpha^2, \alpha^{-2}, \alpha^2\alpha^{-2}\}$.

The quotient graph $S \setminus \Gamma$, then looks like

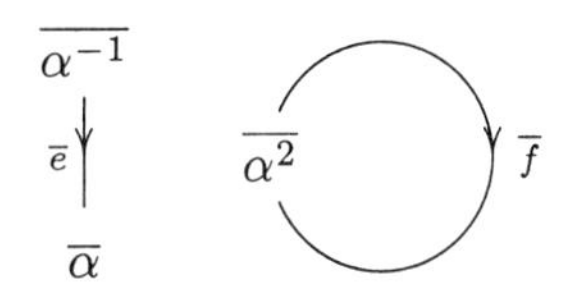

Notice that $S_e = \{\alpha\alpha^{-1}\} = \{\alpha\alpha^{-1}\}\omega, S_f = \{\alpha\alpha^{-1}, \alpha^{-1}\alpha, \alpha^2\alpha^{-2}\} = \{\alpha^2\alpha^{-2}\}\omega$ and so S acts freely on Γ.

In fact we have

Theorem 4.2. *Let S be an inverse semigroup and T a subset of S. Then S acts freely on $\Gamma(S, T)$.*

Proof. Suppose that $p \in S_s$ for some $p, s \in S$. Then $ps = s$ and so $pss^{-1} = ss^{-1}$. Hence $p \geq ss^{-1}$ and $p \in (ss^{-1})\omega$.
Conversely, if $p \in (ss^{-1})\omega$ then $pss^{-1} = ss^{-1}$ and so $ps = s$ and $p \in S_s$. $\square$

4.2. *Graphs of Inverse Semigroups*

An $S-$*transversal* in a partial $S-$act X is a subset Y of X which meets each $S-$orbit exactly once. Hence the composite $Y \subseteq X \to S \setminus X$ is a bijection. Notice that $S \setminus X$ is isomorphic to

$$\dot{\bigcup}_{y \in Y} S/S_y.$$

Lemma 4.3. (cf. [2, Proposition 2.6]) *If X is an $S-$graph and $S \setminus X$ is connected then there exist subsets $Y_0 \subseteq Y \subseteq X$ such that Y is an $S-$transversal in X, Y_0 is a subtree of X, $VY = VY_0$ and for each $e \in EY, \iota e \in VY = VY_0$.*

Proof. Let $\overline{X} = S \setminus X$ and $\overline{x} = S^1 x$ for all $x \in X$. Choose a vertex v_0 of X. By Zorn's Lemma we can choose a maximal subtree Y_0 of X containing v_0 such that the composite $Y_0 \subseteq X \to \overline{X}$ is injective. Let $\overline{Y}_0$ denote the image of Y_0. Then $V\overline{Y}_0 = V\overline{X}$. To see this, suppose that it is not so. Then since $\overline{X}$ is connected, any vertex in $\overline{Y}_0$ is connected to any vertex in $\overline{X} - \overline{Y}_0$ by a path in $\overline{X}$, so some edge $\overline{e}$ of $\overline{X}$ has one vertex $\overline{v}$ in $\overline{Y}_0$ and one vertex in $\overline{X} - \overline{Y}_0$. Here $\overline{v}$ comes from an element v of VY_0 and $\overline{e}$ from an edge e of X; since v lies in the same orbit as a vertex of e, it is a vertex of se for some $s \in S$, and by replacing e with se we may further assume that v is a vertex of e. Let w be the other vertex of e. Notice e, w do not lie in Y_0, since their images do not lie in $\overline{Y}_0$. But $Y_0 \cup \{e, w\}$ contradicts the maximality of Y_0. This proves the claim that $V\overline{Y}_0 = V\overline{X}$.
For each edge $\overline{e}$ in $E\overline{X} - E\overline{Y}_0$, $\overline{\iota}\,\overline{e}$ comes from a unique vertex of Y_0 and as before we can assume $\iota e \in Y_0$. Adjoining the resulting edges to Y_0 gives a subset Y of X such that the composite $Y \subseteq X \to \overline{X}$ is bijective and if $e \in EY$ then $\iota e \in Y$. □

In Lemma 4.3, we call Y a *fundamental $S-$transversal with subtree Y_0*.

Let S be an inverse semigroup and X an $S-$graph such that $S \setminus X$ is connected. Let Y be a fundamental $S-$transversal for X with subtree Y_0. Let $e \in EY$ and consider the $S-$orbits of $\iota e, \tau e$. There is a unique element of $VY, \overline{\iota} e$, which lies in the same $S-$orbit as ιe. Because of the way we construct Y, we can assume that $\iota e = \overline{\iota} e$. In a similar way, there is a unique element of $VY, \overline{\tau} e$, which lies in the same $S-$orbit as τe. This defines functions $\overline{\iota}, \overline{\tau} : EY \to VY$ which makes Y into a graph and it is easy to see that $Y \simeq S \setminus X$.
For each $e \in EY$ we see that τe and $\overline{\tau} e$ lie in the same (transitive) $S-$orbit in EX, so there exists an element $t_e \in S^1$ such that $t_e \overline{\tau} e = \tau e$. If $e \in EY_0$ then $\overline{\tau} e = \tau e$ (by uniqueness) and so we can take $t_e = 1$. Now $S_e \subseteq S_{\iota e}$ and $S_e \subseteq S_{\tau e}$. From Theorem 3.10 we see that $S_{\overline{\tau} e}$ is conjugate to $S_{\tau e}$ and from Theorem 3.8 that if $t_e t_e^{-1}$ is the identity of $S_{\tau e}$ then there is an embedding $S_e \subseteq S_{\tau e} \to t_e^{-1} S_{\tau e} t_e \subseteq S_{\overline{\tau} e}$ given by $x \mapsto t_e^{-1} x t_e$.

We shall refer to this as the *graph of inverse semigroups associated to X* with respect to the fundamental $S-$transversal Y, the maximal subtree Y_0

and the *family of connecting elements* t_e. Notice that it is *sufficient* that $t_e t_e^{-1}$ is the identity of $S_{\tau e}$ in order to guarantee the existence of an embedding $S_e \to S_{\overline{\tau}e}$. At this stage we do not know if it is *necessary*. More generally, by a *graph of inverse semigroups* $(S(-), Y)$, we shall mean a connected graph $(Y, V, E, \overline{\iota}, \overline{\tau})$ together with a function $S(-)$ which assigns to each $v \in V$ an inverse semigroup $S(v)$, and to each edge $e \in E$ a subgroup $S(e)$ of $S(\overline{\iota}e)$ and an inverse semigroup monomorphism $t_e : S(e) \to S(\overline{\tau}e), s \mapsto s^{t_e}$. We shall refer to the semigroups $S(v)$ as the *vertex (inverse) semigroups*, the semigroups $S(e)$ as the *edge (inverse) semigroups* and the maps t_e the *edge functions*.

We present two example of inverse semigroups acting on graphs (in fact trees) and construct the associated graph of inverse semigroups. The first involves the free monogenic inverse semigroup and the second the bicyclic semigroup.

Example 4.1. Let S be the free inverse semigroup on one generator $\{x\}$. From [4,IX.1.1], we see that S is isomorphic to $\{(m, k, n) : m \leq 0 \leq n, m \leq k \leq n, m < n\}$ with multiplication given by $(m, k, n)(m', k', n') = (\min\{m, k + m'\}, k + k', \max\{n, k + n'\})$. Under this isomorphism, x is mapped to $(0, 1, 1)$ and $(m, k, n)^{-1} = (m - k, -k, n - k)$.

Lemma 4.4. *Let S be as above.*

(1) $(m, k, n) \mathcal{R} (p, l, q)$ *if and only if* $m = p$ *and* $n = q$.
(2) $(m, k, n) \mathcal{L} (p, l, q)$ *if and only if* $m - k = p - l$ *and* $n - k = q - l$.
(3) $(m, k, n) \in (p, l, q)^{-1}(p, l, q)S$ *if and only if* $m \leq p - l < q - l \leq n$.
(4) *If* $(m, k, n) \in (p, l, q)^{-1}(p, l, q)S$ *then* $(p, l, q)(m, k, n) = (m+l, k+l, n+l)$.
(5) $(m, k, n) \in S(p, l, q)(p, l, q)^{-1}$ *if and only if* $m - k \leq p < q \leq n - k$.
(6) *If* $(m, k, n) \in S(p, l, q)(p, l, q)^{-1}$ *then* $(m, k, n)(p, l, q) = (m, k + l, n)$.
(7) $(m, k, n) \leq (p, l, q)$ *if and only if* $l = k, p \geq m$ *and* $q \leq n$.

Proof.

(1) $(m, k, n) \mathcal{R} (p, l, q)$ if and only if $(m, k, n)(m, k, n)^{-1} = (p, l, q)(p, l, q)^{-1}$ if and only if $(m, k, n)(m - k, -k, n - k) = (p, l, q)(p - l, -l, q - l)$ if and only if $(m, 0, n) = (p, 0, q)$.
(2) $(m, k, n) \mathcal{L} (p, l, q)$ if and only if $(m, k, n)^{-1}(m, k, n) = (p, l, q)^{-1}(p, l, q)$ if and only if $(m - k, -k, n - k)(m, k, n) = (p - l, -l, q - l)(p, l, q)$ if and only if $(m - k, 0, n - k) = (p - l, 0, q - l)$.

(3) $(p,l,q)^{-1}(p,l,q) = (p-l,-l,q-l)(p,l,q) = (p-l,0,q-l)$. So $(m,k,n) \in (p,l,q)^{-1}(p,l,q)S$ if and only if there exists $(a,b,c) \in S$ with $(m,k,n) = (p-l,0,q-l)(a,b,c) = (\min\{p-l,a\}, b, \max\{q-l,c\})$. But this is true if and only if $m \le p-l$ and $n \ge q-l$ (take $a = m, b = k, c = n$).

(4) By (3), it follows that $m \le p-l$ and $n \ge q-l$ and so $(p,l,q)(m,k,n) = (m+l, k+l, n+l)$.

(5) $(p,l,q)(p,l,q)^{-1} = (p,l,q)(p-l,-l,q-l) = (p,0,q)$. So $(m,l,n) \in S(p,l,q)(p,l,q)^{-1}$ if and only if there exists $(a,b,c) \in S$ with $(m,k,n) = (a,b,c)(p,0,q) = (\min\{a,b+p\}, b, \max\{c,b+q\})$. But this is true if and only if $m \le p+k$ and $n \ge q+k$ (take $a = m, b = k, c = n$).

(6) By (5), it follows that $m \le p+k, n \ge q+k$ and so $(m,k,n)(p,l,q) = (m,k+l,n)$.

(7) $(m,k,n) \le (p,l,q)$ if and only if $(m,k,n) = (m,k,n)(m,k,n)^{-1}(p,l,q)$ if and only if $(m,k,n) = (m,0,n)(p,l,q) = (\min(p,m), l, \max(q,n))$ if and only if $l = k, p \ge m$ and $q \le n$. □

Notice then that the (right) orbit $(m,k,n)S^1$ of an element $(m,k,n) \in S$ is given by

$$(m,k,n)S^1 = \{(m,l,n) : l \ge m\}.$$

Notice then that $(m,k,n)S^1 = (m,m,n)S^1$.
The left orbit $S^1(m,k,n)$ is given by

$$S^1(m,k,n) = \{(m+l, k+l, n+l) : l \le -m\}$$

and so $S^1(0, k-m, n-m) = S^1(m,k,n)$
Now $(m,k,n)\,\mathcal{R}\,(m,k,n)(0,1,1)$ if and only if $k \le n-1$ and so the Schützenberger graph, Γ, of S with respect to $T = \{x\}$ consists of all finite chains of the form

$$(m,m,n) \xrightarrow{x} (m,m+1,n) \xrightarrow{x} \quad \cdots \quad \xrightarrow{x} (m,n,n)$$

for all $m \le 0 \le n, m < n$.
Moreover, since $(m,k,n)\,\mathcal{L}\,(0,k-m,n-m)$ then on writing the $S-$orbit, $(m,k,n)S^1$ as $\overline{(m,k,n)}$, we see that the quotient graph, $S \setminus \Gamma$ consists of the finite chains

$$\overline{(0,0,n)} \longrightarrow \overline{(0,1,n)} \longrightarrow \quad \cdots \quad \longrightarrow \overline{(0,n,n)}$$

for $n \ge 1$.

Consider now the 3$-$element set $V = \{a,b,c\}$ and define a partial action on V from the representation $S \to \mathcal{I}_V$ generated by $x \to \rho_x$ where $\rho_x =$

$\begin{pmatrix} a & c \\ b & a \end{pmatrix}$. Define an S–graph G, as follows

$$c \xrightarrow{x^{-1}e} a \xrightarrow{e} b$$

and note that the quotient graph, $S \setminus G$ is

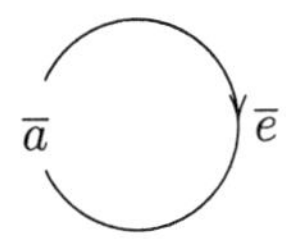

with a fundamental transversal Y

$$a \xrightarrow{e}$$

It is worth noting that the S–graph G is not a complete S–graph.

To construct the associated graph of inverse semigroups, notice that $\iota e = a, \tau e = b, \overline{\tau}(e) = a$, $S_e = \{xx^{-1}\}, S_{\iota e} = \{x^{-1}x, xx^{-1}, xx^{-1}x^{-1}x\} = S_{\overline{\tau}(e)}$, $S_{\tau e} = \{x^2x^{-2}, xx^{-1}\}$ and that xx^{-1} is the identity of $S_{\tau e}$. Hence the graph of inverse semigroups is given by

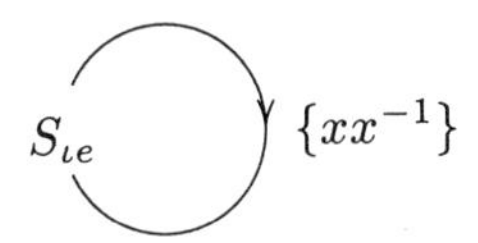

and there is an embedding $\{xx^{-1}\} \to S_{\iota e}$ given by $xx^{-1} \mapsto x^{-1}(xx^{-1})x = x^{-1}x$.

Notice that $x^{-1}x$ is *not* the identity in $S_{\overline{\tau}e}$.

Example 4.2. Let S be the bicyclic semigroup with presentation given by

$$S = \text{Inv}\langle x | xx^{-1}x^{-1}x = x^{-1}x \rangle.$$

The Schützenberger graph of S with respect to $T = \{x\}$ is

$$1 \xrightarrow{x} x \xrightarrow{x} x^2 \xrightarrow{x} \cdots$$

$$x^{-1} \xrightarrow{x} x^{-1}x \xrightarrow{x} x^{-1}x^2 \xrightarrow{x} \cdots$$

$$\vdots \xrightarrow{x} \cdots$$

and the quotient graph $S \setminus \Gamma$ is

$$\overline{1} \longrightarrow \overline{x} \longrightarrow \overline{x^2} \longrightarrow \cdots$$

with corresponding graph of inverse semigroups

$$\{1\}\xrightarrow{\{1\}}\{1\}\xrightarrow{\{1\}}\{1\}\xrightarrow{\{1\}}\cdots$$

Consider now the sets $V = \{v_1, v_0, v_{-1}, v_{-2}, \ldots\}, E = \{e_0, e_{-1}, e_{-2}, \ldots\}$ and define partial actions on V and E from the representations $\rho : S \to \mathcal{I}_V, \varphi : S \to \mathcal{I}_E$ generated by $x \mapsto \rho_x, x \mapsto \varphi_x$ respectively, where $\rho_x = \begin{pmatrix} v_0 & v_{-1} & v_{-2} & \cdots \\ v_1 & v_0 & v_{-1} & \cdots \end{pmatrix}$ and $\varphi_x = \begin{pmatrix} e_{-1} & e_{-2} & \cdots \\ e_0 & e_{-1} & \cdots \end{pmatrix}$. This defines for us an $S-$graph $G = (G, V, E, \iota, \tau)$, with $\iota e_n = v_n, \tau e_n = v_{n+1}$.

$$\cdots\xrightarrow{x^{-2}e_0}v_{-1}\xrightarrow{x^{-1}e_0}v_0\xrightarrow{e_0}v_1$$

and note that the quotient graph, $S \setminus G$ is

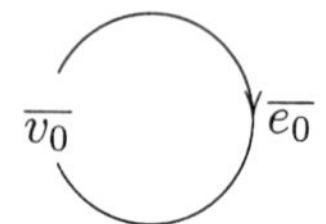

with a fundamental transversal Y

$$v_0\xrightarrow{e_0}$$

To construct the associated graph of inverse semigroups, notice that $xx^{-1} = 1, \iota e_0 = v_0, \tau e_0 = v_1, \overline{\tau} e_0 = v_0$, $S_{e_0} = \{1\}, S_{\iota e_0} = \{1, x^{-1}x\} = S_{\overline{\tau} e_0}, S_{\tau e_0} = \{1\}$ and that xx^{-1} is the identity of $S_{\tau e_0}$. Hence the graph of inverse semigroups is given by

$S_{\iota e_0}$ $\{xx^{-1}\}$

and there is an embedding $\{1\} \to S_{\iota e_0}$ given by $1 \mapsto x^{-1}(1)x = x^{-1}x$.

5. Summary

The contents of this paper are a start at developing a theory of (partial) actions of inverse semigroups on sets along the lines of the group theory approach. It is clear that much more work needs to be done on this and we hope to expand on this work in a future paper [5]. In particular we hope to address the question of a 'structure theorem' for partial actions on trees and consider examples of amalgamated free products of inverse semigroups.

Acknowledgement

The author would like to thank the anonymous referee for his comments and advice on an earlier version of this paper.

References

1. A.H. Clifford and G.B. Preston, *The Algebraic Theory of Semigroups II, American Mathematical Society, Mathematical Surveys 7*, AMS, 1967.
2. Warren Dicks & M.J. Dunwoody, *Groups acting on graphs, Cambridge Studies in Advanced Mathematics 17*, CUP, 1989.
3. J. M. Howie, *Fundamentals of semigroup theory, London Mathematical Society Monographs New Series*, OUP, 1995.
4. M. Petrich, *Inverse semigroups*, Wiley, New York, 1984.
5. J. Renshaw, *Inverse semigroups acting on graphs and trees*, in preparation.
6. A. Yamamura, *HNN Extensions of Inverse Semigroups and Applications*, Internat. J. Algebra Comput. Vol. 7, No. 5 (1997) 605-624.

A MODERN APPROACH TO SOME RESULTS OF STIFFLER

BENJAMIN STEINBERG*
School of Mathematics and Statistics,
Carleton University,
1125 Colonel By Drive,
Ottawa, Ontario K1S 5B6, Canada
E-mail: bsteinbg@math.carleton.ca

We give a modern proof of Stiffler's classical results describing the pseudovarieties of $\mathcal{R}$-trivial semigroups and locally $\mathcal{R}$-trivial semigroups as the wreath product closures of semilattices, respectively semilattices and right zero semigroups. Our proof uses the derived category of a functor developed by the author with B. Tilson. We prove a more general result describing functors between finite categories which are injective on coterminal $\mathcal{R}$-equivalent elements.

1. Introduction and Main Results

In this paper all monoids, semigroups, and categories are finite. Recall that a semigroup is called $\mathcal{R}$-trivial if each principal right ideal has a unique generator. Stiffler showed in the early seventies [13] that the $\mathcal{R}$-trivial semigroups are precisely the divisors of wreath products of the form

$$U_1 \wr U_1 \wr \cdots \wr U_1 \tag{1}$$

where $U_1 = (\{0,1\}, \cdot)$ is the two element semilattice (a divisor is a quotient of a subsemigroup). He also showed that the locally $\mathcal{R}$-trivial semigroups are precisely the divisors of wreath products of the form

$$U_1 \wr U_1 \wr \cdots \wr U_1 \wr R_1 \wr R_1 \wr \cdots \wr R_1 \tag{2}$$

where R_1 is the two element right zero semigroup. Recall that a semigroup is locally $\mathcal{R}$-trivial if all it submonoids are $\mathcal{R}$-trivial.

*The author was supported in part by the FCT and POCTI approved project POCTI/32817/MAT/2000 in participation with the european community fund FEDER and by FCT through *Centro de Matemática da Universidade do Porto.*

In modern terminology, the first result states that the pseudovariety $\mathbf{R}$ of $\mathcal{R}$-trivial semigroups is the smallest pseudovariety closed under semidirect product containing U_1. The second says that the equation

$$\mathbf{LR} = \mathbf{R} * \mathbf{D} \tag{3}$$

is valid where $\mathbf{LR}$ is the pseudovariety of locally $\mathcal{R}$-trivial semigroups, $*$ is the semidirect product operator on pseudovarieties [3] and $\mathbf{D}$ is the pseudovariety of semigroups whose idempotents form a right zero semigroup. The fact that $\mathbf{D}$ is the smallest pseudovariety of semigroups closed under semidirect product containing R_1 is a straightforward exercise [3, 13]. It is also straightforward to verify that the pseudovariety generated by U_1 is the pseudovariety $\mathbf{Sl}$ of all semilattices (idempotent-commutative semigroups).

We remark that Stiffler's first result is equivalent to stating that every $\mathcal{R}$-trivial monoid divides a wreath product of copies of U_1 since U_1 is a monoid and adding an identity does not change whether a semigroup is $\mathcal{R}$-trivial.

Tilson's seminal paper [14] introduced pseudovarieties of categories, generalizing the definition of division to this context. He obtained a strong connection between the semidirect product operator and pseudovarieties of categories. Let $\mathbf{V}$ be a pseudovariety of monoids. Then $\mathbf{gV}$ denotes the pseudovariety of all categories dividing a monoid in $\mathbf{V}$. One denotes by $\ell\mathbf{V}$ the pseudovariety of all categories which are locally in $\mathbf{V}$; a category C is locally in $\mathbf{V}$ if all its endomorphism (or local) monoids $C(c,c)$ belong to $\mathbf{V}$. A pseudovariety of monoids is called local if $\mathbf{gV} = \ell\mathbf{V}$. For instance $\mathbf{Sl}$ is local by a result of Simon [2]. Tilson's Derived Category Theorem says that a monoid M divides a semidirect product $V * N$ with $V \in \mathbf{V}$ if and only if there is a relational morphism $\varphi : M \to N$ with derived category $\mathcal{D}_\varphi \in \mathbf{gV}$ [14]. This was extended to relational morphisms of categories in [12].

Tilson's delay theorem [14] relates the two senses in which we have used the word local.

Theorem 1.1. *A pseudovariety of monoids* $\mathbf{V}$ *is local if and only if*

$$\mathbf{LV} = \mathbf{V} * \mathbf{D}. \tag{4}$$

Since it is easy to verify that a wreath product of semilattices is $\mathcal{R}$-trivial and since every $\mathcal{R}$-trivial monoid can be viewed as a one-object locally $\mathcal{R}$-trivial category, to prove both of Stiffler's theorems it suffices to prove the following theorem which is our main result.

Theorem 1.2. *Let C be a locally $\mathcal{R}$-trivial category. Then C divides an iterated semidirect product of copies of the two element semilattice U_1.*

Our proof uses the theory of derived categories of relational morphisms of categories developed by the author with Tilson in [12]. The proof also makes use of an extension of the results of Rhodes [7, 4, 8, 9, 10] on maximal proper surmorphisms to the setting of categories to prove the result by induction.

Stiffler's result can, in fact, be deduced from the following more general result:

Theorem 1.3. *Let $\varphi : C \to D$ be a (functional) morphism of categories (monoids) whose derived category $\mathcal{D}_\varphi$ is locally $\mathcal{R}$-trivial. Then φ admits a factorization via (functional) morphisms of categories (monoids)*

$$\varphi = \varphi_1 \cdots \varphi_n \tag{5}$$

where each derived category $\mathcal{D}_{\varphi_i}$ divides a semilattice.

However, we prefer to prove Theorem 1.2 directly, since to deduce it from Theorem 1.3 requires the Composition Theorem [12] and the result that $\mathbf{g}(\mathbf{V} * \mathbf{W}) = \mathbf{gV} * \mathbf{gW}$ [11, 12], both of which are non-trivial.

Call a morphism $\varphi : C \to D$ of categories $\mathcal{R}$-faithful if $m, n \in C(c, c')$, $m \mathrel{\mathcal{R}} n$ and $m\varphi = n\varphi$ implies $m = n$. That is, φ is injective when restricted to coterminal $\mathcal{R}$-equivalent elements. We shall see below that $\mathcal{D}_\varphi \in \ell\mathbf{R}$ if and only if φ is $\mathcal{R}$-faithful. In particular, Theorem 1.3 applies to monoid homomorphisms which are injective on $\mathcal{R}$-classes. Since a homomorphism of inverse monoids is injective on $\mathcal{R}$-classes if and only if it is idempotent-pure, this result can be viewed as a generalization of the classical factorization results for inverse monoids [5].

We mention that Stiffler's original proof [13] (see also [3]) involves a classification of the prime transformation semigroups and direct construction of the divisions. A syntactic approach can be found in [1].

2. The Proofs of Theorems 1.2 and 1.3

2.1. *Maximal proper quotients*

In this paper, we shall find it convenient not to distinguish notationally between a category and its arrow set. We shall used $\mathbf{d}, \mathbf{r}$ for the functions that select the domain and range of an arrow.

Let $\varphi : C \to D$ be a quotient morphism [6, 14] of categories. We say that φ is a maximal proper quotient (MPQ) if the associated congruence (φ) is not the identity congruence and contains only the identity congruence. In this case (φ) is called a minimal non-trivial congruence. An MPQ between monoids is often called an MPS (maximal proper surmorphism) [7, 4, 8, 9, 10].

One could study MPQ's by imitating the arguments of [7, 4, 8, 9, 10], but we find it more expedient to deduce the results for categories from the monoid case. To do this, we shall need to make use of the following construction called the consolidation. If C is a category, we can define a monoid $C_{cd} = C \cup \{0, 1\}$ where 0 is a zero, 1 is an identity and all undefined products in C are by fiat 0.

Let $\mathbf{Cat}_Q$ be the category of categories with quotient morphisms. Then $C \mapsto C_{cd}$ is the object part of a functor to the category $\mathbf{Mon}_Q$ of monoids with surjective morphisms. A quotient morphism $\varphi : C \to D$ induces a surjective homomorphism $\varphi_{cd} : C_{cd} \to D_{cd}$ by defining $0\varphi_{cd} = 0$, $1\varphi_{cd} = 1$. Proving that φ_{cd} is a homomorphism uses that φ is a quotient morphism. The key observation is that, for $x, y \in C$, xy is defined if and only if $x\varphi y\varphi$ is defined.

Note that $\varphi : C \to D \in \mathbf{Cat}_Q$ is an MPQ if and only if, for each factorization $\varphi = \varphi_1\varphi_2$ in $\mathbf{Cat}_Q$, exactly one of φ_1 or φ_2 is an isomorphism.

Here is the connection between MPS's and MPQ's.

Proposition 2.1. *Let $\varphi : C \to D$ be a quotient morphism of categories. Then φ is an MPQ if and only if φ_{cd} is an MPS.*

Proof. Suppose (φ) is a congruence on C and $(\sigma) \subsetneq (\varphi_{cd})$. Define an equivalence relation (σ') on C by

$$m_1 \, (\sigma') \, m_2 \iff m_1 \, (\sigma) \, m_2 \tag{6}$$

Since $(\sigma) \subsetneq (\varphi_{cd})$, it follows that $c_1 \, (\sigma') \, c_2$ implies c_1 and c_2 are coterminal. It is then straightforward to verify that (σ') is a congruence on C. Evidently $(\sigma') \subsetneq \varphi$ and $(\sigma'_{cd}) = (\sigma)$. It follows that (φ) is minimal non-trivial if and only if (φ_{cd}) is minimal non-trivial, establishing the result. □

To make use of this results, we need to consider Green's relations $\mathcal{H}$, $\mathcal{R}$, $\mathcal{L}$, and $\mathcal{J}$ on categories (the obvious generalizations of the definitions are left to the reader). We shall also have occasion to use the $\mathcal{J}$-preorder, denoted $\leq_{\mathcal{J}}$. We also extend the notions of regular and null elements and $\mathcal{J}$-classes [4] to the setting of categories. The following proposition is a direct calculation which we omit.

Proposition 2.2. *Let C be a category and $\mathcal{K}$ be one of Green's relations. Suppose $m_1, m_2 \in C$. Then m_1 and m_2 are $\mathcal{K}$-equivalent in C if and only if they are in C_{cd}. A similar result holds for $\leq_{\mathcal{J}}$. Likewise, m is regular in C if and only if it is in C_{cd}.*

Propositions 2.1 and 2.2 allow us to translate results about surjective monoid morphisms to quotient morphisms of categories via the consolidation functor. For instance the following useful lemma [4] carries over to quotient morphisms of categories.

Lemma 2.1. *Let $\varphi : C \to D$ be a quotient morphism of categories and let J be a $\mathcal{J}$-class of D. Then $J\varphi^{-1} = J_1 \cup \ldots \cup J_k$ where the J_i are $\mathcal{J}$-classes of C. If J_i is $\leq_{\mathcal{J}}$-minimal in $J\varphi^{-1}$, then $J_i\varphi = J$. If J is regular, then there is a unique such minimal J_i.*

Let $\varphi : C \to D$ be a quotient morphism of categories. We define a $\mathcal{J}$-class J of C to be $\mathcal{J}$-singular for φ if $\varphi|_{C\setminus J}$ is injective and $x >_{\mathcal{J}} J$ implies $x\varphi \notin J\varphi$. This definition generalizes the notion of $\mathcal{J}$-singularity used in [8], but is stronger than the definition used in [9, 10].

An immediate consequence of Proposition 2.2 is the following proposition.

Proposition 2.3. *Let $\varphi : C \to D$ be a quotient. Then a $\mathcal{J}$-class J of C is $\mathcal{J}$-singular for φ if and only if it is for φ_{cd}.*

Rhodes established the following facts for MPS's $\varphi : M \to N$ of semigroups [7, 4, 8, 9, 10]:

(1) φ has a $\mathcal{J}$-singular $\mathcal{J}$-class;
(2) φ either separates $\mathcal{H}$-classes or is injective on $\mathcal{H}$-classes.

Hence we can deduce the following:

Proposition 2.4. *Let $\varphi : C \to D$ be an MPQ of categories. Then φ has a $\mathcal{J}$-singular $\mathcal{J}$-class. Moreover, φ either separates $\mathcal{H}$-classes or is injective on $\mathcal{H}$-classes.*

In fact, the whole classification scheme of MPS's from [9] can be extended to categories via the consolidation functor. Note though that our definition of $\mathcal{J}$-singular is stronger than the one used in [9] and so, for instance, only Q is $\mathcal{J}$-singular in our sense for class III MPQ's.

2.2. $\mathcal{R}$-faithful morphisms

We define a morphism $\varphi : C \to D$ of categories to be $\mathcal{R}$-faithful if $m_1, m_2 \in C(c_1, c_2)$, $m_1\varphi = m_2\varphi$ and $m_1 \mathcal{R} m_2$ imply $m_1 = m_2$. We now prove some basic facts concerning $\mathcal{R}$-faithful morphisms.

Proposition 2.5. *Suppose $\varphi = \varphi_1\varphi_2$ is a factorization in $\mathbf{Cat}_Q$. Then φ is $\mathcal{R}$-faithful if and only if φ_1 and φ_2 are $\mathcal{R}$-faithful.*

Proof. It is clear that the composition of $\mathcal{R}$-faithful morphisms is $\mathcal{R}$-faithful. If φ is $\mathcal{R}$-faithful, it is clear that φ_1 is $\mathcal{R}$-faithful. Let $\varphi_1 : C \to C'$, $\varphi_2 : C' \to D$. Suppose $x, y \in C'$ are coterminal with $x \mathcal{R} y$ and $x\varphi_2 = y\varphi_2$. Let J be the $\mathcal{J}$-class of x, y. Then, by Lemma 2.1, there is a $\leq_{\mathcal{J}}$-minimal $\mathcal{J}$-class $\widetilde{J}$ of $J\varphi_1^{-1}$ with $\widetilde{J}\varphi_1 = J$. Suppose $xu = y$. Choose $\widetilde{x} \in \widetilde{J}$ with $\widetilde{x}\varphi_1 = x$ and choose $u' \in C$ with $u'\varphi_1 = u$. Since φ_1 is a quotient, $\widetilde{x}u'$ is defined and coterminal with $\widetilde{x}$. Clearly $\widetilde{x}u' \leq_{\mathcal{J}} \widetilde{x}$. Since $\widetilde{x}u'\varphi_1 = xu = y$, we see $\widetilde{x}u' \mathcal{J} \widetilde{x}$ by minimality of $\widetilde{J}$ and so $\widetilde{x}u' \mathcal{R} x$. Since

$$\widetilde{x}u'\varphi = y\varphi_2 = x\varphi_2 = \widetilde{x}\varphi, \tag{7}$$

we see that $\widetilde{x}u' = \widetilde{x}$ by $\mathcal{R}$-faithfulness of φ. Hence $x = y$ and φ_2 is $\mathcal{R}$-faithful. $\square$

We remind the reader of the definition of the derived category of a morphism of categories [12]. If $\varphi : C \to D$ is a morphism of categories, the category $\mathbf{Der}(\varphi)$ can be described as follows:

$$\begin{aligned} \mathrm{Obj}(\mathbf{Der}(\varphi)) &= \{(n, c) \in D \times \mathrm{Obj}(C) \mid n\mathbf{r} = c\varphi\} \\ \mathbf{Der}(\varphi) &= \{(n, m) \in D \times C \mid n\mathbf{r} = m\mathbf{d}\varphi\} \\ (n, m)\mathbf{d} = (n, m\mathbf{d}), &\quad (n, m)\mathbf{r} = (nm\varphi, m\mathbf{r}) \\ (n, m)(nm\varphi, m') &= (n, mm') \end{aligned} \tag{8}$$

The identity arrow at (n, c) is $(n, 1_c)$. One can define a functor $\sigma : \mathbf{Der}(\varphi) \to \mathbf{Set}$ by

$$\begin{aligned} (n, c)\sigma &= \{m \in C \mid m\varphi = n, m\mathbf{r} = c\} \\ (n, m)\sigma &= m' \longmapsto m'm \end{aligned} \tag{9}$$

The derived category of φ is then $\mathcal{D}_\varphi = \mathbf{Der}(\varphi)/(\sigma)$. The (σ)-class of an arrow (n, m) is denoted $[(n, m)]$.

Proposition 2.6. *Let $\varphi : C \to D$ be a morphism of categories. Then φ is $\mathcal{R}$-faithful if and only if $\mathcal{D}_\varphi \in \ell\mathbf{R}$.*

Proof. Suppose first that φ is $\mathcal{R}$-faithful. Let $[(n,m)], [(n,m')] : (n,c) \to (n,c)$ be $\mathcal{R}$-equivalent elements of $\mathcal{D}_\varphi((n,c),(n,c))$. Let $m_L : c_0 \to c \in n\varphi^{-1}$. Suppose $[(n,m)][(n,x)] = [(n,m')]$ and $[(n,m')][(n,y)] = [(n,m)]$. Then

$$\begin{aligned}[(1_{c_0\varphi}, m_L mx)] = [(1_{c_0\varphi}, m_L)][(n,m)][(n,x)] = \\ [(1_{c_0\varphi}, m_L)][(n,m')] = [(1_{c_0\varphi}, m_L m')] \text{ and} \\ [(1_{c_0\varphi}, m_L m'y)] = [(1_{c_0\varphi}, m_L)][(n,m')][(n,y)] = \\ [(1_{c_0\varphi}, m_L)][(n,m)] = [(1_{c_0\varphi}, m_L m)]\end{aligned} \tag{10}$$

Since σ doesn't identify arrows emanating from objects of the form $(1_{c\varphi}, c)$ [12], we see that $m_L mx = m_L m'$ and $m_L m'y = m_L m$. Thus $m_L m \mathrel{\mathcal{R}} m_L m'$. But

$$m_L m\varphi = nm\varphi = n = nm'\varphi = m_L m'\varphi \tag{11}$$

Since $m_L m, m_L m' : c_0 \to c$, we must have $m_L m = m_L m'$ by $\mathcal{R}$-faithfulness. Since m_L was arbitrary in $(n,c)\sigma$, we conclude that $[(n,m)] = [(n,m')]$ and so $\mathcal{D}_\varphi \in \ell\mathbf{R}$.

Conversely, suppose $\mathcal{D}_\varphi \in \ell\mathbf{R}$ and that $n = m\varphi = m'\varphi$ with $m, m' : c \to c'$ and $m \mathrel{\mathcal{R}} m'$. Suppose $mx = m'$ and $m'y = m$. Then $m(xy)^\omega = m$ (where s^ω denotes the idempotent power of s) and $m(xy)^\omega x = m'$. Let $z = (xy)^\omega$. Then $zxy = z$. One easily verifies that $[(n,z)], [(n,zx)], [(n,x)], [(n,y)]$ are arrows of $\mathcal{D}_\varphi((n,c'),(n,c'))$. The equalities

$$[(n,z)][(n,x)] = [(n,zx)], \quad [(n,zx)][(n,y)] = [(n,z)] \tag{12}$$

imply $[(n,z)] \mathrel{\mathcal{R}} [(n,zx)]$ so $[(n,z)] = [(n,zx)]$. Since $m \in (n,c')\sigma$, we see that

$$m = mz = mzx = m', \tag{13}$$

as desired. □

We now prove our main technical result, namely that the derived category of an $\mathcal{R}$-faithful MPQ divides a semilattice. This generalizes the case of $\mathcal{R}$-injective MPS's [8, 10].

Theorem 2.1. *Let $\varphi : C \to D$ be an MPQ of categories. Then φ is $\mathcal{R}$-faithful if and only if $\mathcal{D}_\varphi \in$ **gSl**.*

Proof. Since **gSl** is contained in $\ell\mathbf{R}$, we just need to prove the only if statement. Since **Sl** is local, it suffices to show that each local monoid of $\mathcal{D}_\varphi$ is a semilattice. Since φ is a quotient morphism an object (n,c) is determined by n and so we drop c from the notation. Also $n\sigma = n\varphi^{-1}$ in this context.

Since a monoid is a semilattice if and only if it satisfies the identities $x^2 = x$ and $xy = yx$, we need only show that:

$$(n, m_1), (n, m_2) \in \mathbf{Der}(\varphi)(n, n) \implies \\ [(n, m_1^2)] = [(n, m_1)] \text{ and } [(n, m_1 m_2)] = [(n, m_2 m_1)]. \quad (14)$$

Note that $nm_1\varphi = n = nm_2\varphi$.

Let $J \subseteq C$ be a $\mathcal{J}$-singular $\mathcal{J}$-class for φ (such exists by Proposition 2.4). If $|n\varphi^{-1}| \leq 1$, then (14) clearly holds. So we need only consider the case that $|n\varphi^{-1}| > 1$. Since φ is injective on $C \setminus J$, we conclude that there is at most one element $x \in C \setminus J$ with $x\varphi = n$. Also, by definition of $\mathcal{J}$-singularity, we have that $x \not\geq_{\mathcal{J}} J$. Since $xm_i \leq_{\mathcal{J}} x$ (whence $xm_i \notin J$) and $xm_i\varphi = n(m_i\varphi) = n$ for $i = 1, 2$, we see that

$$xm_i = x, \ i = 1, 2. \quad (15)$$

Suppose $x \neq m \in n\varphi^{-1}$. Then $m \in J$. Since $mm_1\varphi = n(m_1\varphi) = n$, we see that either $mm_1 = x$ or $mm_1 \in J$. In the first case $mm_1^2 = xm_1 = x$ by (15). In the second case, we must have $mm_1 \mathcal{R} m$ (since C is finite). But $mm_1\varphi = n(m_1\varphi) = n$ so $mm_1 = m$ by $\mathcal{R}$-faithfulness. Hence $mm_1^2 = mm_1 = m$. In either case we see that $[(n, m_1^2)] = [(n, m_1)]$. Similarly, either $mm_2 = m$ or $mm_2 = x$.

Suppose $mm_1 = x$. Then $mm_1m_2 = xm_2 = x$. On the other hand, mm_2m_1 is either mm_1 or xm_1. Either way, we see that $mm_2m_1 = x$. If $mm_2 = x$, then we are in the dual situation to the above and so $mm_1m_2 = mm_2m_1$. If $mm_i = m$ for $i = 1, 2$, then clearly $mm_1m_2 = m = mm_2m_1$. So in all cases we obtain $[(n, m_1m_2)] = [(n, m_2m_1)]$, proving (14) and establishing our result. □

2.3. *Proof of Theorem 1.2*

Suppose $C \in \ell\mathbf{R}$. We prove that C divides an iterated semidirect product of copies of U_1 by induction on the size of the longest chain of congruences on C. If the longest chain of congruences on C is of length 1, then the congruence which identifies all coterminal arrows is the identity congruence; that is each hom set of C has at most one element. But then C divides the trivial monoid [14].

Suppose the result has been proven when the longest chain has length n and let $C \in \ell\mathbf{R}$ have a longest chain of congruences of length $n + 1$.

Let (τ) be a minimal non-trivial congruence in such a longest chain. Then $D = C/(\tau) \in \ell\mathbf{R}$ and the longest chain of congruences on D has length n since the lattice of congruences on D is order isomorphic to the lattice of congruences on C containing (τ). Hence, by induction, D divides an iterated semidirect product W of semilattices.

The projection $\tau : C \to D$ is an MPQ by choice of (τ). Since $\mathbf{Der}(\tau)$ divides C via $(n, m) \mapsto m$, we see that $\mathcal{D}_\tau \in \ell\mathbf{R}$ and so τ is $\mathcal{R}$-faithful. Hence, by Theorem 2.1, $\mathcal{D}_\tau \in \mathbf{gSl}$ and so, by the Derived Category Theorem [12], C divides a wreath product $U \wr D$ with U a semilattice. Hence C divides $U \wr W$ as desired.

2.4. *Proof of Theorem 1.3*

As usual, φ factors as $\varphi_{Im}\varphi_F$ with φ_{Im} a quotient and φ_F faithful. Since φ_F is a division $\mathcal{D}_{\varphi_F}$ divides the trivial monoid [12]. By finiteness φ_{Im} factors $\varphi_1 \cdots \varphi_{n-1}$ where $\varphi_1, \ldots, \varphi_{n-1}$ are MPQ's. If φ is a monoid morphism, so are all the φ_i. By Proposition 2.5, the φ_i are all $\mathcal{R}$-faithful. Hence, by Theorem 2.1, $\mathcal{D}_{\varphi_i} \in \mathbf{gSl}$, all i. The result follows by taking $\varphi_n = \varphi_F$.

To prove Theorem 1.2 from Theorem 1.3, one considers the collapsing morphism $\gamma_C : C \to 1$. By the above results $\gamma_C = \varphi_1 \cdots \varphi_n$ with $\mathcal{D}_{\varphi_i} \in \mathbf{gV}$. The Composition Theorem [12] then states that

$$C \cong \mathcal{D}_{\gamma_C} \in \mathbf{gSl} * \cdots * \mathbf{gSl}. \tag{16}$$

But this latter is $\mathbf{g}(\mathbf{Sl} * \cdots * \mathbf{Sl})$ by the results of [11, 12].

References

1. J. Almeida, *Finite Semigroups and Universal Algebra*, World Scientific, 1994.
2. J. A. Brzozowski and I. Simon, *Characterizations of locally testable events*, Discrete Math. **4** (1973), 243–271.
3. S. Eilenberg, Automata, Languages and Machines, Academic Press, New York, Vol B, 1976.
4. K. Krohn, J. Rhodes, and B. Tilson, *Lectures on the algebraic theory of finite semigroups and finite-state machines*, Chapters 1, 5-9 (Chapter 6 with M. A. Arbib) of The Algebraic Theory of Machines, Languages, and Semigroups, (M. A. Arbib, ed.), Academic Press, New York, 1968.
5. M. V. Lawson, Inverse Semigroups: The theory of partial symmetries, World Scientific, Singapore, 1998.
6. S. MacLane, Categories for the Working Mathematician, Springer-Verlag, New York, 1971.
7. J. Rhodes, *A homomorphism theorem for finite semigroups*, Math. Systems Theory **1** (1967), 289–304.

8. J. Rhodes and B. Tilson, *The kernel of monoid morphisms*, J. Pure Appl. Algebra **62** (1989), 227–268.
9. J. Rhodes and P. Weil, *Decomposition techniques for finite semigroups, using categories I*, J. Pure Appl. Algebra **62** (1989), 269–284.
10. J. Rhodes and P. Weil, *Decomposition techniques for finite semigroups, using categories II*, J. Pure Appl. Algebra **62** (1989), 285–312.
11. B. Steinberg, *Semidirect products of categories and applications*, J. Pure Appl. Algebra **142** (1999), 153–182.
12. B. Steinberg and B. Tilson, *Categories as algebras II*, Internat. J. Algebra and Comput. **13** (2003), 627–703.
13. P. Stiffler, *Extension of the fundamental theorem of finite semigroups*, Adv. in Math. **11** (1973), 159–209.
14. B. Tilson, *Categories as algebra: An essential ingredient in the theory of monoids*, J. Pure and Appl. Algebra **48** (1987), 83–198.